Geography and Ethics

This book represents a landmark exploration of the terrains common to geography and moral philosophy, or ethics. Drawing together specially commissioned contributions from distinguished geographers across the UK, North America and Australasia, the place of geography in ethics and of ethics in geography is examined through wide-ranging, thematic chapters.

Geography and Ethics is divided into four sections for the discussion and exploration of ideas, following major thematic emphases in the discipline of geography:

- ethics and space
- ethics and place
- ethics and nature, and
- ethics and knowledge.

Each section points to the rich interplay between geography and ethics.

This collection represents a milestone in literature discussing geography and ethics, benefiting from the close management of James Proctor and David Smith who are two leaders in this emerging subfield of geography.

James D. Proctor is Associate Professor in the Department of Geography, University of California, Santa Barbara and **David M. Smith** is Professor of Geography at Queen Mary and Westfield College, University of London.

Geography and Ethics
Journeys in a moral terrain

**Edited by James D. Proctor and
David M. Smith**

Routledge
Taylor & Francis Group

LONDON AND NEW YORK

First published 1999
by Routledge
4 Park Square, Milton Park, Abingdon, Oxon OX14 4RN
605 Third Avenue, New York, NY 10017

*Routledge is an imprint of the Taylor & Francis Group,
an informa business*

© 1999 Selection and editorial matter James D. Proctor and
David M. Smith; individual chapters, the contributors

Typeset in Galliard by RefineCatch Limited, Bungay, Suffolk

British Library Cataloguing in Publication Data
A catalogue record for this book is available from the British Library

Library of Congress Cataloging in Publication Data
Geography and ethics : journeys in a moral terrain / edited by
   James D. Proctor and David M. Smith.
      p.   cm.
   1. Geography – Moral and ethical aspects.   I. Proctor, James D.,
1957–   .   II. Smith, David Marshall, 1936–   .
   G70.G4433   1999
   910′.01 – dc21   98–47914
                                                              CIP

ISBN 13: 978-0-415-18969-9 (pbk)
ISBN 13: 978-0-415-18968-2 (hbk)

Contents

Illustrations

Figures

Tables

Contributors

Liz Bondi is Senior Lecturer in the Department of Geography, University of Edinburgh, where she has been since 1985. She is founding editor (with Mona Domosh) of *Gender, Place and Culture*, and has published mainly in feminist urban studies. Her recent research concerns gender and gentrification. She is writing a text on gender, sexuality and urban environments with Hazel Christie. She is also interested in and beginning to publish on the links between psychotherapeutic practice and human geography.

Jeremy Crampton is Assistant Professor in the department of Geography and Earth Science at George Mason University. He previously held a post as Lecturer at Portsmouth University, UK. His interests include the relationship of technology and society; particularly the Internet's role in changing place and space. He has published on virtual fieldtrips, the ethics of GIS, and a guide to the World Wide Web for geographers, as well as several websites. He is on the Board of Advisors of the Virtual Geography Department.

Michael R. Curry is Professor in the Department of Geography at the University of California, Los Angeles. Dr Curry holds degrees in liberal arts, philosophy, and geography. His research concerns the development of and interactions among geographic ideas (space, place, nature); geographic technologies (geographic information systems, the written work, the map); the structure of the discipline of geography; and the broader social, cultural, and legal contexts within which the discipline, ideas, and technologies are situated. The author of *The Work in the World: Geographical Practice and the Written Word* (Minnesota, 1996) and *Digital Places: Living with Geographic Information Technologies* (Routledge, 1998), he is currently working on a book on privacy, property, and place.

Alice Dawson is a PhD candidate in the Geography Department at the University of North Carolina at Chapel Hill. Her primary research interests are human/animal relationships and medical geography.

Douglas Deur is a Doctoral Fellow in the Department of Geography and Anthropology at Louisiana State University. A cultural geographer, he holds advanced degrees in both geography and anthropology. His work emphasizes the cultural landscapes, political ecology, and traditional resource management of the indigenous peoples of Northwestern North America, as well as the post-contact environmental history of this region. He has published on such topics as contact period place-naming, the relationship between environmental knowledge and indigenous oral history, and pre-European traditions of estuarine plant cultivation among the Native Americans of the Northwest coast. He is currently completing the production of a multi-author volume which documents traditions of plant cultivation among Northwest coast indigenous peoples and traces the etiology of that region's "non-agricultural" designation within academic discourse.

Brendan Gleeson is a Research Fellow in the Urban Research Program, The Australian National University. His research interests center upon urban social policy, environmental policy and theory, and spatial regulation. His book, *Justice, Society and Nature* (Routledge, 1998) with Nicholas Low, won the 1998 Harold and Margaret Sprout Award of the International Studies Association (USA).

Nuala Gormley has recently completed a doctorate on "Mission and Development: Imagined Spaces for Women" at the University of Edinburgh. She has published articles about conducting fieldwork and volunteering overseas. She is currently involved in policy-related research on social exclusion in Scotland.

Seamus Grimes has been teaching geography (bilingually) for many years in the National University of Ireland, Galway. His PhD from the University of New South Wales looked at networks among Irish immigrants in Sydney. He has published extensively on social and economic aspects of development in the European context, and has focused particularly on the regional development prospects associated with the new information and communication technology. More recently his research is examining the philosophical aspects of development policy including the ideological dimension of Third World population control.

Thomas Herman is completing his PhD in geography in a joint program between San Diego State University and the University of California, Santa Barbara. His dissertation focuses on the local geographies of inner-city children and identifies ways in which children actively engage their environments in reconstructing social relations of the city. Other current research interests include the geography of public education, historical patterns of community development, and creative approaches to qualitative research. In addition, Tom enjoys spending time with his wife, playing with other people's children, and watching Formula One racing.

Sheila Hones teaches American Studies at Keisen University in Tokyo. Her work is concerned with the articulation of particular geographies in English-language texts, and focuses especially on the relationship between socially shared concepts of "the natural" and conventions in figurative language and narrative structure.

R. J. Johnston is a Professor in the School of Geographical Sciences at the University of Bristol; before going there in 1995 he held posts in Australia and New Zealand, was Professor of Geography at the University of Sheffield, and Vice-Chancellor of the University of Essex. His main research interests are in the study of elections, and he has published widely on geographical perspectives on the operation of the British electoral system.

Rob Kitchin is a Lecturer in Human Geography at the National University of Ireland, Maynooth. He has a diverse set of research interests which center on culture, space and power; disability; human spatial behavior; and cyberspace and society. He has recently published his first book *Cyberspace: The World in the Wires* (Wiley, 1998) and has two more books forthcoming: *The Cognitive Understanding of Space* (Johns Hopkins University Press) and *Conducting Research in Human Geography* (with Nick Tate; Longman). He is currently undertaking three "blue-chip" funded research projects through ESRC (geographies of violence), National Science Foundation (geographies of blindness), and the Royal Irish Academy (measuring disabling environments), and is organizing a special section for *Ethics, Place and Environment* concerning geography, disability and ethics.

Nicholas Low is an Associate Professor in the Faculty of Architecture, Building and Planning at the University of Melbourne, Australia. His research includes environmental governance and ethics, and urban theory, policy and planning. His latest book, *Justice, Society and Nature* (Routledge, 1998) with Brendan Gleeson, won the 1998 Harold and Margaret Sprout Award of the International Studies Association (USA).

Doreen J. Mattingly is an Assistant Professor of Geography and Women's Studies at San Diego State University. Her research interests include immigration, local labor markets, community development, and public education. She has forthcoming articles about paid household work (*Urban Geography*) and the politics of public education (*Historical Geography*). When not working, she swims in the ocean, dabbles in arts and crafts, and attends to the beauty and trauma of everyday life.

Caroline Rose Nagel is a post-doctoral fellow at the University of Kentucky. She recently completed her dissertation research on Arab immigrant communities in London, and is currently pursuing an interest in the reconstruction of Beirut, Lebanon.

Gearóid Ó Tuathail (Gerard Toal) is Associate Professor of Geography at Virginia Polytechnic Institute and State University. He is the author of *Critical Geopolitics* (University of Minnesota and Routledge, 1996) and an editor of *The Geopolitics Reader, Rethinking Geopolitics* and *An Unruly World? Globalization, Governance and Geography* (all published by Routledge). He is currently working on a volume entitled *Postmodern Geopolitics* and on a study of the relationship between religion and geopolitics.

James D. Proctor is currently Associate Professor in the Department of Geography at the University of California, Santa Barbara. His intellectual interests concern contemporary geographical thought, environmental ethics and epistemology, and the tensions between scientific and humanistic perspectives on nature. His research has focused on the cultural and ethical dimensions of conflicts surrounding biodiversity protection on local, regional, and international scales. He has also been involved in research and development efforts related to Internet-based means of enhancing undergraduate education. Dr Proctor serves as Editor for the Americas for *Ethics, Place and Environment.*

Paul Roebuck received degrees in philosophy and anthropology from the University of Colorado, and a degree in geography from the University of Minnesota. His academic work is centered on environmental ethics, understanding people's concepts of nature as well as their notions of subjectivity and objectivity. He has worked as a neuro-surgical technician, printers' devil, archaeologist, computer programmer and geography lecturer. He is currently working as a long-range systems planner for Douglas County, Colorado.

Robert Rundstrom (PhD Kansas, 1987), Associate Professor of Geography at the University of Oklahoma, is a cultural geographer/cartographer interested in the representation of geographical ideas in maps, GIS, and landscape. His research focuses on American Indian geographies. He has published on Inuit cartography and toponyms, the social and ethical implications of cross-cultural GIS, and Indian place-making during the 1969 Alcatraz Occupation in *Cartographica, Cartography and GIS, American Indian Culture and Research Journal,* and *Names.* His current project is a book on memories of Indian/non-Indian relations, and how they are expressed in recent monuments and memorials found in the American West.

David M. Smith is Professor of Geography at Queen Mary and Westfield College, University of London. He has published extensively on geographical aspects of inequality, human welfare and social justice. His books include *Human Geography: A Welfare Approach* (Arnold, 1977), *Geography, Inequality and Society* (Cambridge University Press, 1987) and *Geography and Social Justice* (Blackwell, 1994). His research combines theory with case studies set in Eastern Europe, Israel, South Africa and the United States. He is currently working on a book exploring a range of issues at the interface of geography and moral philosophy.

Jeremy Tasch is a PhD candidate in geography at Clark University. He is currently completing his field research on the effects of local autonomy on environmental and resource management in the Russian Far East. Other academic interests include tracing the conceptually and historically varied relations between nature and society; exploring geographic and philosophic notions of "place," and practicing a combination of traditional and computer cartography. Jeremy hopes some day to graduate and change his research focus to somewhere warm.

Yi-Fu Tuan is John K. Wright and Vilas Professor Emeritus of Geography at the University of Wisconsin-Madison. His books include *The Hydrologic Cycle and the Wisdom of God* (1968), *Topophilia* (1974), *Dominance and Affection* (1984), *The Good Life* (1986), *Morality and Imagination* (1989), *Passing Strange and Wonderful* (1993), *Cosmos and Hearth* (1996), and *Escapism* (1998). He is currently working on an experimental-intellectual auto-biography.

Tim Unwin is Reader in Geography and Head of Department at Royal Holloway, University of London. His current research focuses on rural restructuring in Europe, with particular reference to the Baltic States and Iberia. He also has a long established research interest on the links between geography and moral philosophy. His recent books include his edited *A European Geography* (Addison, Wesley Longman, 1998), and *Environmental Management: Readings and Case Studies* (Blackwell, 1997) edited with Lewis Owen. He is Editor of *Ethics, Place and Environment*.

Preface

Although geographers have long touched on questions related to ethics, there has been a significant recent pulse of theory and activity in this area. In the last several years, sessions related to geography and ethics have held a prominent place in numerous geography (and some philosophy) conferences, new research emphasis groups, such as the Values, Ethics, and Justice Specialty Group of the Association of American Geographers, and the Society for Philosophy and Geography, have been formed, and publications by geographers on ethics have grown exponentially – indeed, geographers have even launched a new journal, *Ethics, Place and Environment*, devoted to this overlapping terrain.

The intent of this volume is to reach out to both academics and the broader community interested in what geographers have to say about some of the profound ethical issues of our time. The volume draws together a diverse set of original essays which, individually and collectively, point to the rich interplay between geography and moral philosophy or ethics. Although its origins trace back to a number of intellectual concerns in geography (as reviewed in the Introduction), its immediate antecedent was an initiative known as the Geography/ Ethics Project (GEP), whose participants exchanged essays and ideas during 1996 and 1997, culminating in an all-day workshop held at the Annual Meeting of the Association of American Geographers held in Fort Worth, Texas in April 1997. The notion of this volume arose from that workshop; we subsequently appealed both to GEP participants and geographers throughout the English-speaking world (a regrettable practical constraint) to contribute original essays exploring the place of ethics in geography, of geography in ethics, and/or the ways in which ethics mattered to them intellectually and personally. The result, we trust you will discover, suggests the important voice offered by the geographical perspective on theoretical and practical issues of ethics.

The editors would like to extend appreciation to all geographers whose work has urged us to produce this volume, and especially to its essayists for their hard work on such a relatively short production schedule – our aim was high and our timetable was demanding! We also appreciate the nurturing attitude of Sarah Carty and Sarah Lloyd at Routledge, who gave this volume such strong support throughout its development, and Eric West and two anonymous reviewers, who each took time to provide feedback on the entire volume. David Smith's contribution has been assisted by a Leverhulme Fellowship to support his research on moral thinking in human geography, and by sabbatical leave granted by Queen Mary and Westfield College, University of London, for the year 1997. James Proctor gratefully acknowledges support from National Science Foundation grant No. SBR-9600985. Both of us thank various departmental colleagues for their interest as our research took this new direction.

1 Introduction
Overlapping terrains

James D. Proctor

> "Well?" said the geographer expectantly.
> "Oh, where I live," said the little prince, "it is not very interesting. It is all so small. I have three volcanoes. Two volcanoes are active and the other is extinct. But one never knows."
> "One never knows," said the geographer.
> "I have also a flower."
> "We do not record flowers," said the geographer.
> "Why is that? The flower is the most beautiful thing on my planet!"
> "We do not record them," said the geographer, "because they are ephemeral."
>
> (Antoine de Saint-Exupéry, *The Little Prince*, 1943: 53–54)

Geography? Ethics?

If geography were accurately represented by the man the Little Prince encountered on the sixth planet of his galactic journey, if indeed the one thing the child cared most about meant nothing to the geographer, then this volume would never have come to be. I grew up, as perhaps did many readers, with a sense of geography as the one subject most to be avoided. "A geographer," the man of the sixth planet explained to the Little Prince, "is a scholar who knows the location of all the seas, rivers, towns, mountains, and deserts," or, to quote from a t-shirt I have stuffed in my drawer, "Geography is where it's at." This is not the subject matter most of us would consider to be extremely compelling, intellectually, morally, or otherwise, and thank goodness there is more to say from the perspective of geography than that location counts.

This volume is dedicated to what geographers have to say about ethics – another field of intellectual inquiry ripe with potential for misconstrual. Ethics is often held to be a hopelessly abstract and speculative field, one as impractical as it is incomprehensible, of interest only to scholars paid to think thoughts bearing little connection to reality outside of the ivory tower. As Peter Singer argues in his preface to a recent multi-authored overview of the subject:

> It is vital that ethics not be treated as something remote, to be studied only by scholars locked away in universities. Ethics deals with values, with good

and bad, with right and wrong. We cannot avoid involvement in ethics, for what we do – and what we don't do – is always a possible subject of ethical evaluation. Anyone who thinks about what he or she ought to do is, consciously or unconsciously, involved in ethics.

(Singer 1993: v)

If ethics is not necessarily limited to intellectual abstraction on the one hand, neither is it necessarily dominated by moral evangelism on the other. The very term "ethics" conjures up for many the specter of strident declarations of right and wrong, of facile moral judgment, or even worse, of cloaking the realms of power in moral drapery, as charged by Marx in his famous dismissal of morality as ideology (Wood 1993).

This volume speaks to the possibility of creating a space for ethics somewhere other than that inhabited by the out-of-touch scholar and the "in-your-face" evangelist. Its motivation lies in the important work geographers are doing that explores ethics from the diverse perspectives that constitute contemporary geography (for recent reviews, see Smith 1997, 1998a; Proctor 1998a). The subtitle, *Journeys in a Moral Terrain*, suggests both the geographical grounding of these essays and their inherent pluralism: there is no one journey, no final word possible on the relationship between ethics and the geography of our lives. Our hope here is that, rather than closing exploration of this overlapping terrain, something of the richness and relevance of geography and ethics emerges as an inspiration for further work in this area.

The purpose of this brief introduction is to sketch a space for geography and ethics that minimally avoids the misunderstood identities of each, and ideally suggests something to the reader of their intimate relation. I will begin with a clarification of ethics, then propose a conceptual framework for geography and ethics that addresses the dual nature of geographical practice as ontological project (the realities geographers seek to elucidate) and epistemological process (the means of knowledge generation by which geographers represent these realities). I close with an overview of the structure, emphasis, and limitations of this volume and its eighteen constituent essays.

Ethics

Most people place the term ethics in the same category as values and morality; some clarification and differentiation of the three may be helpful at the outset. The term *values* is the kind of word everyone understands but few have examined carefully. Its common usage often involves idealistic and static/atomistic connotations – namely, that values "guide" actions (a form of idealistic reductionism: ideas determine practice) and that values are things (rather than processes) that exist primarily at the level of individual persons. Its reputation in social research ranges from its outright ostracism as the polluter of sound factual knowledge to its elevation in status as the ultimate determinant of what people believe to be facts (Outhwaite 1993). Nonetheless, in spite of its problematic tendencies, the term values points to a whole realm of concerns that somehow never get mentioned in scholarly discourse (as feminists have long reminded us), often due to their highly personal and political implications, and of course their assumed polarity with "facts." For this reason alone I find the term a useful corrective in an intellectual climate that largely exists in denial of this realm.

Morality is often used in a repressive Victorian sense in Western society, implying sexual taboos and the like; but this is a severely myopic definition. The Latin root of morality is also found in the word mores, meaning the manner, customs, or conduct of a particular society. Morality thus refers in a very general sense to standards of conduct by which human action is judged right or wrong in an absolute sense, or better or worse in a relative sense. Yet right/wrong or better/ worse decisions affect a wide range of scales in our lives, running for example from deliberations over capital punishment to decisions concerning how one should dress for the day. The difficulty in weeding out relatively trivial decisions from the weightier ones leads some philosophers such as Feldman (1978) to consider morality to be a difficult term to bound, though most of us would accept that matters of prudence and etiquette are different (if only in degree) from moral matters.

Morality concerns the *normative* sphere of human existence and practice, a term which (as with values) has been used in a pejorative sense in much twentieth-century social science. Under the influence of positivism, the realm of the normative has been unfavorably contrasted with positive knowledge, knowledge gained via dispassionate empirical observation in the spirit of the natural sciences (Wacquant 1993). This judgment follows from the original Comtean spirit of positivism, a theory that knowledge has evolved through earlier theological and metaphysical stages – both tangled up in unprovable normative speculation – to the contemporary positive stage. Yet, thankfully, some recent accounts (most significantly, Sayer and Storper 1997) call for a "normative turn" in social theory and social science, not so much to discard so-called "positive knowledge" as to shed intellectual light on the values that inform the object and the process of social inquiry.

What, then, is *ethics*? In science, ethics typically involves reflection upon moral questions that arise in research, publication, and other professional activities (e.g. NAS Committee on Science 1995); yet philosophical usage is broader than this prevailing scientific interpretation. Ethics, also known as moral philosophy, is commonly understood as systematic intellectual reflection on morality in general, or specific moral concerns in particular. The former can be called theoretical ethics and the latter applied ethics, though the two are closely related. One realm of applied ethics that has garnered considerable attention outside of philosophy focuses on professional conduct; thus the moral questions asked in the fields of science, law, medicine, and business are common examples of ethical inquiry.

Another distinction is typically drawn between descriptive ethics, normative ethics, and metaethics (though only the latter two are represented in philosophical literature). The aim of descriptive ethics is to characterize existing moral schemes; this has been an important feature of, for instance, cultural anthropology, which in so doing has raised the problem of relativism (Benedict 1934; Geertz 1989). Normative ethics is devoted to constructing a suitable moral basis to inform human conduct; contemporary examples include Rawls' theory of justice (Rawls 1971) and, in a quite different approach, the contrasting ethics of care proposed by feminists building upon psychologist Carol Gilligan's pioneering work (1982). Metaethics, in distinction, is more an examination of the characteristics of ethical reasoning or systems of ethics. A classic metaethical problem, as exemplified in David Hume's is–ought dichotomy (Hume 1978), concerns the relationship between facts (descriptive statements) and values (normative statements); this problem has been a major concern of, for instance, twentieth-century social theory (O'Neill 1993).

Much work in Western ethics is thus derived from the way in which moral philosophy has developed. For instance, one major theme to which many theoretical discussions – primarily normative but also metaethical – have returned involves the relationship between the *right* and the *good*. While the right corresponds to a particular act or intent, the good implies rather the end or justification for a particular act or intent. These terms are of primary significance in Western ethics in that they correspond to the two major classes of moral theories: teleological theories such as utilitarianism, where the good is the primary concern, and deontological theories, where the right becomes a more paramount concern (for introductory discussion, see Davis 1993: 206ff.; Goodin 1993: 241). Recent developments in Western ethics are many (see, for instance, the online Ethics Updates site at http://ethics.acusd.edu): important examples include feminist and postmodernist/post-structuralist approaches, which have critically reexamined, though in important ways also extended, this heritage (e.g. Benhabib 1992; Bauman 1994). The result is that those interested in ethical reflection have, perhaps more so than in any previous era, a greatly enriched conceptual vocabulary to draw upon.

This heritage of thought on ethics may sound intellectually formidable; yet ethics is too important to be left to the moral philosopher. Perhaps the most important step in doing ethics simply involves asking questions such as "How is it that people say this is a bad thing?", or "Why do I feel I am right in doing this?" This is something we can all do, whether or not we are versed in virtue theory or Kantian deontology. Many areas of our lives – our jobs, our hobbies, our family and social relationships – are treated as ethically unproblematic, and thus gain *de facto* moral legitimacy (i.e. a thing is *right* because it *is*) precisely because we do not ask the question. In its best sense, then, ethics becomes a practice of consistent (hopefully not neurotic!) moral reflection, turned both inward and outward.

The geographical perspective: a framework

Geography and ethics: process and substance

The confluence of geography and ethics represents no radical recent turn of events: one need only go back to Immanuel Kant to find "moral geography" proposed as a major subdivision of the discipline, focusing on "diverse customs and characteristics of people of different regions" (May 1970: 263). Yet Kant would never have imagined the range and depth of philosophical questions geographers have explored in the last century (Johnston 1986; Livingstone 1992; Buttimer 1993), and geographical engagement with philosophical issues touching on ethics (e.g. social justice and related concepts – see Smith 1994; Hay 1995) has grown tremendously in recent decades.

A glimpse of contemporary work by geographers on ethics is impressive. Two examples of recent monographs include David Harvey's *Justice, Nature and the Geography of Difference* (1996), which offers a materialist and geographically-situated grounding of environmental and political values and attempts a rapprochement of social justice and environmental concern; and Robert David Sack's ambitious *Homo Geographicus* (1997), which grounds human existence in geography (hence the title) and ultimately moves toward a geographical framework for morality. Or, consider a recent issue of *Environment and Planning D: Society and Space* (volume 15 no. 1, 1997) devoted explicitly to the re-engagement of social science in general, and geography in particular, with moral theory. As examples of the important contributions this issue makes, Andrew Sayer and Michael Storper argue convincingly for a greater level of normative self-criticality among geographers and social scientists, given the moral propositions they often unreflectively deploy (Sayer and Storper 1997), and Sarah Whatmore revisits mainstream approaches to social and environmental ethics, offering a feminist and geographically-informed relational theory of ethics based on a reconfiguration of the self (Whatmore 1997). Indeed, an entire journal, *Ethics, Place and Environment*, has recently been launched by an international body of geographers, devoted broadly to geographical and environmental dimensions of ethics.

How did this burst of geographical scholarship on ethics arise? One clear antecedent is a broader interest dating back several decades among geographers on values (Buttimer 1974), in large part a response to the professed value-neutrality of the burgeoning quantitative approach in geography, with its emphasis on objectivist spatial analysis (Billinge, Gregory and Martin 1984; Cosgrove 1989). Values-based concerns among geographers have underscored diverse political struggles in the discipline, including those calling for greater relevance in research (Mitchell and Draper 1982), more explicitly critical theoretical approaches (e.g. Peet and Thrift 1989), and the inclusion of women in general, and feminist perspectives in particular, in the practice and substantive emphasis of geography (Rose 1993).

Two paths distinguish geographical engagement with ethics: the first attends to the process of doing geography and is broadly similar to professional ethics; the second the substance of geographical inquiry and is more akin to theoretical ethics. These paths are intimately related, as the former represents the context out of which the content, the result, of substantive ethics emerges. In this paired approach, geographers point to a manner of being properly reflexive in the moral statements they make about the world without getting lost in this reflexivity to the point that they cannot speak anything of substance.

This twofold approach suggests one possible conceptual framework for a geographical perspective on ethics. As with other academic disciplines, geography is in large part a knowledge-building enterprise consisting of two major components: its *ontological project* and its *epistemological process*. Geography's ontological project is, simply, to make sense of those aspects of reality (thus "ontology," a term referring to being or reality) historically engaged in geographical analysis. Much of geography's ontological project is bound up in specific metaphors used to organize reality; for convenience I will adopt the common threesome of *space, place,* and *nature* as the interweaving metaphors informing the geographical imagination (e.g. Gregory 1994: 217). Space is the metaphor underlying a good deal of geography's ontological project, including emphases as disparate as spatial science and Marxist critique. The metaphor of place is prominent in more humanistic and interpretive work in geography; it speaks of a reality as lived and understood by active human subjects. The metaphor of nature underlies physical geography and geography of the society–nature tradition. Though these three metaphors are by no means comprehensive, they do suggest the different ways in which geography proceeds in its project of making sense of reality.

Geography accomplishes this ontological project via an epistemological process; knowledge of space, place, and nature do not arise from thin air. This is the manner in which professional and substantive ethics in geography are connected, as process and product, context and content, are not comprehensible outside of the other. Yet the epistemological process of geography is far broader than what is typically subsumed under the category of "professional ethics." Minimally, this process involves a set of *guiding concepts* implemented via research and analytical *techniques* to generate knowledge, which has a certain form of *representation* and leads to specific social and other *implications*. Guiding concepts include the metaphors of reality discussed above, which play an important general role in the constitution and reconstitution of geography's identity and thus provide a delimited range of appropriate inquiry in geographical research. Guiding concepts also include philosophical commitments as to how knowledge is to be produced and what kind of knowledge is worth producing, other important components of the constitution of geography. Research and analytical techniques are more specific and include methods of data collection and analysis, such as qualitative interviews, field reconnaissance, Geographic Information Systems (GIS)-based spatial modeling, and so forth. Representation of research results by geographers commonly includes mapping and writing, though other forms of representation are possible as well. Implications, whether intentional or unintentional, follow from the production of geographical knowledge; these may touch upon social, environmental, political, intellectual, and/or other worlds.

Ethics and geography's ontological project

The metaphor of space provides perhaps the most familiar entry of geographers into substantive questions of ethics. Indeed, one of the strongest areas of attention among geographers has concerned spatial dimensions of social justice (Harvey 1973, 1993; Smith 1994; Gleeson 1996). This work builds on geo-graphical analyses of spatial exclusion and control (Ogborn and Philo 1994; Sibley 1995), and considers questions such as geographical perspectives on some of its major philosophical figures (Clark 1986), professional and personal responsibilities to spatially distant and less powerful others (Corbridge 1993, 1998), immigration and social justice (Black 1996), and territorial justice (Boyne and Powell 1991).

Work by geographers on social justice is not, however, limited to its spatial dimensions. Geographers are, for instance, devoting increasing attention to environmental racism and justice, bridging the social justice paradigm to the metaphor of nature. Though contributions by geographers are barely evident in recent anthologies (e.g. Bryant 1995; Westra and Wenz 1995), an upswing of book-length publications (Pulido 1996; Low and Gleeson 1998), articles in mainstream journals (Bowen *et al.* 1995), and indeed whole issues of geo-graphical journals (see, for instance, *Antipode* 28(2), *Urban Geography* 17(5)), attest to its burgeoning significance.

As another example of this interweaving of metaphors, David Smith has recently posed the question, "How far should we care?" (Smith 1998b), in an effort to work through the dual perspectives of ethics as spatial justice, where principles of indifference and universality are prioritized, and ethics as care, a relationally-based ethics where one's families, communities, and other social groups of relational significance are the primary emphasis, where ethics and par-tiality, morality and passion, are not polar opposites. Smith's question clearly considers on the plane of ethics what many others have considered on the plane of epistemology: the tension between the objectivist, rationalistic metaphor of space, and the explicitly perspectival, embodied metaphor of place (Tuan 1977; Buttimer and Seamon 1980; Entrikin 1991; Sack 1992).

Place is, of course, already a significant category in the works of Sibley and others noted above. It is perhaps best exemplified, however, in work on "moral geographies," which could loosely be translated as thick descriptions of the moral features of place. To call this work "descriptive ethics" is missing something, however, as place-based ethical inquiry may be closer to the mark of understand-ing human morality than its placeless equivalents commonly in abundance in more abstract normative and metaethical inquiry (Walzer 1994; O'Neill 1996: 68). Indeed, though geographical work in moral geographies and other questions of ethics has shied away from an explicitly normative and/or metaethical focus, the fact that geographers have attended to questions such as universalist versus particularist ethics suggests the relative ease with which place-based geographical analysis lends itself to addressing these more abstract issues (e.g. Corbridge 1993).

The concept of place itself has been invoked by geographers in order to reflect critically on the problematic objectification of subjective community or regional values (Entrikin 1991: 60–83), as well as to ground the moral context of production and consumption in advanced industrial societies (Sack 1992: 177–205). Indeed, the moral realm is deeply implicated in the work of many humanist geographers on place – of which the example of Yi-Fu Tuan is perhaps most prominent (Tuan 1974, 1989, 1993). But the sheer range of recent work on moral geographies makes the important collective point that the diverse places geographers study are inescapably normative, that normativity is not so much something to be added on to place as to be teased out of it. Some instances of this work include the explorations of Jackson and others on moral order in the city (Jackson 1984; Driver 1988), "moral locations" of nineteenth-century Portsmouth (Ogborn and Philo 1994), the moral geography of reformatories (Ploszajska 1994), the moral geography of the Norfolk Broadlands (Matless 1994), the moral discourse of climate (Livingstone 1991), and the "moral geography of the everyday" (Birdsall 1996). Though the term has had some use outside of geography (e.g. Shapiro 1994; cf. Slater 1997 for a related geographical perspective), it would be a gross overstatement to suggest that, by means of moral geographies, geographers have made their indelible mark on how ethics ought to be encountered. Though Kant would perhaps be perplexed at this outcome, geographers are rather used to intellectual anonymity; the question is whether the important voice geographers have to add on the ethics of place will be heard outside of the discipline.

The metaphor of nature (understood as biophysical environment) is evident in much of what was presented above, but as a primary focus of ethical interest among geographers it has not enjoyed such diffuse attention as social justice and moral geographies. One important reason is that the vast majority of work by geographers under this metaphorical trajectory is largely physical and life science-based, and as such rarely if ever entertains questions of human ethics (one important recent exception being a forum on ethics in environmental science in *Annals of the Association of American Geographers* (88: 2)). Is this lack of attention by physical geographers to ethics justified? At the level of their immediate topics of interest, perhaps: fluvial geomorphology and microclimatology involve processes that have important human impacts and arise in part from human drivers, but in and of themselves there is arguably little ground for ethical reflection. Yet the historical process by which science decoupled from explicit attention to morality is well-rehearsed elsewhere, and as such suggests that this immediate detachment of physical geography from ethics is as much a particular historical result as some inevitable corollary of its subject-matter.

Nonetheless, there has been a rising interest among geographers in environmental ethics (Proctor 1998c). In addition to the literature cited above on environmental racism and justice, there is ample supplemental evidence of this interest. The inaugural issue of *Philosophy and Geography*, for instance, was devoted to environmental ethics (Light and Smith 1997). Whole books are now arising which engage with questions of nature and morality in significant ways (Simmons 1993; Harvey 1996).

Work in this area is predictably diffuse, though not at all limited to the recent past, as suggested for instance in the writings of Reclus (Clark 1997). Some geographers have situated questions of environmental ethics in the context of culturally-based ideas of nature (Simmons 1993), while others have discussed the spatial scale dependency of optimal formulations of environmental ethics (Reed and Slaymaker 1993), and still others have critically reviewed the values underlying environmental movements (O'Riordan 1981; Lewis 1992), at times rejecting them in favor of less sociopolitically naive alternatives (Pepper 1993). Some have looked at environmental ethics from a cross-cultural perspective (Wescoat 1997), while others have engaged with the modernist and anti-modernist underpinnings of Western environmental thought (Gandy 1997). Indeed, the diverse linkages geographers have drawn between social theory and environmental ideology and ethics (Proctor 1995; Gandy 1996) are broadly suggestive of the important contributions geographers can make.

As suggested above, perhaps the most interesting substantive work by geographers on ethics transcends the boundaries between the metaphors of space, place, and nature. Indeed, the key contribution geographers have to make arises from the diverse metaphors of reality they invoke; hence critical tensions between universals and difference, justice and care, can be thoughtfully entertained by geographers, given the solid establishment of the discipline upon the metaphors of space and place. This strength in metaphorical diversity is also evident in the contribution geographers can make to environmental ethics; here, for instance, the problem of how to resolve conflicts between social and natural goods can be meaningfully addressed, as geographers have a foot planted in both nature and culture. The diversity of geographical imaginations cast upon this world thus offers an important point of beginning for geographers to make a real contribution to moral discourse.

Ethics and geography's epistemological process

Geographical knowledge does not arise in a vacuum. The statements geographers make about space, place and nature come out of a particular process, of which four sequential steps were noted above. The first step, guiding concepts, draws upon the metaphors that inform geography's ontological project, as well as intrinsic or extrinsic epistemological rules (e.g. universalizability or the lack thereof) that govern the application of these metaphors to knowledge-building. This discussion is well-rehearsed in the literature: the critique of positivism over the last several decades, for instance, is in large part a critique of how particular ontological and epistemological assumptions associated with positivism have constrained the kinds and implications of knowledge arising from geographical research (Gregory 1979). Though this critical literature does not go by the self-ascription of "ethics," nonetheless its reasoned normative pronouncements are of similar intent. Further inquiry into the ways in which basic ontological and epistemological assumptions shape geographical research in ethically significant ways is needed.

One of the most familiar areas of ethical inquiry in geography involves research and analytical techniques, ranging from cartography (Harley 1991; Monmonier 1991; Rundstrom 1993) to remote sensing and geographic information systems (Wasowski 1991; Lake 1993; Curry 1994; Crampton 1995). The act of research itself, and the consideration of the role of the researcher vis-à-vis the research subject(s), has also been a popular subject of inquiry (e.g. Eyles and Smith 1988; England 1994). Another area where important work has been done concerns how geographical knowledge is represented, in realms ranging from cartography (see above) to academic publication (Brunn 1989; Curry 1991) to education (Havelberg 1990; Kirby 1991; Smith 1995). Less work has considered implications of geographical research, though explicit attention has been paid to areas with direct social significance, such as planning (Entrikin 1994), and it should be noted that some of the most provocative publications by geographers have taken seriously the implications of geographical research as a starting-point for reconfiguring geography (e.g. Kropotkin 1885; Harvey 1974). Indeed, ethical issues become more focused as one moves from a particular geographical concept to its technical implementation and finally to its application. For instance, conceiving space as an isotropic surface appears innocent enough until one builds a GIS upon this naive assumption for the purpose of, say, specifying social service facility location. This example suggests also the interrelation of ethical issues across the continuum of geography's epistemological process, and points out the severe limitations in a "professional ethics" circumscribed solely to questions of research data and publication (though see Brunn 1998; Hay 1998, for recent theoretically and historically rich accounts of professional ethics in geography).

The essays

What follows includes eighteen original essays contributed by geographers on ethics. The volume is structured into the themes of space, place, nature, and knowledge following the framework above; each section is preceded by a short introductory summary of the component essays and some major issues they jointly raise. The volume ends with a conclusion by David M. Smith which offers further reflections on these four themes, linking them to related issues in geography and ethics.

Yet, as already noted, what is interesting about geographical work is often the ways in which these themes are joined – indeed, some would argue that this is a defining feature of geography. The inherent multidimensionality of the essays comprising this volume indeed presented the editors with an organizational challenge: do we categorize them *a posteriori* according to the prevalent theme they address, or do we simply present them as they are, without some imposed (and admittedly modernist) Procrustean structure? We have chosen the former approach as a manner of providing some conceptual clarification on the potentially limitless ways in which geography and ethics interweave, as well as to suggest the flavor and diversity of each of these themes as exemplified in their representative essays. Some clearly focus on their given theme; others invoke it only implicitly. Some themes (e.g. the production of geographical knowledge) are "tighter" than others. We assume the reader will not expect an overly tidy match between essay and theme: we gave each author free rein to explore the confluence of geography and ethics, and thus all stand on their own as well as contribute to a given thematic conversation. The real work of real geographers is far too many-layered to collapse entirely onto such a simple rubric.

There are other commonalities among essays the reader will detect that are not emphasized in this fourfold structure. Some overarching commonalities combine geographical themes: hence, for example, the tension between space and place comes out as a tension between universalism and particularism, thin and thick moralities, justice and caring. Other commonalities speak more to the broad relationship between ethics, reality and knowledge. For instance, the well-known "is–ought" problem surfaces in discussions of ethnicity and morality, the natural as good, and knowledge as power; and some essays speak explicitly of the onto-logical and epistemological embeddedness of ethics. Some commonalities involve method: many essays adopt a case method of argumentation, whereas others proceed more in the abstract. And, of course, there is a strong resonance among all essays as to the geographical embeddedness of ethics, an argument made implicitly or explicitly that geography matters in finding clarifications of, or solutions to, ethical questions. The most accurate organizational motif for these essays would thus probably be some sort of analogue to hyperspace, in which each essay were linked to essays related to it in all the ways noted above; this approach, however, is clearly far more suited to electronic than hard-copy publication.

The volume's diverse essays speak not only to a multitude of ways to consider ethics from geographical perspectives; they also speak to some flexibility in what ethics is all about. Here, several key tensions are important. One tension – particularly exemplified in comparing the essays on knowledge and ethics with the other essays – concerns the difference between ethics as "thinking about caring" and ethics as "caring": in the first sense (much as ethics was defined above), being ethical involves intellectual reflection on moral matters, and in the second, being ethical involves doing the right (at least the best possible) thing. The net effect of these essays is that both are honored as key in any authentic project of ethics: an excess of thinking-without-doing, or an excess of caring-without-thinking, would otherwise result. Another key tension (highlighted especially in the important essay by Ó Tuathail) considers the valence of any project labeled "ethics": are its moral implications ultimately positive or, viewed in a far more negative sense, can projects of this nature impose some political project, veiled in moral guise, upon others? Though most will probably agree that the kinds of moral reflection exemplified in these essays are by and large positive in their implications, any project of ethics such as that comprising this volume that is mindless of its potential coercive power is dangerous.

A final, important, note. There are many fine authors and ideas that made this volume; there are many more that did not. We regret, for instance, the non-participation of physical geographers in our volume, yet there is no necessary reason why they should be excluded (Proctor 1998b); and though we made a point of encouraging balance of gender, seniority, subspecialty, philosophical predisposition, and other differences among us, still many sectors are inadequately represented. The editors particularly regret the absence of a chapter with an explicitly feminist stance, but take comfort from the reflection of feminist perspectives in some of the essays as well as in the Conclusion, and from the focus on feminist ethics in other recent geographical publications. Ethics in geography is simply too rich at this moment to be fully captured in one volume, for the simple reason that geography is far richer than that suggested by its representative on that distant planet visited by the Little Prince on his journey. If this volume is a testimony to that fact, it is also in its finitude an indication of the rich work yet to be written as geographers continue their journeys on this moral terrain.

Acknowledgments

I acknowledge the helpful comments of Caroline Nagel, David M. Smith, and two anonymous reviewers on an earlier draft. I also acknowledge kind permission from the Royal Geographical Society (with the Institute of British Geographers) to use portions of a recently-published review article on geography and ethics in this essay (Proctor 1998a).

References

Bauman, Z. (1994) *Postmodern Ethics*, Oxford: Blackwell Publishers.
Benedict, R. (1934) "A defense of moral relativism", *Journal of General Psychology* 10: 59–82.
Benhabib, S. (1992) *Situating the Self: Gender, Community and Postmodernism in Contemporary Ethics*, Cambridge: Polity Press.
Billinge, M., Gregory, D. and Martin, R. L. (1984) *Recollections of a Revolution: Geography as Spatial Science*, New York: St Martin's Press.
Birdsall, S. S. (1996) "Regard, respect, and responsibility: Sketches for a moral geography of the everyday", *Annals of the Association of American Geographers* 86: 619–629.
Black, R. (1996) "Immigration and social justice: Towards a progressive European immigration policy?", *Transactions of the Institute of British Geographers* 21: 64–75.
Bowen, W. M., Salling, M. J., Haynes, K. E. and Cyran, E. J. (1995) "Toward environmental justice: Spatial equity in Ohio and Cleveland", *Annals of the Association of American Geographers* 85: 641–663.
Boyne, G. and Powell, M. (1991) "Territorial justice: A review of theory and evidence", *Political Geography Quarterly* 10: 263–281.
Brunn, S. D. (1989) "Editorial: Ethics in word and deed", *Annals of the Association of American Geographers* 79: iii–iv.
—— (1998) "Issues of social relevance raised by presidents of the Association of

American Geographers: The first fifty years", *Ethics, Place, and Environment* 1: 77–92.

Bryant, B. (ed.) (1995) *Environmental Justice: Issues, Policies, and Solutions*, Washington, DC: Island Press.

Buttimer, A. (1974) "Values in geography", Association of American Geographers Resource Paper no. 24.

—— (1993) *Geography and the human spirit*, Baltimore: The Johns Hopkins University Press.

Buttimer, A. and Seamon, D. (1980) *The Human Experience of Space and Place*, New York: St Martin's Press.

Clark, G. L. (1986) "Making moral landscapes: John Rawls' original position", *Political Geography Quarterly Supplement* 5: 147–162.

Clark, J. (1997) "The dialectical social geography of Elisée Reclus" in A. Light and J. M. Smith (eds) *Philosophy and Geography I: Space, Place, and Environmental Ethics*, Lanham, Maryland: Rowman and Littlefield Publishers Inc.

Corbridge, S. (1993) "Marxisms, modernities and moralities: Development praxis and the claims of distant strangers", *Environment and Planning D: Society and Space* 11: 449–472.

—— (1998) "Development ethics: Distance, difference, plausibility", *Ethics, Place, and Environment* 1: 35–54.

Cosgrove, D. (1989) "Models, description and imagination in geography" in B. MacMillan (ed.) *Remodelling Geography*, Oxford: Blackwell Publishers.

Crampton, J. (1995) "The ethics of GIS", *Cartography and Geographic Information Systems* 22: 84–89.

Curry, M. R. (1991) "On the possibility of ethics in geography: Writing, citing, and the construction of intellectual property", *Progress in Human Geography* 15: 125–147.

—— (1994) "Geographic information systems and the inevitability of ethical inconsistency" in J. Pickles (ed.) *Ground Truth*, New York: Guilford Press.

Davis, N. A. (1993) "Contemporary deontology" in P. Singer (ed.) *A Companion to Ethics*, Oxford: Blackwell Publishers.

de Saint-Exupéry, A. (1943) *The Little Prince*, New York: Harcourt, Brace and World, Inc.

Driver, F. (1988) "Moral geographies: Social science and the urban environment in mid-nineteenth century England", *Transactions of the Institute of British Geographers* 13: 275–287.

England, K. V. L. (1994) "Getting personal: reflexivity, positionality, and feminist research", *Professional Geographer* 46: 80–89.

Entrikin, J. N. (1991) *The Betweenness of Place: Towards a Geography of Modernity*, Baltimore: The Johns Hopkins University Press.

—— (1994) "Moral geographies: The planner in place", *Geography Research Forum* 14: 113–119.

Eyles, J. and Smith, D. M. (eds) (1988) *Qualitative Methods in Human Geography*, Oxford: Polity Press.

Feldman, F. (1978) *Introductory Ethics*, Englewood Cliffs, New Jersey: Prentice-Hall.

Gandy, M. (1996) "Crumbling land: The postmodernity debate and the analysis of environmental problems", *Progress in Human Geography* 20: 23–40.

—— (1997) "Ecology, modernity, and the intellectual legacy of the Frankfurt School", in A. Light and J. M. Smith (eds) *Philosophy and Geography I: Space, Place,* and *Environmental Ethics*, Lanham, Maryland: Rowman and Littlefield Publishers Inc.

Geertz, C. (1989) "Anti anti-relativism" in M. Krausz (ed.) *Relativism*, Notre Dame, Indiana: University of Notre Dame Press.

Gilligan, C. (1982) *In a Different Voice*, Cambridge, Mass.: Harvard University Press.

Gleeson, B. (1996) "Justifying justice", *Area* 28: 229–234.

Goodin, R. E. (1993) "Utility and the good" in P. Singer (ed.) *A Companion to Ethics*, Oxford: Blackwell Publishers.

Gregory, D. (1979) *Ideology, Science, and Human Geography*, New York: St Martin's Press.

—— (1994) "Geographical imagination" in R. J. Johnston, D. Gregory and D. M. Smith (eds) *The Dictionary of Human Geography*, Oxford: Blackwell Publishers.

Harley, J. B. (1991) "Can there be a cartographic ethics?", *Cartographic Perspectives* 10: 9–16.

Harvey, D. (1973) *Social Justice and the City*, London: Edwin Arnold.

—— (1974) "What kind of geography for what kind of public policy?", *Transactions of the Institute of British Geographers* 63: 18–24.

—— (1993) "Class relations, social justice and the politics of difference" in M. Keith and S. Pile (eds) *Place and the Politics of Identity*, New York: Routledge.

—— (1996) *Justice, Nature and the Geography of Difference*, Oxford: Blackwell Publishers.

Havelberg, G. (1990) "Ethics as an educational aim in geography teaching", *Geographie und Schule* 12: 5–15.

Hay, A. M. (1995) "Concepts of equity, fairness and justice in geographical studies", *Transactions of the Institute of British Geographers* 20: 500–508.

Hay, I. (1998) "Making moral imaginations. Research ethics, pedagogy, and professional human geography", *Ethics, Place, and Environment* 1: 55–76.

Hume, D. (1978) *A Treatise of Human Nature*, Oxford: Oxford University Press.

Jackson, P. (1984) "Social disorganization and moral order in the city", *Transactions of the Institute of British Geographers* 9: 168–180.

Johnston, R. J. (1986) *Philosophy and Human Geography: An Introduction to Contemporary Approaches*, London: Edward Arnold.

Kirby, A. (1991) "On ethics and power in higher education", *Journal of Geography in Higher Education* 15: 75–77.

Kropotkin, P. (1885) "What geography ought to be", *The Nineteenth Century* 18: 940–956.

Lake, R. W. (1993) "Planning and applied geography: positivism, ethics, and geographic information systems", *Progress in Human Geography* 17: 404–413.

Lewis, M. W. (1992) *Green Delusions: An Environmentalist Critique of Radical Environmentalism*, Durham: Duke University Press.

Light, A., and Smith, J. M. (eds) (1997) *Philosophy and Geography I: Space, Place, and Environmental Ethics*, Lanham, Maryland: Rowman and Littlefield Publishers Inc.

Livingstone, D. (1991) "The moral discourse of climate: Historical considerations on race, place and virtue", *Journal of Historical Geography* 17: 413–434.

—— (1992) *The Geographical Tradition: Episodes in the History of a Contested Discipline*, Oxford: Blackwell Publishers.

Low, N. and Gleeson, B. (1998) *Justice, Society and Nature: An Exploration of Political Ecology*, London: Routledge.

Matless, D. (1994) "Moral geography in broadland", *Ecumene* 1: 127–156.

May, J. A. (1970) *Kant's Concept of Geography and its Relation to Recent Geographical Thought*, Toronto: University of Toronto Press.

Mitchell, B. and Draper, D. (1982) *Relevance and Ethics in Geography*, London: Longman.

Monmonier, M. (1991) "Ethics and map design. Six strategies for confronting the traditional one-map solution", *Cartographic Perspectives* 10: 3–8.

NAS Committee on Science, E., and Public Policy (1995) *On Being a Scientist: Responsible Conduct in Research*, Washington, DC: National Academy Press.

O'Neill, J. (1993) "Ethics" in W. Outhwaite and T. Bottomore (eds) *The Blackwell Dictionary of Twentieth-Century Social Thought*, Oxford: Blackwell Publishers.

O'Neill, O. (1996) *Toward Justice and Virtue: A Constructive Account of Practical Reasoning*, Cambridge: Cambridge University Press.

O'Riordan, T. (1981) *Environmentalism*, London: Pion.

Ogborn, M. and Philo, C. (1994) "Soldiers, sailors and moral locations in nineteenth-century Portsmouth", *Area* 26: 221–231.

Outhwaite, W. (1993) "Values" in W. Outhwaite and T. Bottomore (eds) *The Blackwell Dictionary of Twentieth-Century Social Thought*, Oxford: Blackwell Publishers.

Peet, R. and Thrift, N. (eds) (1989) *New Models in Geography: The Political-Economy Perspective*, London: Unwin Hyman.

Pepper, D. (1993) *Eco-Socialism: From Deep Ecology to Social Justice*, London: Routledge.

Ploszajska, T. (1994) "Moral landscapes and manipulated spaces: Gender, class and space in Victorian reformatory schools", *Journal of Historical Geography* 20: 413–429.

Proctor, J. D. (1995) "Whose nature? The contested moral terrain of ancient forests", in W. Cronon (ed.) *Uncommon Ground: Toward Reinventing Nature*, New York: W. W. Norton.

—— (1998a) "Ethics in geography: Giving moral form to the geographical imagination", *Area* 30: 8–18.

—— (1998b) "Expanding the scope of science and ethics: A response to Harman, Harrington and Cerveny's 'Balancing scientific and ethical values in environmental science'", *Annals of the Association of American Geographers* 88: 290–296.

—— (1998c) "Geography, paradox, and environmental ethics", *Progress in Human Geography* 22: 234–255.

Pulido, L. (1996) *Environmentalism and Economic Justice: Two Chicano Struggles in the Southwest*, Tucson: University of Arizona Press.

Rawls, J. (1971) *A Theory of Justice*, Cambridge, Mass.: Harvard University Press.

Reed, M. G. and Slaymaker, O. (1993) "Ethics and sustainability: a preliminary perspective", *Environment and Planning A* 25: 723–739.

Rose, G. (1993) *Feminism and Geography: The Limits of Geographical Knowledge*, Minneapolis: University of Minnesota Press.

Rundstrom, R. A. (1993) "The role of ethics, mapping, and the meaning of place in relations between Indians and whites in the United States", *Cartographica* 30: 21–28.

Sack, R. D. (1992) *Place, Modernity, and the Consumer's World: A Relational Framework for Geographical Analysis*, Baltimore: The Johns Hopkins University Press.

—— (1997) *Homo Geographicus: A Framework for Action, Awareness, and Moral Concern*, Baltimore: The Johns Hopkins University Press.

Sayer, A. and Storper, M. (1997) "Guest editorial essay – Ethics unbound: For a normative turn in social theory", *Environment and Planning D: Society and Space* 15: 1–17.

Shapiro, M. J. (1994) "Moral geographies and the ethics of post-sovereignty", *Public Culture* 6: 479–502.

Sibley, D. (1995) *Geographies of Exclusion: Society and Difference in the West*, London: Routledge.

Simmons, I. G. (1993) *Interpreting Nature: Cultural Constructions of the Environment*, London: Routledge.

Singer, P. (1993) "Introduction" in P. Singer (ed.) *A Companion to Ethics*, Oxford: Blackwell Publishers.

Slater, D. (1997) "Spatialities of power and postmodern ethics: Rethinking geopolitical encounters", *Environment and Planning D: Society and Space* 15: 55–72.

Smith, D. M. (1994) *Geography and Social Justice*, Oxford: Blackwell Publishers.

—— (1995) "Moral teaching in geography", *Journal of Geography in Higher Education* 19: 271–283.

—— (1997) "Geography and ethics: A moral turn?", *Progress in Human Geography* 21: 583–590.

—— (1998a) "Geography and moral philosophy: Some common ground", *Ethics, Place, and Environment* 1: 7–34.

—— (1998b) "How far should we care? On the spatial scope of beneficence", *Progress in Human Geography* 22: 15–38.

Tuan, Y.-F. (1974) *Topophilia: A Study of Environmental Perception, Attitudes, and Values*, Englewood Cliffs, NJ: Prentice-Hall.

—— (1977) *Space and Place: The Perspective of Experience*, Minneapolis: University of Minnesota Press.

—— (1989) *Morality and Imagination: Paradoxes of Progress*, Madison: The University of Wisconsin Press.

—— (1993) *Passing Strange and Wonderful: Aesthetics, Nature, and Culture*, Washington, DC: Island Press.

Wacquant, L. J. D. (1993) "Positivism" in W. Outhwaite and T. Bottomore (eds) *The Blackwell Dictionary of Twentieth-Century Social Thought*, Oxford: Blackwell Publishers.

Walzer, M. (1994) *Thick and Thin: Moral Argument at Home and Abroad*, Notre Dame, Ind.: University of Notre Dame Press.

Wasowski, R. J. (1991) "Some ethical aspects of international satellite remote sensing", *Photogrammetric Engineering and Remote Sensing* 57: 41–48.

Wescoat, J. L., Jr. (1997) "Muslim contributions to geography and environmental ethics: The challenges of comparison and pluralism" in A. Light and J. M. Smith (eds) *Philosophy and Geography I: Space, Place, and Environmental Ethics*, Lanham, Maryland: Rowman and Littlefield Publishers Inc.

Westra, L., Wenz, P. S. (eds) (1995) *Faces of Environmental Racism: Confronting Issues of Global Justice*, Lanham, Maryland: Rowman and Littlefield Publishers, Inc.

Whatmore, S. (1997) "Dissecting the autonomous self: Hybrid cartographies for a relational ethics", *Environment and Planning D: Society and Space* 15: 37–53.

Wood, A. (1993) "Marx against morality" in P. Singer (ed.) *A Companion to Ethics*, Oxford: Blackwell Publishers.

Part 1
Ethics and space

Space is at once the most obvious and subtle of geographical concerns. It is implicated in ethics in many ways, including discussions of unevenness of rights or goods, in the moral ramifications of particular configurations of spatial relations, and ultimately in our very conceptualization of space itself. Thus, to arrange a selection of essays under the bald heading of 'ethics and space' may raise more expectations than the resolutions offered. However, our contributors open up some salient issues, if well short of the comprehensive and fully coherent treatment which further work may eventually provide.

Paul Roebuck's broad essay on geography and ethics leads into the section's emphasis on universalist versus particularist concepts of space by contrasting objectivist and relational conceptions of knowledge, as a fundamental dualism opened up within the discipline of geography. His essay is thus placed at the beginning of the volume because it alludes to connections between this section and the final section on ethics and knowledge. Roebuck refers to the frameworks which humans impose on their world, to try to make sense of it. The Enlightenment framework sought the universal, in contrast to which he sees the counter-Enlightenment move towards expressionism as a way of clarifying the meaning of life in the process of living. In recognizing the authenticity of different ways of life without a retreat into relativism, he stresses the importance of critique (of the self as well as of others), based on an understanding of the lives of others in their own terms. This involves a "perspicuous contrast," allowing us to make distinctions sensitive to the perspectives of others. He thus suggests a possible route through the uncritical particularism encouraged by recognition of the fact of geographical diversity in ways of life, which can be defended from a moral point of view.

Nicholas Low and Brendan Gleeson follow with a discussion of one aspect of morality which is subject to obvious geographical diversity: human rights. They point to some of the implications of the emergence of the discourse of rights from spatially situated struggles against injustice. Recognition and specification of rights has proceeded from individual protection against the actions of other individuals and of the state, to the right to welfare under the uneven distribution of power exemplified by class relations of capitalism, and on to the right to environmental quality which has been espoused more recently. The authors relate this progression to an expanding conception of the human self, from atomism to social interdependence and to ecological embeddedness. However, the waning power and autonomy of the nation state, in the face of globalization, is threatening the spatial framework within which rights (and obligations) have usually been implemented in the context of citizenship. A new form of rights consistent with the emerging placeless and spaceless world of global capital is therefore required, if the rights people actually experience are not to become even more unevenly distributed geographically.

Ron Johnston moves the discussion on to the political dimension, exploring the notion of fairness (in the sense of an egalitarian conception of social justice) involved in the design and implementation of democratic systems. He points to the difficulties which the geographical setting raises for the application of democratic principles. The bounding of spaces which determine where individuals can vote (within a national territory) means that geography is actively implicated, and manipulated, in conflicts over electoral power and fairness in obtaining political representation. The notion of fairness in this context may be applied to individuals, communities, minorities and political parties. But meeting more than one of these can lead to conflict, as is illustrated from British experience. Some intriguing issues are raised when the context is broadened to the international scale, as in the case of the European Union. The ongoing process of globalization has important implications for international regulatory regimes, the activities of which may be judged by different criteria of fairness: to nation states, to population groups, and so on. The conclusion that fairness in general is extremely difficult to achieve, and that fairness based on geographical building blocks is almost impossible, underlines the demanding ethical role which space can play when introduced into political philosophy.

One of the most obvious spatial expressions of the threat to human rights and social justice is what is usually referred to as uneven development. Seamus Grimes provides an introduction to some issues in the growing field of development ethics. He points to the failure of the dominant market-led model of development to involve the great majority of humankind in the benefits of growing wealth creation. The gap between rich and poor is increasing globally (sometimes translated geographically into north and south), and at more local spatial scales. He identifies shortcomings in the economistic approach of mainstream development analysis, and in the political-economy and neo-Malthusian perspectives which have challenged the prevailing orthodoxy. He argues for a greater awareness of the ethical dimension of development theory and practice.

The final essay in Part 1 takes the discussion into cyberspace, with some of the implications of modern means of communication over distance. In addressing the ethics of the Internet, Jeremy Crampton compares the competing scenarios of liberation (e.g. through the promotion of a wider spatial scope of connectivity) and of hegemony (e.g. enabling the rich and powerful to extend more effective surveillance over others). Access to the Internet is shown to be extremely uneven geographically, reflecting other disparities between core and periphery at a world scale, as well as inequalities at the regional and local levels. There are also differentiated societal degrees of access, reflecting gender, race, income, education and so on. All this raises important issues of social justice, which tend to be overlooked by both technophiles and politicians invoking access to the net as some kind of panacea. The author also explores issues of professional ethics associated with Internet communication (linking on to Part 4 of this book).

2 Meaning and geography

Paul Roebuck

What is the place of ethics in geography, of geography in ethics and why does ethics matter to us intellectually and personally? To explore these questions I will contrast objectivist and relativist concepts of science and ethics, consider what these perspectives have to do with meaning and morality, and conclude with how geography helps us negotiate the intellectual terrain of ethics and geography.

Geographers study processes, both human-caused and natural, writing themselves on the face of the earth. Geography encompasses human/nature interactions and all the possible worlds and spaces that humans create in actuality and imagination (Lukermann 1961). Ethics is people's character (ethos) and conduct. When we think about ethics we focus on the values and norms by which people measure their lives and in which they find meaning. We ask how we (or others) should live, mindful of ends, means and intentions (see Midgley 1993).

For the most part, Western moral philosophy views ethics top-down, as an expression of theory. It tries to justify systematized sets of rules of right and wrong through logic and consistency or coherence, looking for a single universal objective foundation for moral judgment. This objectivism almost inevitably slides into ethnocentrism as some privileged understanding of rationality is falsely legitimated by claiming for it an unwarranted universality. However, ethics, as a social practice, lived from the bottom-up, goes beyond codified rules to touch our sources of meaning, clarifying our understanding of ourselves and our lives, both individually and in the many overlapping collectivities of which we are a part.

Geographers, studying both theory and practice, investigate the place of rules and meaning in the processes that are changing the earth, the worlds that people create and the spaces they inhabit. Incorporating ethics into geography we not only account for objective phenomena, but also explicate and critique the world views underlying our own and other's beliefs and practices. Geography is partially constitutive of ethical problems and must be accounted for in ethical reasoning. We can take geography's context-sensitive insights about place, space, scale, circulation, interaction, distance, proximity, and multiculturalism and use them to guide us as we struggle with environmental and social problems, helping to clarify issues, ascertain moral considerability and give perspicacious understanding of others' ways-of-life (see Lynn 1998). To avoid ethnocentric prejudice we must move beyond objectivist explanations to understanding and interpretation that makes sense of agents by contrasting their self-understanding with our own. In authentically coming to know others we expand our horizons and come to deeper understanding of ourselves. This also helps us understand science's "situatedness" – the importance of geographical and historical perspective in understanding natural and human phenomena (Geertz 1983; Taylor 1985b; Marcus and Fischer 1986; Barnes and Duncan 1991).

Objectivism, relativism, pluralism

Geography, history, anthropology, travel, trade, exploration, art, literature, long-distance communication, conquest and colonization reveal a huge variety in humanity's forms-of-life. The simple fact of diversity, however, is no disproof of the possibility that there are some beliefs and practices better to have than others because they are truer or more justified. How do we justify judging some values to be truer, better, etc.?

Western moral philosophy has long sought universal norms that would apply forever, to everyone, in all situations. It seeks a single objective foundation for knowledge or against which all are judged (e.g. Plato's *Theaeteus*, Descartes' *Meditations*, Kant's *Critiques*, Rawls' *Theory of Justice*, Habermas' theory of communicative action). Such moral universalism insists we judge others solely by our own criteria, resulting in ethnocentric projection of our values onto others.

From the beginning of European expansion, a countervailing force, Isaiah Berlin's (1973) "Counter-Enlightenment," questioned this universalism, raising considerations of pluralism, authentic self-expression, and autonomous freedom. Las Casas, Vico, Herder, and others, responded to observed cultural variety asserting that moral truth and justifiability, if they exist at all, are relative to culturally and geo-historically contingent factors. The danger with this kind of pluralism is that it may slide into cultural relativism when we judge the validity of beliefs and practices of others solely from the point of view of the agent, or describe what she does only in her own terms or those of her society, place and time (Winch 1964). This would make social science un-illuminating as agents can be mistaken and our scientific accounts should go beyond common-sense to make agents' activities clearer than they are to themselves.

Modernity is marked by an extreme fear of relativism that Bernstein (1983: 18) calls Cartesian Anxiety. It takes many forms – religious, metaphysical, epistemological and moral. At the heart of the objectivist's vision, and what makes sense of his or her passion, is the belief that there must be some permanent, secure constraints to which we can appeal. In contrast, relativism casts us adrift:

> At its most profound level the relativist's message is that there are no such basic constraints except those that we invent or temporally (and temporarily) accept. Relativists are suspicious of their opponents because, the relativists claim, all species of objectivism almost inevitably turn into vulgar or sophisti-

cated forms of ethnocentrism in which some privileged understanding of rationality is falsely legitimated by claiming for it an unwarranted universality. The primary reason why the agony between objectivists and relativists has become so intense today is the growing apprehension that there may be nothing – not God, reason, philosophy, science, or poetry – that answers to and satisfies our longing for ultimate constraints, for a stable and reliable rock upon which we can secure our thought and action.

(Bernstein 1983: 19)

The solution to this tension lies not in searching for firmer foundations, but in finding modes of expression that give us clearer insight. Our understanding differs between competing philosophies of science – empiricism, pragmatism, idealism, realism – all have different "takes" on what constitutes theory, knowledge, and the contents of the universe. Choosing one perspective shapes possible questions, acceptable answers, and comes freighted with metaphysical assumptions about the world and human/nature interactions. We cannot authentically embrace any particular position without acknowledging our prejudices and adopting a particular perspective relative to others. We are turned back to geo-historically situated knowledge – not knowledge of the essence of the world in isolation – but rather understanding in a context of places, periods, narratives, vocabularies, and metaphysical assumptions. This knowledge is scientific, but also relative, aware of history and the views of others. It is value-laden, practical and ethical, part of an ongoing discourse in critical and hermeneutic (interpretivist) studies. Perhaps the best way to understand this shifting ground of subjectivity and objectivity is through the notion of frameworks.

Frameworks

> The imposition of meaning on life is the major end and primary condition of human existence.
>
> (Weber 1919)

Our values, norms and sources of meaning are part of conceptual and cultural constructions called "horizons" by Gadamer and Nietzsche, "worldviews" by Dilthey, "genre de vie" by Vidal, "forms-of-life" by Wittgenstein, "epistemes" by Foucault, "constellations of absolute presuppositions" by Collingwood and "paradigms" by Kuhn. I use the term "framework" to characterize constructions human beings impose on their world to make sense of it. Frameworks are active processes expressed in language and forms-of-life. People use frameworks of feeling and understanding to define the world, its organization, processes and direction. These constructions define how people judge their lives and determine how full or empty their lives are. Frameworks are our sources of identity – what Charles Taylor (1989) calls "the sources of the self." At different times in history and in different places in the world, frameworks have been based on such things as the belief in an hierarchical chain of being in the universe, the call of God made clear in revelation, the guidance of dreams obtained on a spirit quest, or the space of glory in the memory and song of the tribe.

In the West, the mainstream of Enlightenment thought looked on nature and society as having only instrumental significance – potential means to the satisfaction of human desire and nothing more (e.g. Bacon, Hobbes, Descartes, Locke, Bentham). These views seem natural to those of us in the Enlightenment tradition and alienate us from non-Western cultures because of how thoroughly the Enlightenment altered Western understanding of ontology, epistemology, identity, and morality.

With the conceptual revolution of the Renaissance and Enlightenment, the world was transformed from a meaningful order – a sign from God – to an inert, manipulable thing. Orderliness that the Greeks and their intellectual heirs took to be in the world, came to be thought of as orderliness in our minds that we project onto a contingent earth. Notions of subjectivity and objectivity changed. Formerly, meaning had been thought to be in-the-world and in things such that material objects could be sacred or embody meaning. For example, groves or mountains could be holy. In the Enlightenment, however, meaning came to be thought of as only for a subject. With the Reformation, humanity's relationship with the Divine was internalized. It ceased to be meaningful to talk about subjects other than humans as having meaning, value or rights. Only reasoning, communicating humans could be subjects. There could be no meaning in the world in-itself. Only humans confer value and have intrinsic value – that is, value in-itself and not merely as means toward (human) instrumental ends.

People came to define themselves no longer in relation to a cosmic order, but as subjects who possessed their own picture of the world within them, their own purposes and drives. With this new notion of subjectivity went an objectification of the world. The old view of the world as a cosmic order to which people were essentially related was replaced by a domain of neutral facts, mapped by tracing correlations and manipulated in fulfillment of human purposes. This vision of an objectified, contingent world was valued as a confirmation of the new identity before it was valued as the basis of mastery over nature, which only later manifested itself as technological progress (Taylor 1975). Objectification extended beyond external nature to englobe human life, resulting in a certain vision of humanity: an associationist psychology, utilitarian ethics, atomistic politics of social engineering and ultimately a mechanistic social science to go with the mechanistic natural science (Berlin 1979; Taylor 1985a; Toulmin 1990).

Ethical notions of identity changed in the Enlightenment and through the Romantic period. Self, that had been defined in relationship to family, guild, community or hierarchy, changed to an atomistic, individualized ego, and then changed again to a self-determining autonomous subject. Notions of the "good life" changed from public service to the community or contemplation of Ideal Forms beyond the self, to private, autonomous self-fulfillment and self-realization, freely chosen according to one's own lights.

Uniformitarianism

The Enlightenment framework would have us search for a universal norm, a particular vocabulary, and one set of intellectual categories to use as a foundation and from which to judge others. It holds that all reality and all the branches of our knowledge of it, form a rational coherent whole, and there is ultimate consistency between human ends. This constitutes a sort of philosophical monism:

> The central doctrines of the progressive French (Enlightenment) thinkers, whatever their disagreements among themselves, rested on the belief, rooted in the ancient doctrine of natural law, that human nature was fundamentally the same in all time and places; that local and historical variations were unimportant compared with the constant central core in terms of which human beings could be defined as a species . . . that there were universal human goals; that a logically connected structure of laws and generalizations susceptible of demonstration and verification could be constructed and replace the chaotic amalgam of ignorance, mental laziness, guesswork, superstition, prejudice, dogma, fantasy, and above all, the "interested error" maintained by the rulers of mankind and largely responsible for the blunders, vices and misfortunes of humanity.
>
> (Berlin 1979: 1)

Enlightenment thinkers believed that since Newtonian methods had worked for certain purposes in physics, they could be applied to ethics, politics and human relations with the expected result that human social institutions could be rationalized for the greater good:

> Once this had been effected, it would sweep away irrational and oppressive legal systems and economic policies the replacement of which by the rule of reason would rescue men from political and moral injustice and misery and set them on the path of wisdom, happiness and virtue.
>
> (Berlin 1979: 2)

Moderns sought universal knowledge of things true everywhere, for all time and of everyone. One common strain of this is what Lovejoy and Boas (1935) called uniformitarianism – the idea that humans develop everywhere the same way, that their development is uniform, and moves through predictable stages. Uniformitarians believe all societies are arrayed along a single path of development toward the same intellectual, moral and cultural ends. Typically, uniformitarians believe in progress. Unenlightened beliefs are taken to be the things that are responsible for holding back human progress. The Enlightenment took as its project the progress of humanity – its liberation from delusions using empirical science and instrumental thought.

Uniformitarians often distinguish between folk conceptions, which they regard as error-ridden and deluding, and another deeper variety of discourse – usually science, which consists of systematically rendered concepts drawn, for example, from linguistic or psychoanalytic theory. Thus in Hegel's phenomenology of mind, Freud's psychoanalysis, and Marx's critique of ideology, reflection on things universal to humanity provides for the experience of emancipation by means of critical insight into relationships of power – power that derives its strength, in part, from the fact that these relationships have not been seen through. Potentially, our thoughts and actions liberate us from coercive illusions. The hope is, by applying critical insights universally, humanity is enabled to progress.

Alienation

With the increasing variety and pace of change in world views we moderns question all frameworks. The frameworks from which our predecessors took their meaning have lost their foundations. We must seek meaning from our inner resources, yet when we look inside ourselves we find uncertainty because we have been cut off from everything that once supplied the resources we are seeking. This is what Weber meant by disenchantment – the dissipation of our sense of the cosmos as a meaningful order:

> The fate of our times is characterized by rationalization and intellectualization and, above all, by the disenchantment of the world. Precisely the ultimate and most sublime values have retreated from public life either into the transcendental realm of mystic life or into the brotherliness of direct and personal human relations. It is not accidental that our greatest art is intimate and not monumental.
>
> (Weber 1919)

Hegel (1966: 215) called Christianity the agent of disenchantment: "Christianity has depopulated Valhalla, hewn down the sacred groves, and rooted out the phantasy of the people as shameful superstition, as a diabolical poison."

Ultimately, modernity disenchanted Christianity as well. This disenchantment altered the framework in which people had lived their spiritual lives. Nietzsche, in his "God is dead" passage used the term "horizon" to characterize such constructions: "How could we drink up the sea? Who gave us the sponge to wipe away the whole horizon?" (Nietzsche 1887: para. 125).

Though modernity questions all frameworks, we nonetheless cannot live our lives without some framework. To have no source of meaning, no values, and no context within which to define our world would render us inhuman or insane. Though frameworks change as societies change, we cannot dispense with them altogether. Like language, they are a fundamental constituent of who we are, how we think and why we act. We come to full expression of ourselves through our frameworks and cannot exist without them. Any social science which purports to understand what we are about must take account of them and the anxiety that occurs when they clash with others' world views.

The counter-Enlightenment

Various counter-Enlightenment movements arose in the late eighteenth and early nineteenth centuries protesting alienation caused by the objectification of humanity and the division from nature which the mainstream view entailed. They offered alternative notions of humanity whose dominant images were of expressive and morally-free agents. These movements toward expressivism (Berlin 1965; Taylor 1975) and moral freedom are intricated in the Western tradition's current notions of language, art, history and culture. The focus on morally free agency derives from Protestantism but was given particular emphasis in Kant's philosophy, which continues to shape our attitudes about freedom, subjectivity, moral considerability of others (including nature), rationality, sovereignty, and nationalism. Expressivists saw human life possessing expressive unity analogous to a work of art, where every aspect found its proper meaning in relation to all others and thought each life unique. Expressivism serves as the foundation for contemporary comparative anthropology, philology, ethnography, and human geography.

Though the Enlightenment had been opposed by the Scholastics and the Church, neither of these questioned the basic premise of uniformitarianism. The counter-Enlightenment, which arose in reaction to but was partially an extension of the Enlightenment, did call into question this basic dogma. The primary challenge came in the ideas of nationalism, pluralism and expressivism. The sources of those concepts in Western thought lie with the counter-Enlightenment thinkers. The most influential was Johann Gottfried von Herder. "All regionalists, all defenders of the local against the universal, all champions of deeply rooted forms of life . . . owe something, whether they know it or not, to the doctrines which Herder . . . introduced into European thought" (Berlin 1976: 176).

Nationalism, pluralism, and expressivism, concepts given an original twist with Herder, are counter to the Enlightenment's philosophical monism, and shape contemporary notions of language, culture, self, and human/nature interaction. Nationalism is the belief that people can realize themselves fully only when they belong to an identifiable culture with roots in places, tradition, language, custom, and common historical memories. Pluralism recognizes a potentially infinite variety of systems of values, all equally ultimate and incommensurable with one another. It renders incoherent the belief in a universally valid nomological path to human fulfillment sought by all people at all places and times. Expressivism is the notion that all people's works "are above all voices speaking" – forms of expression which convey a total view of life.

Herder, along with Rousseau and Hamann, are the originators of the Expressivist theory. They viewed human life as "self-expression." They meant by expression something akin to the following: We try to live a "good" life. We try to live up to our capacities and express ourselves authentically. Through expression, we clarify for ourselves what our values are, and thereby what and who we are. Our lives realize an essence or form. The idea of who I am is not fully determinate beforehand. It is only made determinate in being fulfilled, in the sense that sometimes I don't know what I think until I am able to articulate it or act it out – to express it. Living our lives expresses our purposes, allows us to realize them and can clarify for us what our purposes are. Expression, therefore, is the fulfillment of life and can clarify its meaning. In living a "good" life, I fulfill my humanity but also clarify what my humanity is about. As clarification, my life-form is not just the fulfillment of purpose but the embodiment of meaning – the expression of an idea, e.g. of fidelity, honesty, or bravery. This is how we define our values: the ways in which I express my bravery in my social practice clarifies and simultaneously redefines for me and members of my community what bravery is for us. This connection between meaning and being breaks with the Cartesian dualism of subjectivity and objectivity that rigorously separated meaning and being.

Herder's insight was that for peoples and for individuals, my humanity is unique, it is not equivalent to yours and the unique quality can only be revealed in my life itself. "Each man has his own measure, as it were an accord peculiar to him of all his feelings to each other" (Herder 1877–1913: VIII. 1). Differences between people take on a moral importance so we can ask if a form-of-life is an authentic expression of an individual or a people. Are they living up to their potential? This is what is meant by self-realization. This powerful insight fired the imagination of Kant, Goethe, Schiller, the Humboldts, Hegel, and Ritter and is still an essential part of how we understand subjectivity today. Through the internalization of Romantic ideas it has become part of the modern framework. And finally, notions of nationalism, pluralism and expressivism are key components of contemporary human and regional geographic understanding.

Beyond ethnocentrism

Social science can no longer ignore other's meaning. This is true not just as an exculpation of imperialist/colonialist excesses or science's sometimes racist treatment of non-Enlightenment-tradition cultures, but on epistemological grounds. Our theories are adequate for explaining human societies only if they provide an accurate account of what people are doing in the context of their cultural and linguistic practices. Following the objectivist natural scientific model and theorizing people's ways-of-life without understanding their subjective self-descriptions, leads to failure because such theories cannot be validated. Taylor (1985b), suggests:

> Suppose we are trying to give an account of a society very different from our own, say a primitive society. The society has (what we would call) religious and magical practices. To understand them . . . we come to grasp how they use the key words in which they praise and blame, describe what they yearn for or seek, what they abhor and fear, and so on. Understanding their religious practices would require that we come to understand what they see themselves as doing when they are carrying out the ritual we have provisionally identified as a "sacrifice", what they seek after in the state we may provisionally identify as "blessedness" or "union with the spirits." (Our

provisional identifications, of course, just place their actions/states in relation to our religious traditions, or to ones familiar to us. If we stick with
these, we may fall into the most distorted ethnocentric readings.) We have no
way of knowing that we have managed to penetrate this world in this way
short of finding that we are able to use their key words in the same way
they do.

(Taylor 1985b: 120–121)

Other societies may be incomprehensible in terms of our own frameworks and we
must strive for a perspective that explains what they do and shows it to make sense
to us under their description. I am not suggesting that we should solely adopt the
point of view of the other, assuming all societies are relative and cannot be compared. Taylor calls this kind of cultural relativism the "incorrigibility thesis"
because describing cultures solely in their own terms rules out accounts which
show them up as wrong or confused – they are incapable of being corrected or
critiqued, i.e. incorrigible. Our discourses should explain what the agent is doing
and they should improve upon common-sense understanding. When judging
others, how can we avoid making cross-cultural study an exercise in ethnocentric
prejudice?

The answer lies in confronting others' frameworks while authentically being
open to change in our own viewpoints. This is close to Gadamer's (1993) "fusion
of horizons." Openness to change in theoretical stance and self-understanding of
the researcher is missing from the grand totalizing theories of science – e.g.
psychoanalysis, evolutionism, or sociobiology.

When we moderns consider indigenous knowledge of nature, for example,
something we provisionally represent as a "rain dance," the challenge to be open
to change in our own viewpoint is difficult in light of how integral our natural-
istic, atomistic, and instrumental views of nature are to our own world views. Our
investigation should challenge both our language of self-understanding and
theirs, maintaining science's own, proper, critical role. We seek a language of
clearly understood contrast that compares our framework and the other's framework as alternative possibilities in relation to some human constants at work in
both. This would not be akin to Frazer's (1890) portrayal of "primitive" magic as
largely mistaken incipient science or a failed attempt to master the environment.
Nor would it be the cultural relativism of the incorrigibility thesis (Winch 1964;
Beattie 1970; and much contemporary ethnography) that the "rain dance" is a
very different activity which has no corresponding practice in our society and
which can only be seen as the integration of meaning through symbolic acts rather
than an attempt to change the world. The incorrigibility thesis only appears to
escape ethnocentric projection because it implicitly assumes that all cultures make
a disjunction between understanding (and controlling the world) on the one
hand, and integrating meaning through expressive acts on the other. They do not.
That disjunction was made in Western culture at the beginning of the modern era.
The Platonist view was that to know the truth is to love it. Knowledge of nature
and being in harmony with it, in touch with our sources of meaning, were

formerly connected in the Western tradition. This is no longer true in the modern space of neutral facts to which we are only contingently related. Taylor (1985b: 128–129) suggests we understand activities like rain dances in the context of a framework in which the disjunction of proto-technology or expressive activity is not made. Identifying these two possibilities amounts to finding contrasting language that enables us to give an account of the procedures of both societies in terms of the same cluster of possibilities.

Finding a language of perspicuous contrast avoids the pitfall of ethnocentric projection of our values onto others. It also avoids the problem of mindless relativism in which no comparisons can be made between societies. Social science requires that we come to a deeper understanding and that we be able to critique society – our own and others'. Others are not uncriticizable, yet they must be criticized on relevant grounds. In the context of ability to manipulate nature, the segregation perspective has proved most effective, as the technological achieve-ments of the Western tradition attest. However the West is alienated and in that context the fusion perspective proves more effective in putting people in harmony with the world. Overcoming the alienation of segregation was the impetus for the Romantic movement and underlies much of current environmentalism's attempts at reconciling humanity's actions with nature's "needs."

We are always in danger of seeing our ways of being as the only conceivable ones. The language of perspicuous contrast allows us to make finer distinctions, sensitive to the other's perspective. In finding this language we redescribe what we and others are doing. If done while authentically open to change in our normative viewpoints, such activity provides the flexibility to alter our own self-understanding and avoid ethnocentricity. Potentially, understanding is increased, the range of possible human expression is enlarged, we gain new tools with which to approach human problems and achieve social science as cultural critique.

Thus, our geographies are not just accounts of natural phenomena, but also are explications and critiques of the ethics underlying our own and others' beliefs and practices. Geography's insights into people's finite and situated perspectives on the world can inform and enlarge Western moral philosophy, expanding its horizons, challenging prejudice, and advancing the dialogic conversations we use to understand the places and spaces that humans create in the context of lived-experience.

As we work to understand nature, ourselves, and others, we produce a "new" world. This is why we can productively study geography and history, other cultures or past times. It also gives an ethical dimension to our constructions. Through our criticisms we shape the lived-world in our geographical imagin-ations and reflective understanding. By applying the insights learned from cultural comparisons and critiques, while maintaining sensitivity to the situatedness of geographic circumstances, we have the potential to reshape the frameworks of ourselves and others with whom we communicate and interact.

References

Barnes, T. and J. Duncan (1991) *Writing Worlds*, New York: Routledge.

Beattie, J. (1970) "On understanding ritual" in B. Wilson (ed.) *Rationality*, New York: Harper & Row.

Berlin, I. (1965) "Herder and the Enlightenment" in E. Wasserman (ed.) *Aspects of the Eighteenth Century*, Baltimore: The Johns Hopkins University Press.

—— (1973) "Counter-Enlightenment" in P. Wiener (ed.) *Dictionary of the History of Ideas*, New York: Charles Scribner's Sons.

—— (1976) *Vico and Herder*, London: Chatto and Windus.

—— (1979) *Against the Current*, London: Penguin Books.

Bernstein, R. (1983) *Beyond Objectivism and Relativism*, Philadelphia: University of Pennsylvania Press.

Frazer, J. (1890) [1911] *The Golden Bough*, London: Macmillan.

Gadamer, H. (1993) *Truth and Method*, trans. J. Weinsheimer and D. Marshall, New York: Continuum.

Geertz, C. (1983) *Local Knowledge*, New York: Basic Books.

Glacken, C. (1967) *Traces on the Rhodian Shore*, Berkeley: University of California Press.

Hegel, G. (1966) *Hegels Theologische Jugendschriften, nach den Handschriften der Koniglichen Bibliothek*, Berlin: Hrsg von Herman Nohl.

Herder, J. (1877–1913) *Vom Erkennen und Empfinden der Menschlichen Seele, von Bernhard Suphan, Volume VIII*, p. 199, Berlin: Weidmann.

Lovejoy, A. and Boas, G. (1935) *Primitivism and Related Ideas in Antiquity*, Baltimore: The Johns Hopkins University Press.

Lukermann, F. (1961) "The Concept of Location in Classical Geography", *Annals of the Association of American Geographers* 51: 194–210.

Lynn, W. S. (1998) *Geoethics: Geography, Ethics, and Moral Understanding*, Dissertation, Department of Geography, University of Minnesota.

Marcus, G. and Fischer, M. (1986) *Anthropology As Cultural Critique*, Chicago: University of Chicago Press.

Midgley, M. (1993) *Can't We Make Moral Judgments?*, New York: St Martin's Press.

Nietzsche, F. (1887) *The Gay Science*, trans. G. Taylor, in C. Taylor *The Sources of the Self*, Cambridge, Mass.: Harvard University Press, p. 17.

Taylor, C. (1975) *Hegel*, Cambridge: Cambridge University Press.

—— (1985a) *Philosophy and the Human Sciences Philosophical Papers 1*, Cambridge: Cambridge University Press.

—— (1985b) *Philosophy and the Human Sciences Philosophical Papers 2*, Cambridge: Cambridge University Press.

—— (1989) *The Sources of the Self*, Cambridge, Mass.: Harvard University Press.

Toulmin, S. (1990) *Cosmopolis*, New York: Free Press.

Weber, M. (1919) [1949] "Science as a Vocation" in *The Methodology of the Social Sciences*, trans. E. Shils and H. Finch, Glencoe, IL: Free Press.

Winch, P. (1964) "Understanding a primitive society", *American Philosophical Quarterly* 1: 307–324.

3 Geography, justice and the limits of rights

Nicholas Low and Brendan Gleeson

> There is nothing which so generally strikes the imagination, and engages the affections of mankind, as the right of property, or that sole and despotic dominion which one man claims and exercises over the external things of the world, in total exclusion of the right of any other individual in the universe.
>
> (Blackstone 1766)

Introduction

Rights are not transcendental universal norms for behaviour handed down from on high but human discoveries about our human condition made in the course of human conflict. While this is a condition we all share as members of the human species and indeed as animals, the discovery takes place in specific places and times and is embedded within different cultures and ways of thinking.

We argue that the discourse of human rights has emerged from the spatially situated struggle against injustice. Rights should be understood as protections against power in society. We interpret power in terms of the system elaborated in the critical realist philosophy of Roy Bhaskar (1993). Rights, in our view, are protections against what Bhaskar terms power$_2$ or 'master-slave relations' – making space for power$_1$ or human agency. The sources of social power in the world are many. Mann (1986: 28) for example, groups these into four main clusters: ideological, economic, military and political. Many different combinations of these clusters are possible: fascism, forms of corporatism, Stalinist, Maoist or Dengist socialism (consider the work of Schmitter 1974, and Winkler 1976 on corporatism, and the discussions which followed). Our critique is directed principally at a particular constellation of power, that of global capitalism.

The claim of critical realism is that our understanding of ethical systems is closely related to our understanding of the way the world outside us works, that is to say to an understanding of the causal mechanisms operating in the real world. An ethic implies an ontology. We do not argue that capitalism is the only, or in many localities and in many instances, even the dominant cause of social events. Yet almost everywhere in the world today its power is felt, sometimes as pressure (the IMF 'rescue' of Mexico and Indonesia), sometimes as explosive force (the Gulf War). Less obviously, capitalism inserts itself into local cultures and structures of power, creating a multiplicity of forms under a single logic.

The ethic within capitalism contains the message of human emancipation, freedom and rights: the language of rights formulated eloquently by Kant (1996) [1797] and Locke (1970) [1690]. But this ethic of emancipation is contradicted by the structures of economic and social power that capitalism erects: structures of exploitation and domination fuelled by human labour and the natural world. In contrast to the institutional picture of 'reality' painted by capitalism, we posit a reality comprising vulnerable human persons, socially related and embedded in ecological systems with finite capacity to support human collective activity. The transcendent quality of ethics lies in this reality which has both a material and a spiritual dimension. It is this reality from which the idea of rights draws its power.

Rights embody the need for protection of the human person in nature. Our view of rights is negative in the sense that rights are a defence and a protection. Rights do not *make* people behave in a certain way, they make space for people to behave in the way they want in accordance with their desires and needs. They are a shield rather than a sword. What people do within their rights is not the subject of the rights discourse. Likewise, with regard to environmental rights, nature has its own business to get on with. We humans cannot ourselves produce a lovely environment in any but the most trivial sense but we can stop interfering with nature's capacity to provide us with the environment which nurtures us and is part of us. So in this sense we try and see through our cultural predilection for the positive.

In the first section of this chapter we first posit that rights have emerged historically as protectors of the person from the structures of power in which persons are embedded; second, that the stratified structure of rights is derived from the stratified constitution of the self. Third, we pose the question of whether the discourse of rights is meeting its limits. In the following section we take up this question, arguing that there are two aspects of the limits of the rights discourse. One is the fading autonomy of the nation state which provided the vehicle for the protections afforded by human rights. The second is the dualistic tendency in human rights discourse which breaks down when confronted with the idea of injustice to the non-human world. We conclude, however, that the idea of rights will continue to have a place in the global struggles necessary for the constitution of humane governance.

The ontology of rights

We first become conscious in an inchoate way of the violation of our humanity. We are able to imagine ourselves in the position of others and thus conceive of the violation of humanity in general. This consciousness arises as we also become aware that things might be different. And so we are forced to think about and define justice as we become conscious of our position as unjust, and as we refuse to accept the situation as it is.

The history of human rights is a history of struggle enacted over the spaces and territories of the globe. Nash (1989) has recognised this struggle in the widening circle of beings to whom rights have come to apply. But he does not really explain the connection between human struggle, human rights and the nature of the self. The 'legislation', so to speak, of a single foundation for human rights can only diminish the scope of this ongoing struggle – in this we agree with Bauman (1993). Rather we see successive 'levels' of rights emerging from moments in the process of struggle in which humanity is engaged. Conceptions of justice are articulated at different levels as the offspring of the more elemental struggle against injustice, the struggle to 'absent', that is to take away, the constraints on the full development of our humanity (Held 1987: 271; Bhaskar 1993: 208; Young 1990).

Certain phases have been defined in the struggle against injustice: against feudal authority, against capitalist oppression, against colonial rule, and against ecological destruction (Eckersley 1996: 220; citing Marks 1981). From these struggles emerge different 'generations of rights'. But 'generations', suggesting a certain chronological continuity or sequence, is not quite the right word. Such continuity can only be seen from a particular, and usually Eurocentric, viewpoint. Struggles against the oppression of traditional forms of class, ethnic or gender structure continue today as Falk (1995: 63–70) illustrates. There is a historical element but it is not necessarily chronologically coherent so, in order to link the emergence of rights to the struggle against injustice, we propose three main 'moments' which give rise to a layered structure of three 'levels' of rights.

The first moment comprises the struggles for the rights of individuals to be protected against direct violations committed by economic actors, states and political institutions. Thus the contract and the rule of law became sanctified protections for 'individuals' against other 'individuals', and against the arbitrary despotism of the state. From the middle of the nineteenth century in the newly industrialized parts of the world it became evident that such a formulation of rights was insufficient to protect people and their communities from serious harm. A second moment of the struggle for rights, therefore, addresses the abuse of personal and social integrity arising from class structural power. A person's position in a class structure, it was perceived, prevented him or her from enjoying the bodily protections offered by individual rights of the first level. The second level of rights is often expressed in terms of the right to the satisfaction of basic needs (Shue 1980; Doyal and Gough 1991: 94; Galtung 1994).

The combination of first and second levels has spread to encompass the critique of other structures of power – of race and of gender – and of international relations. Both race and gender may be inscribed in traditional and capitalist power structures as well as any combination of the two. The anti-colonial struggle produced a volume of work on the right to national sovereignty and self-determination under the aegis of the United Nations which accorded a formal equality to constituent states (e.g. United Nations, 1988). In reality, of course, highly uneven power structures persist in the international sphere. The UN has been rather powerless to enforce its own charter of rights. The term 'global apartheid' was coined in the 1970s by Gernot Kohler to describe the emerging world order in which the developed world of the 'north' adopts different standards for its own behaviour from that it applies to the developing 'south' (Kohler, 1982).[1] The second level also encompassed the right to safe, healthy and pleasing local environments. Thus the second moment of struggle led to the institutional innovations of the welfare state, public health and town planning.

A third moment in the struggle for rights conceives of the right to ecological integrity, a third level. Both the Brundtland Commission and the Rio Declaration on Environment and Development are informed by the idea of ecological rights: 'All human beings have the fundamental right to an environment adequate for their health and wellbeing' (World Commission on Environment and Development 1987, Annex 1; see also Norton 1982a and b; Eckersley 1992; Merchant 1992, 1996). We need hardly expand upon the many well known examples of ecological destruction (deforestation, chemical and nuclear disasters, atmospheric pollution, anthropogenic climate change, species depletion, etc.) which, though spatially differentiated, affect the integrity of the whole planet. The third level asserts that the human species has a right to an undamaged and ecologically benign environment in perpetuity.

In the course of history, laws and institutions have been created to protect rights. In so far as rights refer to the reality of the human condition, they are epistemic markers indicating what we know to be true of human vulnerability and its need for protection (see Turner 1993). In so far as rights are constitutionally enshrined they are discursive 'pegs' necessary, but often far from sufficient, to hold in place institutions protecting the vulnerable human person from structures of social power.

The rights to freedom of speech and assembly, for example, are restraints upon the power of the state which are voluntarily upheld by the state in order to maintain its legitimacy. Such institutions enable people (those with the power to use their rights) to develop their autonomy in freedom, limited only by the need for the protection of others. The protection offered by the first level of rights is for the person against other persons (and groups) and the power of the state. The second level offers protection against structures of power in civil society mediated by the market. The third level affords protection against the structural power of nature mediated by its human exploitation. All three levels are necessary to human emancipation.[2]

As rights become redefined, so the conception of autonomy and of 'moral considerability' develops and changes. The idea of moral considerability arises primarily from an understanding of who we are, not from what we should do. In the discourse of rights we can trace an expanding conception of the human self. The first level of rights conceives of the self as an independent atom. Yet even this image is fraught with contradiction for there is no such thing in nature as an *independent* atom. The second level conceives of the self as connected with other humans whose welfare is in some sense our own. The third level expands the self to include the non-human world (see Mathews 1991; Plumwood 1993). The image of the ecological self includes the welfare of non-human nature as in some sense our own.

With Bhaskar, we do not view these images of the self as incommensurable. Rather we say that a layered structure of protection is made necessary by a layered structure of real power within which the vulnerable person is embedded. This layered structure of power is also *constitutive* of the human person: the 'self'. So we regard the dialectic of rights as a struggle to elucidate the structured reality of the person: first as an individual in a world of competing individuals, second as an individual shaped by and shaping social power structures, third as an individual in the context of natural power structures (see Bhaskar 1993: 171). To depersonalize the discussion for a moment, every item in the word processor on which this chapter is being typed is composed of atoms and molecules, but we are not saying much about the computer by describing its atomic structure. The set of microchips and a collection of other pieces of hardware, together with sets of instructions about how it should react to the stimuli of having its keys pressed and its mouse moved and clicked, comprise the computer. This we see before us on the table and interact with quite usefully. A further layer in the reality of this computer is introduced when we connect the computer to networks of computers and engage an Internet, a web. So what is our word processor? Is it a collection of atoms, a set of parts, a node in a web? Obviously it is all three simultaneously. The powers of the computer are derived through its participation in this layered reality. Of course the identification of three levels falls far short of the real multi-tiered nature of reality, but three ontological levels will suffice to illustrate our view of a reality which exists independent of our ontologies.

So it is with the human person. The particular interpretation of utilitarianism which forms the ethical-ontological underpinning of the market orthodoxy overemphasizes a single level of the person, the 'individual' as a kit of organs plus a consciousness (or subjectivity). This view pervades the common sense of our time. Persons as 'customers' or 'consumers' have replaced persons as 'citizens' in the language of modern policy-making. But this narrowed perspective leads to an overwhelming sense of helplessness and alienation which is self-reinforcing, as Charles Taylor discusses:

> A fragmented society is one whose members find it harder and harder to identify with their political society as a community. This lack of identification may reflect an atomistic outlook, in which people come to see society purely instrumentally. But it also helps to entrench atomism, because the absence of effective common action throws people back on themselves.
>
> (Taylor 1991: 117)

Taylor's concern can be extended to the third level of rights. A fragmented society makes it harder for people to identify with the natural world which is seen in the radically alienated form of commodities for the exclusive use of humans. Rainforests become paper or chopsticks. Food animals become so much packaged meat on the supermarket racks. Humans are encouraged by a vast advertising and information industry to ignore both how such items are produced and the effects of their production.

In such an alienated world we need to ask: has the discourse of rights now reached its limits? If rights are protections against structural power, what part if any will rights play in the political struggles necessary to modify that power? Can the existing structures of social power be adapted in such a way as to enshrine the protections of the three levels of rights, or has the whole idea of rights reached the limit of its usefulness in the pursuit of human emancipation?

There are two main reasons for believing that the discourse of rights as part of an emancipatory praxis is reaching its limits. The first is that human rights have up to now always been pursued in the institutional context of the nation state. The state provides the necessary framework for the institutions which give real force to the discursive peg. Today the power and autonomy of the nation state is waning and the framework for a global substitute which could provide what Falk (1995) calls 'humane governance' seems a distant prospect. The second reason is that the idea of rights itself is a dualistic device which distinguishes between those beings who have rights and those who do not. Plumwood (1995, 1997) has argued cogently that such distinctions are imbued with the notion of human supremacy which has its roots in patriarchy. Eckersley (1996), on the other hand, takes a more optimistic view of the future of the rights discourse.

The institutional limits of rights

Let us return to the historical emergence of rights. Though we may live for the future, we live in the past (Bhaskar 1993: 54). As far back as the mid-nineteenth century, economists and capitalists were extolling the virtues of the market as a liberating, democratic force. The laissez-faire discourse correctly pointed to the dynamic spatial character of the market as a set of restless, expanding forces that were already integrating regions and countries within a common political economic system and, in the process, rapidly transforming long embedded sociocultural relations. However, whilst agreeing with the economists' view of the market as a radically transformative force, Marx was quick to pour scorn on the moral claims of capitalism:

> It claims to have obtained political freedom for everybody; to have loosed the chains which fettered civil society; to have linked together different worlds; to have created trade promoting friendship between the peoples; *to have created pure morality* and a pleasant culture; to have given the people civilized needs in place of their crude wants, and the means of satisfying them.
>
> (Marx 1977: 86, emphasis added)

Marx's synopsis of the 'free industry' rhetoric appears to anticipate with astonishing precision the common prospectus advanced by contemporary neo-liberal institutions and ideologues for the global economy. For Marx, industrial capital imposed a 'pure morality' on the landscapes and social systems over which it was victorious – by this he meant simply the formal structure of rights that were necessary even to brutal (i.e. laissez-faire) forms of capitalist democracy. Even in the mid-nineteenth century, some states realized the emptiness of such 'pure morality', though, as now, political resistance to the capitalization of nature through markets seemed difficult, if not futile (Marx 1977: 88).

But Marx was not prescient. Capitalism unfettered is today simply regressing to the form it took in Marx's day. Nor is there anything new about the neo-liberal rhetoric which is mobilized in justification. The set of layered rights to which we refer is a historically-specific feature of the liberal democratic state that has provided the political architecture for capitalism. These layers of rights developed first in antagonism to the previous mode of production, feudalism, and later in response to the socio-political contradictions of capitalism as manifested over the last two centuries. The 'liberalism' of laissez-faire saw human rights roughly 'traded for economic growth' (Macpherson 1985: 47), and 'order' was only restored through the bitter struggles which marked the eventual emergence of the second level of social rights.

Feminists have raised analogous objections to the discourse of rights, showing how notions of impartiality and formal equality conceal deeper structural inequalities between men and women (Young 1990; Plumwood 1995; Fraser 1997). The meta-critique at issue is captured in the well-known shibboleth: 'there is nothing so unequal as the equal treatment of unequals'. Masculinism and capitalism together shape an ethical and normative system which hypostasizes the individual, self-maximizing success, insecurity, competition, consumerism, narrow rationality, hierarchy and commodification (Plumwood 1995: 142–143).

The first level of (liberal democratic) rights was not conceived in a moral vacuum, but actually supplanted a complex system of traditional rights which rested on the mutual obligations between subaltern and superordinate classes anchored in law and tradition. The transition to liberal democracy – most vividly marked by the French Revolution – involved not the birth of rights, but rather the triumph of a new set of political ethics over an older order of rights. The same process is now being repeated as capitalism expands into new territories. It is no accident that the right to private property – the alienation of nature from the human collective – was (and remains) the first political goal of the nascent bourgeoisie. Throughout newly capitalist Asia and post-socialist Europe, the first task has been that of cadastral survey in order to privatize the land. The recent financial crisis in Asia is symptomatic of the familiar disease of capitalism: over-accumulation of capital in the hands of the dominant class, backed as always in early capitalism by coercive force. With the failure of nepotistic regimes to redistribute wealth, and the consequent failure to develop mass markets, locally accumulated capital must devalue or move out.

Such economic contradictions within capitalism may eventually speed up the democratization of authoritarian regimes and establish a minimum level of human rights as they did in developed Western nations early this century, and more recently in Eastern Europe and Latin America. But today the problem lies deeper. Just as the forms of democracy are spreading world-wide, so its substance is diminishing – not just for the developing but also for the developed world. While the collapse of the centrally planned economies has revealed devastated environ-ments, it is looking increasingly unlikely that this devastation will be repaired by a benign, ecologically modernized capitalism. In astonishingly prescient work Richard Falk and his colleagues in the World Order Models Project have long recognized the unfolding inhumanity and ecological destructiveness of the world capitalist order (Falk 1975; Mendlovitz 1975; Kim 1984). Held (1995) has documented the diminished autonomy of the nation state within this order. Martin and Schumann (1997) provide a vividly critical account of the effects of globalization today.

Economic globalization disembeds, deterritorializes and internationalizes nation states which become mere 'competition states' equipped with goals of short term economic survival analogous to those of corporations operating in the global market (Cerny 1996; Falk 1997). In these circumstances, the pursuit of second level rights is severely limited. While democracy and human rights become the universally adopted norm, they also become increasingly empty processes and hollow formalities. In societies exposed to the world market's systemic con-straints, formal democracy costs nothing for those who hold substantial economic power. On the contrary it reduces social frictions and thus economic transaction costs. But the claims to political participation on the part of the people are abrogated by deregulation.

While the struggle for second level rights will intensify during the next century in the world's most populous regions, especially in India, China and East Asia, a new discourse is opening up which brings to light a contradiction between the second and third levels of rights: third level rights of the developed world versus the second level rights of the developing world. Put bluntly, the standard of living of the developed world (generated and sustained by the capitalist system of pro-duction as currently constituted) cannot be generalized without destroying the planet's nature to such an extent that human life on earth is jeopardized. To illustrate: 500 million people can drive cars but 6 billion driving cars will destroy the atmosphere (see Altvater 1993, 1994, 1997).

We have argued elsewhere that the global expansion of second level rights without both transformation of economic power and forms of global governance is simply inconsistent with ecological sustainability (Low and Gleeson 1998). The most likely scenario is that while formal democracy and rights will become entrenched, in practice real second level and third level rights will be extremely unevenly distributed. Neo-liberal reforms speed up globalization while paying scant attention to second level rights in the developing world (beyond some lukewarm lip service to 'trickle down' effects). Deregulatory initiatives overseen by the World Trade Organization and the OECD have dismantled part of the public regulatory capacity necessary to control environmental degradation. Environmental and social movements and NGOs have gained momentum and, indeed, have now become part of some new structures of political governance at regional, national and even at global scale. But the power of these bodies to intervene in processes of environmental degradation is as yet very limited.

Injustice to nature

Out of the discourse of third level rights there now emerges the question of the rights of non-human nature. There are those who have sought to extend the rational rights of liberalism to the non-human, to break down the demarcation between human and non-human (Singer 1975, 1979; Regan 1983). But it is difficult to see how such an extension can change the way in which people relate to the rest of nature in a globalizing economy predicated on an instrumental attitude to all things non-human. Extending the protection of rights to animals does not go very far towards protecting ecosystems. Are the world's great forests, now in imminent danger of extinction, to be denied the protection of 'rights'? To say that forests are vulnerable and deserve protection, have the right to protection, are in need of protection, is to refer to the bases of our common sense of justice. Of course forests, like animals and non-articulate humans (babies, the intellectually disabled) cannot assert their rights, but, as Regan has argued with respect to animals, that is no reason to assume their rights away (in particular Regan 1997).

Eckersley (1996) has argued that the ecocentric vision can inform an expansion of the rights discourse to engage third level vulnerabilities:

> By widening the circle of moral considerability, humans, both individually and collectively, have a moral responsibility to live their lives in ways that permit the flourishing and well-being of both human and non-human life. This more inclusive notion of autonomy would necessarily involve the 'reading down' or realignment of a range of 'liberal freedoms' in ways that are consistent with ecological sustainability and biological diversity.
>
> (ibid.: 223)

Plumwood (1993), however, observes that the problem lies deeply embedded in the ethics in which social esteem is distributed according to dualisms such as mind and body, reason and nature, with the latter pole always being less worthy of rights. As Plumwood notes, such dualisms have a long history. Plumwood points to the way in which liberal democracy essentializes the rational egoism of the 'individual' self, the self whose world is contained within the boundaries of the skin and whose sole vantage point is a pinpoint in the vicinity of the brain. This strange image of the self allows for no stratification and thus its interests are defined without connection or care for the world beyond. Self and Other become a duality in which the interests of the Other can never become any part of those of the Self. 'Freedom comes to be interpreted in elitist and masculinist terms as lack of relationship and denial of responsibility, and self-determination as the rational mastery of external life conditions' (Plumwood 1995: 151). To continue:

> In the fin de siècle neo-liberal orgy of the Nasty Nineties, security is interpreted not as a collective good resting in social trust, social 'capital' and the mutual provision of satisfying, basic life standards for all, but in the

individualist terms of punitive law enforcement as social warfare waged against the criminalized Other.

(ibid.)

The Promethean 'compact' with nature which has underwritten capitalist modernization was drawn up on disastrously unequal terms. As Blackstone put it so aptly two centuries ago, at the core of liberal ethics lie property rights, 'that sole and despotic dominion . . . over the external things of the world'. The awful reality of this bargain has been concealed from society by the atomized nature of capitalist political ethics, which has discouraged an open social discourse (within humanity) on the question of justice and rights between humanity and non-human nature.

If we continue to conceive of society as structured in a Humean way as consti-tuted only by the accidental collision of so many billiard balls (Hume's image), then we cannot protect ourselves from the power of society, or from the power of nature, or from the two in combination. The need for protection from society is now well enough understood, and, in truth, even in today's reactionary world, in most developed nations the 'social' rights which protect people from social power remain stubbornly entrenched. But there is, as yet, no order of human rights based upon an understanding of the structural power of nature, the power of the non-human world to react against human exploitation by withdrawing the bene-fits (summed up in ideas of global integrity and biodiversity) that humans have come to depend upon (recall Engels' observations on 'the revenge of nature'). Certainly Leninist dictatorships took no heed of Engels' concerns.

There is, however, an increasingly powerful set of theoretical and institutional discourses which are attempting to prevent capitalist modernization from provok-ing the 'revenge of nature'. Both 'free market environmentalism' and 'ecological modernization' perspectives express a common aspiration to a non-conflictual relationship between nature and its exploitation via the market mechanism; in short, they articulate an inchoate 'third level' of 'ecological rights' to entrench a new compact between humanity and nature. Our purpose in pointing to this new level of 'ecological rights' is not to reify capitalist modernity's externalized (and anthropocentric) view of nature, as reflected in the historical and contemporary forms of conventional ethical systems; on the contrary, we seek to problematize this emergent level of protection itself.

Beyond liberal rights

The discourse of rights is steeped in contradiction but this should not lead us to dismiss its importance. Rights are protections for the stratified self against polit-ical and social power. The need to assert rights will remain and indeed grow. However, as we have argued, the liberal discourse of rights cannot provide these dual protections. There is a need to articulate a new form of rights within a broader dismantling of the sources of unequal social power. Moreover, dem-ocracy which has been the vehicle of rights has always developed within a *place*.

Rights have evolved in different ways, with different emphases in different places at different times. Now the places of democracy are fading in significance before a placeless and to some degree spaceless capitalism. Today we must contemplate global ethics. Can there be, in some sense, a place which is the globe, a place within which we are all citizens rather than just consumers? How can the contradiction between spatially situated ethics and their global application be overcome?

Falk (1987, 1995) has argued that institutional transformation at every level must take place if the world is to move towards humane governance enshrining first, second and third level rights. But this transformation must be accompanied by a cultural transformation with significance at the personal level:

> I think the beginning of a response is to acknowledge that only a miracle will get us out of the present trap . . . To rest our hopes on a miracle means, it seems to me, most essentially that we believe in the credible rebirth of some kind of religious civilization in our midst. This rebirth would have to transform the relationship between self, society and Nature in a direction that transcends the materialist consensus long dominating our value system.
>
> (Falk 1987: 292)

The idea of humane governance of course demands a new universalism, not to replace the currently fashionable particularism but to augment it at a new level. Those who deny that such a synthesis is possible or desirable might refer to Arran Gare's critique of postmodernism (Gare 1995).

Dismayed by the results of both religious fundamentalism and communism, Falk later argues for a 'politics of bounded conviction' involving a 'broad scanning of normative horizons', and a discursive politics arising not within the state system but in an emergent global civil society (Falk 1995: 44). Such a politics is also broadly supported in the work of Dryzek (1990, 1994, 1996) and Habermas (1990). We have elsewhere argued for a politics of enlarged thought (see Low and Gleeson 1997, 1998), recalling Arendt (1977) and Benhabib (1992).

We have surely learned during the twentieth century that the Jacobin model of violent social revolution (the politics of the French Revolution adopted by Lenin) transfers power but does not on the whole transform it. Transformation takes place from within. It is a process in which transformative agency avails itself of power structures in order to change them (Bhaskar 1993). Within this politics the discourse of rights will surely continue to have a place. But it will have to shed the implication that rights exclude or demarcate privilege, whether between genders, nations or species.

Acknowledgements

The authors would like to acknowledge that the initial inspiration for this chapter came from an unpublished paper by Professor Elmar Altvater. Helpful comments on the text were also received from the editors, James Proctor and David Smith, as well as from Seamus Grimes and Gearóid Ó Tuathail whose work appears in this volume. Responsibility for the finished text, however, rests with the authors.

Notes

1 Scare quotes are necessary here since the image of a division between a developed north and an undeveloped south is far from geographically accurate as the examples of, say, Algeria or Ethiopia in the north and Australia in the south demonstrate.
2 Undue concentration on second level to the exclusion of first and third level rights led to appalling consequences under socialist dictatorships. Undue concentration on first level, to the partial exclusion of second and third level rights, have produced dismal social and environmental consequences in the USA.

References

Altvater, E. (1993) The Future of the Market, An essay on the regulation of money and nature after the collapse of 'actually existing socialism', London: Verso.

—— (1994) 'Ecological and Economic Modalities of Time and Space' in O'Connor, M. (ed.) *Is Capitalism Sustainable? Political Economy and Political Ecology*, New York: Guilford Press, pp. 76–90.

—— (1997) 'Restructuring the Space of Democracy, The effects of capitalist globalization and of the ecological crisis on the form and substance of democracy', paper given at the Conference on Environmental Justice: Global Ethics for the 21st Century, The University of Melbourne, Australia, 1–3 October.

Arendt, H. (1977) *The Life of the Mind vol. 1: Thinking*, New York and London: Harcourt Brace and Jovanovich.

Bauman, Z. (1993) *Postmodern Ethics*, Oxford: Blackwell Publishers.

Benhabib, S. (1992) *Situating the Self, Gender, Community and Postmodernism in Contemporary Ethics*, Cambridge: Polity Press.

Bhaskar, R. (1993) *Dialectic, The Pulse of Freedom*, London: Verso.

Blackstone, W. (1766) *Commentaries on the Laws of England: Volume II of the Rights of Things*, 1979 edition, Chicago: University of Chicago Press.

Cerny, P. (1996) 'Globalization and other stories: the search for a new paradigm for international relations', *International Affairs* 15/4: 617–637.

Doyal, L. and Gough, I. (1991) *A Theory of Human Need*, London: Macmillan.

Dryzek, J. (1990) *Discursive Democracy: Politics, Policy and Political Science*, Cambridge: Cambridge University Press.

—— (1994) 'Ecology and discursive democracy: beyond liberal capitalism and the administrative state' in O'Connor, M. (ed.) *Is Capitalism Sustainable? Political Economy and the Politics of Ecology*, New York and London: Guilford Press.

—— (1996) 'Political and ecological communication' in Mathews, F. (ed.) *Ecology and Democracy*, London and Portland, USA: Frank Cass.

Eckersley, R. (1992) *Environmentalism and Political Theory, Towards an Ecocentric Approach*, London: UCL Press.

Eckersley, R. (1996) 'Greening liberal democracy: the rights discourse revisited' in Doherty, B. and De Geus, M. (eds) *Democracy and Green Political Thought: Sustainability, Rights and Citizenship*, London: Routledge, pp. 212–238.

Falk, R. (1975) *A Study of Future Worlds*, New York: Free Press.
—— (1987) *The Promise of World Order, Essays in International Relations*, Philadelphia: Temple University Press.
—— (1995) *On Humane Governance*, Cambridge: Polity Press.
—— (1997) 'State of siege: will globalism win out?' *International Affairs* 73/1: 123–136.
Fraser, N. (1997) *Justice Interruptus: Critical Reflections on the 'Postsocialist' Condition*, New York: Routledge.
Galtung, J. (1994) *Human Rights in Another Key*, Cambridge: Polity Press.
Gare, A. (1995) *Postmodernism and the Environmental Crisis*, London: Routledge.
Habermas, J. (1990) *Moral Consciousness and Communicative Action*, Cambridge: Polity Press.
Held, D. (1987) *Models of Democracy*, Cambridge: Polity Press.
—— (1995) *Democracy and the Global Order, From the Modern State to Cosmopolitan Governance*, Cambridge: Polity Press.
Kant, I. (1996) (1797) *The Metaphysics of Morals* trans. M. Gregor, Cambridge: Cambridge University Press.
Kim, S. S. (1984) *The Quest for a Just World Order*, Boulder: Westview Press.
Kohler, G. (1982) 'Global apartheid' in Falk, R., Kim, S. S. and Mendlovitz, S. H. (eds) *Toward a Just World Order*, Boulder: Westview Press, pp. 315–325.
Locke, J. (1970) [1690] *Two Treatises of Government*, Laslett, P. (ed.) Cambridge: Cambridge University Press.
Low N. P. and Gleeson, B. J. (1997) 'Justice in and to the environment, ethical uncertainties and political practices', *Environment and Planning A* 29: 21–42.
Low, N. P. and Gleeson, B. J. (1998) *Justice, Society and Nature, An Exploration of Political Ecology*, London and New York: Routledge.
Macpherson, C. B. (1985) *The Rise and Fall and Fall of Economic Justice*, Oxford.
Mann, M. (1987) *The Sources of Social Power, Vol. 1, A History of Power from the Beginning to AD 1760*, Cambridge: Cambridge University Press.
Marks, S. (1981) 'Emerging human rights: a new generation for the 1980s?', *Rutgers Law Review* 33: 435.
Martin, H.-P. and Schumann, H. (1997) *The Global Trap: Globalization and the Assault on Prosperity and Democracy*, London: Zed Books.
Marx, K. (1977), *Economic and Philosophic Manuscripts of 1844*, Moscow: Progress Publishers.
Mathews, F. (1991) *The Ecological Self*, London: Routledge.
Mendlovitz, S. H. (ed.) (1975) *On the Creation of a Just World Order*, New York: Free Press.
Merchant, C. (1992) *Radical Ecology: the Search for a Livable World*, New York: Routledge.
—— (1996) *Earthcare: Women and the Environment*, London: Routledge.
Nash, R. F. (1989) *The Rights of Nature, A History of Environmental Ethics*, Madison: University of Wisconsin Press.
Norton, B. (1982a) 'Environmental ethics and non-human rights', *Environmental Ethics* 4: 17–36.
—— (1982b) 'Environmental ethics and the rights of future generations', *Environmental Ethics* 4: 319–337.
Plumwood, V. (1993) *Feminism and the Mastery of Nature*, London: Routledge.

Plumwood, V. (1995) 'Has democracy failed ecology? An ecofeminist perspective', *Environmental Politics* 4: 134–168.

—— (1997) 'From rights to recognition? Ecojustice and non-humans', paper given at the Conference on Environmental Justice: Global Ethics for the 21st Century, The University of Melbourne, Australia, 1–3 October.

Regan, T. (1983) *The Case for Animal Rights*, Berkeley: University of California Press.

—— (1997) 'Mapping Human Rights', paper given at the Conference on Environmental Justice: Global Ethics for the 21st Century, The University of Melbourne, Australia, 1–3 October.

Schmitter, P. C. (1974) 'Still the century of corporatism', *The Review of Politics* 36/1: 85–131.

Shue, H. (1980) *Basic Rights, Subsistence, Affluence and U.S. Foreign Policy*, Princeton: Princeton University Press.

Singer, P. (1975) *Animal Liberation*, New York: Avon Books.

—— (1979) *Practical Ethics*, Cambridge: Cambridge University Press.

Taylor, C. (1991) *The Ethics of Authenticity*, Cambridge, Mass.: Harvard University Press.

Turner, B. S. (1993) 'Outline of a theory of human rights', *Sociology* 27/3: 489–512.

United Nations (1988) *Human Rights: A Compilation of International Instruments*, New York: United Nations Publications.

Winkler, J. T. (1976) 'Corporatism', *Archives Européennes de Sociologie* 17/1: 100–136.

World Commission on Environment and Development (1987) *Our Common Future*, Australian Edition, Melbourne, Oxford: Oxford University Press.

Young, I. M. (1990) *Justice and the Politics of Difference*, Princeton, NJ: Princeton University Press.

4 Geography, fairness, and liberal democracy

R. J. Johnston

Justice, fairness and equality are commonly-used terms in discussions regarding the design and operation of electoral systems. All three have deep roots in considerations of ethics. Justice, according to Rawls (1958: 164), involves two principles:

> first, each person participating in a practice, or affected by it, has an equal right to the most extensive liberty compatible with a like liberty for all; and second, inequalities are arbitrary unless it is reasonable to expect that they will work out for everyone's advantage.

Fairness follows from this:

> A practice will strike the parties (to a contract or joint activity) as fair if none feels that, by participating in it, they or any of the others are taken advantage of.

A fair set of rules can be considered just, therefore, if people are not treated unequally, unless that inequality can be justified as being in the greater good of all. This is the basis of Rawls' (1971: 11) celebrated definition of 'justice as fairness':

> the principles that free and rational persons concerned to further their own interests would accept in an initial position of equality as defining the fundamental terms of their association.

Inequalities are only just if they produce compensating benefits for all: 'It may be expedient but it is not just that some should have less in order that others may prosper' (Rawls 1971: 15).

Equality, according to Arneson (1993), has two aspects: equality of condition (or of life prospects) and equality of democratic citizenship. Democracy is usually interpreted (as in the OED definition) as 'a system of government by the whole population, usually by elected representatives,' and sometimes phrased as 'rule of the people, by the people, and for the people'. It is a contested concept, however, whose meaning has varied substantially over time and space (Barber 1987: 119, refers to it as 'one of the most cherished and at the same moment most contested of political ideals'); according to Arblaster (1996: 9) 'At the root of all definitions of democracy, however refined and complex, lies the idea of popular power, of a situation in which power, and perhaps authority too, rests with the people,' either to direct governments or to limit their degrees of freedom. One of democracy's basic and continuing core ideas is 'equal political rights for all' (p. 25), though political equality is very difficult to achieve without economic and social equality too (see Marshall and Bottomore 1996). Gutmann (1993: 412) extends this definition by arguing that: 'democracy represents fair terms of a social contract among people who share a territory but do not agree upon a single conception of the good. On this common contractarian view, democracy consists of a fair moral compromise.'

Reaching that compromise involves geographical considerations in most societies, because most democratic systems are organized as spatial systems. In this chapter, I explore the notions of fairness (and hence justice and equality) involved in the design and operation of democratic systems, pointing up their geographical components and indicating the difficulties of applying democratic principles in geographical settings.

On democracy

Unanimity on how issues should be resolved is likely to be rare in large-scale societies, which require means of ensuring a system of majority decision-making which does not permanently disadvantage any minorities. Methods of identifying views and representing them have to be established, so that all citizens are treated equally in decision-making processes. Held (1996) identified four main types of democratic decision-making system – classical, republican, liberal and direct. All entail citizen participation in elections held to influence how their state is governed, by whom, and to what ends (Birch 1993). Elections are central to most contemporary democratic operations, and electoral systems must be evaluated against criteria of equality and fairness.

From classical Greece onwards, major questions addressed in all democracies have been 'which people?' should participate, and 'in what ways?' In Athens, for example, the entire citizenry formed the sovereign assembly, which met almost weekly, but a Council of 500 undertook preparatory work for those meetings and a Committee of 50 guided the Council (Held 1996: 21–22). Means of choosing Council and Committee members included direct elections, selection by lot, and straightforward rotation. As societies expanded in scale, elected bodies became central to democratic practice because the exercise of power on a day-to-day basis had to be delegated to a representative body, and it became a *sine qua non* that 'The opportunity to vote in periodic competitive elections is the minimum condition that a governmental system must satisfy to qualify as democratic' (Birch 1993: 80).

Three sets of institutional guarantees are necessary to such a liberal democracy according to Dahl (1978):

1   in the formulation of preferences, involving the freedoms to form and join organizations, of expression, of information, to compete for votes and to stand for public office;
2   in signifying preferences, through free and fair elections; and
3   in the equal weighting of preferences, and their relationship to policy-making.

On the last of these, he argued that 'all procedures for making binding decisions must be evaluated according to *the criterion of political equality*' (Dahl 1979; his emphasis). Each citizen must have an equal vote; otherwise equally valid claims are not given equal shares of power and some claims are considered more valid than others.

The right to vote for the body or bodies allocated power to rule within the state apparatus has thus become associated with the concepts of equality and fairness: citizens are entitled not only to vote for representatives to participate in their government but also to equality in that voting process. Any system which denies them such equality is thus deemed 'unfair'.

Almost all liberal democratic electoral systems have major geographical components, sets of bounded spaces within the national territory which determine where individuals can vote, and who will represent them. Many different systems are used (the IDEA 1997 refers to hundreds currently in use: see also Dummett 1997 and Farrell 1997). Whichever is chosen can have a 'profound effect' on political life (IDEA 1997: 1), because the system shapes 'the rules of the game under which democracy is practiced' (IDEA 1997: 7). Electoral systems are relatively easy to manipulate for political gain, however, and 'the choice of electoral system can effectively determine who is elected and which party gains power'. Thus geography is actively implicated and manipulated in conflicts over electoral power and fairness in obtaining political representation.

Fairness in liberal democracies and the design of electoral systems

The concept of fairness is used in a variety of ways in this context, all relevant to appreciation of the role of geography in securing justice within a liberal democracy. Four fairness domains are identified here, with particular reference to the election of legislatures. Each is examined in turn before the complications which emerge when two or more are combined are introduced.

Fairness to individuals

All citizens should be equally powerful: each person's vote should be worth the same as everybody else's, and unequal power to influence the outcome of an election violates the rights of those whose votes are worth less than others' and whose preferences consequently carry less weight.

Such fairness could be delivered if election to a legislature involved a single contest across the whole of the state's territory and if fairness to individuals was the sole domain being applied. This is rarely the case, however, though it is used in elections for executive positions, such as state presidents (though not for the President of the USA). Only a small number of countries, including Israel, Namibia and the Netherlands, elect their legislatures in that way. Most have legislators as representatives from sub-national constituencies – contiguous blocks of territory in virtually every case – on the rationale that this creates a link between the legislator and that segment of the electorate whom he or she represents, and to whom they can turn for advice and assistance.

Ensuring that election from spatially-defined constituencies does not violate the fairness to individuals criterion involves securing similarity in constituency size. The ratio of electors to legislators should be the same in all constituencies. The relevant United Kingdom legislation requires each single-member constituency to have the same number of electors 'as far as is practicable', for example, and in the United States the Fourteenth Amendment was deployed in a series of legal decisions beginning in the 1960s (known as the reapportionment revolution) which required states to create Congressional Districts and other constituencies with equal numbers of electors in order to ensure that all voters were treated equally.

Fairness to communities

A society is not an aggregation of isolated, atomized individuals but an amalgamation of separate communities with their own cultures and interests. Each community should be fairly represented in the democratic decision-making fora, otherwise power is not equitably distributed among society's major component parts. Community is another term, like democracy, however, that is both 'much cherished and much contested', and one of the problems involved in ensuring fairness to communities is defining them (see Bell and Newby 1978), except in federal states, like Australia, Canada and the USA, where long-established sub-national territories form the major communities to which fairness is due under the relevant constitution.

Many communities (or interest groups) are not spatially-identifiable within a national territory, but the 'fairness to communities' criterion in the design of electoral systems usually assumes that communities can be defined as occupying contiguous blocks of territory. Representation of such communities was the basis of the Parliamentary democracy which evolved from the thirteenth century onwards in the United Kingdom, until the move towards universal adult franchise in the nineteenth and twentieth centuries brought fairness to individuals increasingly to the fore. Parliaments comprised the representatives of the (very small) enfranchised populations in the shires (the traditional units of local government) and boroughs (the urban areas with royal charters for holding markets and fairs). Each elected a set number of representatives.[1] The three major nineteenth-century reforms (in 1832, 1867 and 1885) involved political conflicts not only over extending the franchise but also over the relative representation of town and country, of the interests of landowners and manufacturers, of property and of trade; fairness to communities was stressed rather than fairness to (the enfranchised) individuals.

Fairness to minorities

Few national societies are culturally homogeneous and most contain one or more separate minorities defined by criteria such as ethnicity, language and religion (Mikesell and Murphy 1991). National cohesion and stability depends on these minorities being fully and equitably involved in the overall society, otherwise civil order could be threatened if one or more groups considered themselves unequally treated.[2]

In the United States, fairness to minorities includes the Supreme Court-determined requirement that members of defined minorities (i.e. those recognized by the Constitution, normally taken to be ethnic minorities) have a level of representation from each state commensurate with their relative size. On this basis, if, for example, black voters comprise 25 percent of a state's electorate, 25 per cent of the state's congressional districts should have a majority of black residents, which has led to the legally-acceptable practice of racial gerrymandering.[3] Other countries have designed systems to ensure minority representation. In New Zealand, for example, four seats were originally set aside in the House of Representatives for the Maori population in the nineteenth century – although this was done when voting was based on a property qualification, and all Maori property was communal: there are now five. (See also Mathur 1997, on an ingenious system used in Mauritius.)

Ensuring that minorities are represented in a legislature (or have the potential to be represented there), commensurate in numbers with their relative size within the population, may meet a minimum definition of fairness in this domain, but their political power may not be consistent with that (unless they hold the balance of power in the legislature): they will probably remain a minority with little political influence and therefore feel alienated from the political system. In Northern Ireland, for example, an electoral system which ensured that the so-called 'nationalist' (Roman Catholic) minority won representation relative to its size meant that the majority (the 'Unionists') always had a majority of seats in the provincial legislature and the latter's unwillingness to recognize the minority's claims meant that they felt powerless, despite the apparent 'fairness' of the electoral system (see Arblaster 1996: 68–69).

Fairness to political parties

Political parties are at the centre of contemporary liberal democracies: they provide stability for a government within a legislature, which can rely upon the (usual) support of those parties which underpin it and so guarantee that its legislative proposals are accepted; and they provide continuity of support among the electorate, through a portion of the population who can normally be counted upon to vote for the same party at each election. They are the focus of efforts to mobilize the electorate and to organize government.

Fairness to parties as a criterion to be implemented in the design and operation of an electoral system is usually interpreted as proportional representation: a party should be allocated the same proportion of seats in the legislature as it won votes in the relevant election. In the United Kingdom, for example, the experience of the Liberal Democrat party (and its various predecessors) in winning between 15 and 25 per cent of the votes cast in general elections from 1974 on, but never winning even 10 per cent of the seats in the House of Commons, is widely considered 'unfair' (the maximum was 7 per cent of the seats in 1997, when the party won 17 per cent of the votes). For this reason, many (not only supporters of such discriminated-against parties) argue for adoption of an electoral system which results in proportional representation for parties (either by explicit design or because of properties which tend to favour that outcome).[4]

Electoral systems can be designed to meet 'fairness to parties' criterion.[5] But, as with 'fairness to minorities', proportional representation does not necessarily lead to 'proportional power' (indeed, it will almost certainly not do so: Johnston and Pattie 1997). A party may be proportionally represented but have neither power nor influence unless it is either the majority party in the legislature or in a coalition with other parties which form a stable majority.[6] This is particularly important in legislative assemblies, whose major roles include the passing of laws and the creation and maintenance of governments: it is somewhat less important in deliberative assemblies, which debate issues and policies but have few law-making powers, and so do not require permanent majority governments, either single-party or coalition.[7] Proportional representation and proportional power are particularly desirable in legislative bodies.

Consider a legislature of 100 members elected by proportional representation, in which four parties have the following number of seats (I am indebted to Laver 1997, for this example):

A	B	C	D
26	26	26	22

If it were a legislative assembly, with 51 votes needed to pass any measure, then as no party has that majority of the votes, either a permanent coalition (a government) has to be created or an *ad hoc* majority has to be assembled for each measure. Three potential party groups meet the criterion of a 'minimal winning coalition' – a group of parties with sufficient votes to pass a measure, with all parties necessary to that majority:[8] AB, AC, and BC. Party D is not necessary to a single minimal winning coalition despite having over one-fifth of the seats in the legislature: any group of parties of which it was a member and which had a majority of the votes would still have a majority even if D were excluded. So D is proportionally represented but has no power there. The 'fairness to parties' criterion has been met in terms of the size of its representation, but its allocation of power within the assembly is extremely 'unfair'. It has sufficient 'voice', relative to the portion of the electorate which voted for it, to make the case sustained by its supporters and so seek to influence the other parties, but no 'exit' power to affect decision-making, and so has to accept whatever the others decide. In the terminology of Hirschman's (1970) classic model, the only course open to it and its supporters is. 'loyalty', acceptance of the majority view; it is excluded from that majority unless it agrees with two of the other parties, who need take no account of what party D thinks.

This is not an isolated example of the potential unfairness on one criterion (proportional power) of an electoral system which meets one of the fairness criteria (proportional representation). Indeed, proportional representation is very rarely equated with proportional power (see Johnston 1998). Nor is it always the smaller parties which are 'unfairly' treated.[9] The distribution of power within a legislature where no party has a majority can be very volatile with changes in the allocation of seats, and small variations in the configuration of votes across the parties can have major impacts on their relative power.

Meeting more than one fairness criterion: the British case

Few electoral systems are designed to meet just one of the four criteria outlined above, and most are at least implicitly supposed to be 'fair' on at least two. Achieving that goal is extremely difficult, however.

The British electoral system illustrates this. Fairness to individuals has become the dominant criterion in defining the single-member constituencies used to elect MPs.[10] The initial legislation establishing the procedure for creating constituencies by independent Boundary Commissions required those bodies to ensure 'fairness to communities' by matching local government and constituency boundaries, as far as possible, and to meet the 'fairness to individuals' criterion by ensuring that no constituency had an electorate which deviated from the national average by more than 25 per cent. The Commissions informed Parliament that meeting both was not feasible: they could not ensure fairness to communities within the (wide) degrees of freedom set by the fairness to individuals criterion. Parliament's response was that fairness to communities was the more important, and the fairness to individuals criterion was modified to say that all constituencies should have the same electorates 'as far as is practicable', having regard for the primary requirement of fitting constituencies within the local government template.

Changes to the local government map and in the distribution of electors mean that over time the constituencies (or at least some of them) no longer meet the criteria. The original legislation required the Commissions to review all constituencies every five to seven years, and recommend changes to bring them in line with the criteria. The Commissions did this in 1954, seven years after their original report in 1947, and created considerable consternation among MPs and their parties because, although 398 of the 625 constituencies were unchanged, many felt that frequent change to constituency boundaries would disrupt the work of parties in mobilizing voter support and weaken the links between electors and their representative. Parliament changed the rules in two ways: first, the period between reviews was lengthened to 10–15 years;[11] and second, when proposing changes the Commissions were to take account of 'local ties' and the disruption that normally follows the creation of new constituencies. As the Home Secretary of the time put it, the presumption was 'against making changes unless there is a very strong case for them'. 'Fairness to individuals' was, to some extent, subordinated to both 'fairness to communities' and 'fairness to parties'.

This erosion of the 'fairness to individuals' criterion was continued in 1982 when four senior Labour Party officers sought to have the Boundary Commission for England's Third Periodic Report referred back by the courts before it was presented to Parliament, on the grounds that it had not paid sufficient attention to that criterion. This was rejected on the grounds that the legislation, although it said that all constituencies should have equal electorates, as far as practicable, was not an invitation for the Commissions to undertake 'an exercise in accounting'. Instead they were given a number of criteria and wide freedom to exercise their judgement – which could only be challenged if it was clear that they had not undertaken the task conscientiously (Rossiter, Johnston and Pattie 1999).

'Fairness to individuals' has become increasingly important in the operation of the British electoral system (by default if not design), because the Commissions now recommend the great majority of constituencies within 10 per cent of the national quota. This does not necessarily imply that each individual's vote is of equal weight, however. In an electoral system entirely dominated by the political parties, many individuals' votes are of little consequence, because they are very unlikely to have an impact on the outcome in their constituency. Labour Party supporters living in a constituency where Conservative voters predominate realize that their vote is very unlikely to affect the result there: if they bother to turn out, their vote will in effect be 'wasted'. Similarly, Labour Party supporters living in constituencies where their party has a large majority also realize that their votes are unlikely to affect the outcome: if they do turn out, their votes will be 'surplus to requirements'. A party wants all of its supporters' votes to count (as they would do under proportional representation), not to be wasted in its opponents' safe seats or to be piled up creating large majorities in its own safe seats. The most efficient resolution is for its votes to be so distributed across the constituencies that it wins as many as possible by a comfortable majority, but no more, and that where it loses, it has virtually no supporters. This can be an intended outcome of the definition of constituencies by Boundary Commissions because parties are invited to make representations to the Commissions, which invariably seek to ensure that boundaries are drawn to their benefit (Rossiter, Johnston and Pattie 1997a). Alternatively, a party may so mobilize its electoral support that its votes are distributed much more efficiently than its opponents'. This was done very successfully by the Labour Party at the 1997 general election. It targeted its campaign resources very carefully on the marginal seats,[12] and the outcome was that with 44 per cent of the votes cast it won 63 per cent of the seats, compared to 53 per cent of seats won by the Conservative Party with a similar vote percentage in 1979. Furthermore, so efficient was its strategy that if there was a uniform movement of support away from Labour to the Conservatives across all constituencies at the next general election, with each party getting the same percentage of the votes cast (37), Labour would still have an 82-seat lead over its opponent (Johnston *et al.*, 1998).

The 1997 British general election was the first fought in a new set of relatively equal-sized constituencies, so that 'fairness to individuals' was largely ensured.[13] But the result was 'unfairness to political parties'. Hence the pressure in parts of the British polity for electoral reform, to introduce a proportional representation system in which every vote counts, wherever it is cast. This undoubtedly means multi-member constituencies, in part if not entirely,[14] which will mean breaking the MP–constituency link,[15] and probably lead to a further deterioration in fair representation of communities, unless the communities are large.[16] Carefully designed, multi-member constituencies can ensure both 'fairness for individuals' and 'fairness for minorities' – especially if minorities have their own political parties. But, like all other electoral systems, they are open to partisan manipulation (Gallagher 1992).

Globalization and fairness in the international arena

This discussion has focused so far on political power within liberal democracies. But the world economy has become increasingly integrated over recent decades and regulation of that new reality calls for institutions which have a territorial scope beyond that of any individual state.

States are the only actors in the international arena recognized as having the sovereign power to negotiate on behalf of the populations living within their territories, so they are key players in multinational and global regulatory agencies. Creation of those agencies raises the same issues of fairness addressed above. Should every state have the same number of votes, which might be interpreted as enacting 'fairness to communities' (where the territorial state is equated with a community), for example, or should a state's number of votes be commensurate with its population, so ensuring 'fairness to individuals'?

This question was addressed in the creation of the institutions associated with what was the European Economic Community and is now the European Union; it is revisited whenever the Union incorporates more members. It was resolved in the construction of both the European Parliament and the European Council of Ministers (the executive body) by allocating seats and votes to each member country relative to its population, but because populations varied so substantially, from Luxembourg (0.4 million) to Germany (81.4 million) in 1997, the allocations deviated substantially from proportionality, especially for the smaller countries. With only six member countries (1958–1972), for example, Luxembourg had one vote in the Council of Ministers and Germany had four. Enlargement to nine members in 1973 saw Luxembourg's representation doubled and Germany's increased to 10, out of a total of 58 Council votes. Those numbers have not changed since (although Germany's population increased by some 20 million with the incorporation of East and West Germany in 1991) but the

Council has been expanded and the 15 members from 1995 onwards had a total of 87 votes. In the original six, Germany had 23.5 per cent of the votes and 32.2 per cent of the population, whereas Luxembourg had 5.9 and 0.2 per cent respectively: in the current 15-member Union, Germany's figures are 11.5 and 22.0 respectively, and Luxembourg's are 2.3 and 0.1. (The figures are from Felsenthal and Machover 1997.)

This distribution of votes in the Council of Ministers does not meet the 'fairness to individuals' criterion, and it is hard to construct an argument that it meets the 'fairness to (national) communities' criterion either. Reflecting this, the Council has evolved a voting procedure for certain key issues (known as Qualified Majority Voting) which requires a majority of approximately 71 per cent to be cast for a measure to be implemented: an alternative interpretation is that a minority of countries with just under 30 per cent of the votes can block measures supported by a majority. (The QMV majorities in the five phases of the EU's growth have been 12/17, 41/58, 45/63, 54/76, and 62/87 or 0.706, 0.707, 0.714, 0.711, and 0.713.)

As with proportional representation in legislative assemblies, however, even if the allocation of votes in the EU Council of Ministers was adjudged fair to 'communities' – in this case countries – it may not also meet the criterion of 'fairness to individuals'. Felsenthal and Machover (1997) showed that, with the exception of Luxembourg, the power of each country with regard to QMV voting is largely commensurate with its number of votes, but not with its population. Whether it is adjudged fair overall, therefore, depends on which of the criteria is considered salient, and asks a crucial question regarding the construction of international agencies: how important is a country's population in determining its voting strength?

'Fairness to individuals' is not the only criterion applied in the debates over voting strength in international agencies. In the debates over the Montreal Protocols implementing the international agreements to decrease CFC emissions and so reduce the threat to the ozone layer, for example, the United States argued that they should only come into effect when they had been ratified by at least 11 countries which were together responsible for two-thirds of the world's production of CFCs, which allocated veto power to itself and a small number of other 'developed countries'. A 14-country committee was established to oversee implementation of the Protocols; the USA insisted on seven members being 'developed countries'; that it be allocated a permanent place; and that a two-thirds majority decision-making procedure be adopted, thereby giving the 'developed countries' effective veto power.[17] American interests were protected far in excess of any 'fair representation', unless 'fairness by wealth' is taken as an important criterion.

Other agencies created to regulate environmental use illustrate the problems of ensuring fairness – indeed in defining fairness in such contexts. These involve major debates about the 'ownership' of much of the environment – especially the atmosphere and the oceans. International law recognizes two types of 'territory' beyond that claimed by sovereign states: *res extra commercium (rec)*, which is territory that cannot be claimed by any state but which all may 'exploit' (i.e. it is a commons which all can graze); and *res communis humanitatis (rch)*, which is recognized as the common property of all humankind, and should be managed accordingly. Many countries wished to apply rch principles in the evolving component of the International Law of the Sea dealing with the economic exploitation of the ocean bed and proposed that it be governed by an authority with 36 countries elected as members, which would distribute the royalty income obtained to all countries (see Prescott 1985). The United States declined to ratify the treaty, however, on grounds which included: the authority would not be so constituted as to accord the US, and similar countries, a 'role that fairly reflects and protects their interests'; and amendments could be passed even if the US disagreed, because it was given no power of veto (Dubs 1986: 113–114). American intransigence, and that of other 'developed countries', blocked implementation of the agreement, and a much weaker one was eventually ratified by sufficient UN members in 1997.

As globalization continues apace, and as the resolution of environmental problems becomes increasingly acute, so the nature of international regulatory regimes will become a focus of important debate. If the global economy is to be regulated for the greater good of all, and if the global environment is to be protected and sustained for future generations, questions of fairness in the construction of international regimes will take centre stage. (On regime theory, see Vogler 1996.) How can systems be constructed which are 'fair', and what does 'fairness' mean in this context? Is it the same as 'fairness' in the construction of representative liberal democracies within individual states?

Conclusion

Winston Churchill once told the British House of Commons that:

> Many forms of government have been tried, and will be tried in this world of sin and woe. No one pretends that democracy is perfect or all-wise. Indeed, it has been said that democracy is the worst form of Government except all those other forms that have been tried from time to time.[18]

As this chapter has illustrated, that 'least worst' system of government may be readily defined in terms of basic principles (or freedoms), but it is much more difficult to implement. Creating a system which is 'fair' on one criterion is not always straightforward; creating one which is fair on two or more may be extremely difficult, if not impossible.

The crux of these difficulties is the relationship between representation and power. Politics in any democracy involves contests over power, over the right to focus decision-making on policies which are almost certain to disadvantage some people, however much they may advantage others. Thus even if it is possible to construct a system which is fair on predetermined criteria in its allocation of representation to people, communities, minorities and interest groups (political parties), it may well not be fair in the power which they get – indeed, it will almost certainly be unfair to some (or more fair to some than others). Fairness per se is extremely difficult to achieve: fairness based on geographical building blocks is almost impossible – though, of course, some systems are fairer than others.

Acknowledgements

I am grateful to Jim Proctor, David Smith and Albert Weale for valuable comments on drafts of this chapter, while absolving them from any responsibility for its final contents.

Notes

1 As the franchise was extended in the nineteenth century, many boroughs were too small to justify separate representation. Most lost their separate status as Parliamentary boroughs and were merged with their surrounding Parliamentary counties. Within some counties, however, and especially in Scotland and Wales several boroughs were grouped together to elect one MP, even though they were not contiguous, on the grounds that urban and rural areas had different interests (i.e. were different types of community) and so should be separately represented: the last constituency of this type was abolished only in the 1950s.

2 Fairness to women is an issue in some countries – although in almost all countries women form a majority of the population and electorate, after many decades of conflict over the enfranchisement of women. It remains the case that in most countries women are substantially under-represented in the various legislatures, and some commentators argue that particular electoral systems (notably those with multi-member constituencies) are more likely to lead to the election of substantial numbers of women candidates (even the nomination of substantial numbers, given that most political parties are male-dominated) than are others (such as the first-past-the-post system employed in the USA and the UK). The main issue is cultural, however, relating to the treatment of women within society: because there is little, if any, difference in the geographical distribution of women and men within a society, the unequal treatment of women within a political system cannot be addressed through geographical solutions.

3 A recent Supreme Court decision had modified this requirement, stating that black majority districts should not be created if in so doing this subordinates 'traditional districting principles, including but not limited to compactness, con-tiguity, respect for political subdivisions or communities defined by actual shared interests, to racial considerations' (*Miller v. Johnson*: 132 L Ed 2d 780).

4 The single transferable vote (STV) system is frequently promoted by proponents of electoral systems that will be fairer to parties; it was not designed with proportional representation in mind, but will normally produce a proportional outcome – certainly much more so than the plurality-majority systems favoured in most English-speaking countries.

5 There are, however, may technical details regarding the choice not only of the system but also its basic parameters – such as the size (number of elected members) of a constituency – which influence how well the fairness criterion is met: for full details see Gallagher (1992).

6 There is a massive literature on government formation through coalition: see, for example, Laver and Shepsle (1996)

7 The European Parliament is a deliberative assembly, and opponents of

proportional representation for legislative assemblies in the current British government (including the Prime Minister, Tony Blair, and the Home Secretary, Jack Straw) have agreed to the use of a proportional representation system for the elections to that Parliament in June 1999, along with the first elections to the Scottish Parliament and the Welsh Assembly (in May 1999). PR was used for the first elections to the new Northern Ireland Assembly in July 1998, and there are suggestions that it will be introduced for local government elections in Scotland in 2001.

8 Coalitions which are more than 'minimal winning' are sometimes formed, notably in periods of 'national emergency', such as war. The more parties there are in a coalition, however, the harder it is to sustain, hence the importance of minimal winning coalitions which contain the smallest number of parties consistent with achieving a majority on every vote.

9 Though they usually are in the allocation of seats: Taagepera and Shugart (1989).

10 On that trend, see Rossiter, Johnston and Pattie (1999).

11 It was changed to 8–12 years in 1992, by a government which wished to ensure that new constituencies were in place before the next general election, believing that it would profit electorally if they were.

12 On targeted campaigns, see Denver and Hands (1997).

13 One 'unfair' aspect was the relative over-representation of Scotland and Wales relative to England and Northern Ireland, a differential sometimes defended as 'fairness to minorities' (the Scots and Welsh – but not the Northern Irish, the smallest minority of the three).

14 One of the most popular schemes propounded is the additional member system, a variant of the German hybrid mixed member system, which has a proportion of the members elected for single-member constituencies and the remainder from multi-member units, with the number of the latter determined so that pro-portional representation is achieved overall. This was adopted in New Zealand, by popular referendum, in 1993.

15 Though the Irish system of multi-member constituencies has not precluded strong links between individual MPs and electors in particular parts of their constituencies.

16 In the USA and Australia, of course, single-member constituencies ensure fairness for individuals in elections to the House of Representatives, whereas multi-member constituencies for the Senate ensure fairness of representation for the communities – with communities defined as the states.

17 The EU has similar procedures. On certain 'special' economic issues, the majority must not only meet the QMV criterion but must include 'at least two member-states each of which produces at least one-tenth of the total value of the coal and steel produced by the EU' (Felsenthal and Machover 1997: 34).

18 On 11 November 1947. The quotation is reproduced here from the *Oxford Dictionary of Quotations*, new edition, 1979.

References

Arblaster, A. (1996) *Democracy* Buckingham: Open University Press.

Arneson, R. J. (1993) 'Equality' in R. E. Goodin and P. E. Pettit (eds) *A Companion to Contemporary Political Philosophy*, Oxford: Blackwell Publishers, pp. 489–507.

Barber, B. R. (1987) 'Democracy' in D. Miller, J. Coleman, W. Connolly and A. Ryan (eds) *The Blackwell Encyclopaedia of Political Thought*, Oxford: Blackwell Publishers, 114–119.

Bell, C. and Newby, H. (1978) 'Community, communion, class and community action: the social sources of the new urban politics', in D. T. Herbert and R. J. Johnston, (eds) *Social Areas in Cities: Processes, Patterns and Problems.* Chichester: John Wiley, pp. 283–302.

Birch, A. H. (1993) *The Concepts and Theories of Modern Democracy,* London: Routledge.

Dahl, R. A. (1978) 'Democracy as polyarchy', in R. D. Gastil (ed.) *Freedom in the World: Political Rights and Civil Liberties,* Boston: G. K. Hall, pp. 134–146.

—— (1979) 'Procedural democracy', in P. Laslett and J. S. Fishkin (eds) *Philosophy, Politics and Society,* fifth series, Oxford: Blackwell Publishers, 97–133.

Denver, D. T. and Hands, G. (1997) *Modern Constituency Campaigning,* London: Frank Cass.

Dubs, M. (1986) 'Minerals of the deep sea: myth and reality', in G. Pontecorvo (ed.) *The New Order of the Oceans,* New York: Columbia University Press, pp. 85–124.

Dummett, M. (1997) *Principles of Electoral Reform,* Oxford: Oxford University Press.

Farrell, D. M. (1997) *Comparing Electoral Systems.* Hemel Hempstead: Harvester-Wheatsheaf.

Felsenthal, D. S. and Machover, M. (1997) 'The weighted voting rule in the EU's Council of Ministers, 1958–1995: intentions and outcomes', *Electoral Studies,* 16: 33–47.

Gallagher, M. (1992) 'Comparing proportional representation electoral systems: quotas, thresholds, paradoxes and majorities', *British Journal of Political Science,* 22: 469–496.

Gutmann, A. (1993) 'Democracy', in R. E. Goodin and P. E. Pettit (eds) *A Companion to Contemporary Political Philosophy,* Oxford: Blackwell Publishers, pp. 411–421.

Held, D. (1996) *Models of Democracy* (second edition), Cambridge: Polity Press.

Hirschman, A. O. (1970) *Exit, Voice and Loyalty,* Cambridge, Mass.: Harvard University Press.

IDEA (1997) *The International IDEA Handbook of Electoral System Design,* Stockholm: International Institute for Democracy and Electoral Assistance.

Johnston, R. J. (1995) 'The conflict over qualified majority voting in the European Union Council of Ministers: an analysis of the UK negotiating stance using power indices', *British Journal of Political Science* 25: 245–253.

—— (1998) 'Proportional representation and a fair electoral system for the United Kingdom', *Journal of Legislative Studies* 4: 128–148.

Johnston, R. J. and Pattie, C. J. (1997) 'Electoral reform without constitutional reform: questions raised by the proposed referendum on proportional representation for the UK', *The Political Quarterly* 68: 379–387.

Johnston, R. J., Rossiter, D. J., Pattie, C. J., Dorling, D. F. L., MacAllister, I. and Tunstall, H. (1998) 'Anatomy of a landslide: the constituency system and the 1997 general election', *Parliamentary Affairs* 51:131–148.

Laver, M. (1997) *Private Desires, Political Action: An Invitation to the Politics of Rational Choice,* London: Sage Publications.

Laver, M. and Shepsle, K. A. (1996) *Making and Breaking Governments: Cabinets and Legislatures in Parliamentary Democracies,* Cambridge: Cambridge University Press.

Marshall, T. H. and Bottomore, T. (1996) *Citizenship and Social Class,* London: Pluto Press.

Mathur, R. (1997) 'Parliamentary representation of minority communities: the Mauritian experience', *Africa Today*, 44: 61–82.

Mikesell, M. W. and Murphy, A. B. (1991) 'A framework for comparative study of minority-group aspirations', *Annals of the Association of American Geographers* 81: 581–604.

Morrill, R. L. (1973) 'Ideal and reality in reapportionment', *Annals of the Association of American Geographers* 63: 463–477.

Paddison, R. (1976) 'Spatial bias and redistricting in proportional representation electoral systems: a case-study of the Republic of Ireland', *Tijdschrift voor Economische en Sociale Geographie* 67: 230–241.

Prescott, J. R. V. (1985) *The Maritime Political Boundaries of the World*, London: Methuen.

Rawls, J. (1958) 'Justice as fairness', *Philosophical Review* 67: 164–194.

—— (1971) *A Theory of Justice*, Oxford: Oxford University Press.

Rossiter, D. J., Johnston, R. J. and Pattie, C. J. (1997a) 'Estimating the partisan impact of redistricting in Britain', *British Journal of Political Science* 27: 319–331.

—— (1997b) 'Redistricting and electoral bias in Great Britain', *British Journal of Political Science* 27: 466–472.

—— (1999) *The Boundary Commissions: Redrawing the United Kingdom's Map of Parliamentary Constituencies*, Manchester: Manchester University Press.

Taagepera, R. and Shugart, M. S. (1989) *Seats and Votes: The Effects and Determinants of Electoral Systems*, New Haven: Yale University Press.

Vogler, J. (1996) *The Global Commons: A Regime Analysis*, Chichester: John Wiley.

5 Exploring the ethics of development

Seamus Grimes

Introduction

The primary concern of this chapter is the failure of the predominant market-led model of development to include the majority of humankind in its benefits. From stating the extent of this failure it reviews various traditions within geography and the social sciences generally which try to account for widespread exclusion. Beginning with economistic analyses, it also examines the important contribution of the political economy tradition. Neo-Malthusian perspectives are also critically reviewed, while the increasing questioning of the nature of development by poststructuralists is seen as a more positive, though limited, contribution.

While most of these analyses of development have emerged from some form of moral reasoning and most have shared a concern with improving the lot of humanity, the growing literature dealing specifically with an ethical evaluation of development thinking is also examined. Attempts to incorporate an ethical dimension by means of concepts such as human flourishing, transcendence, solidarity and the common good are reviewed. The practical difficulties of operationalizing these philosophical concepts in a policy arena are also considered.

Development and underdevelopment

While many geographers and social scientists have long been preoccupied with the failure of development models within the developed world to bring about greater participation by society in the fruits of wealth creation, a significant part of development literature has focused for obvious reasons on the enormous and growing inequality between the rich north and the poor south. The extent of the gap involved is revealed by the fact that the share of world income for the richest 20 per cent rose from 70 per cent in 1960 to 85 per cent in 1991, while the share of the poorest 20 per cent fell from 2.3 per cent to 1.4 per cent. In 1960 the top fifth of the world's population made 30 times more income than the bottom fifth and by 1989 this gap had expanded to 60 times. The extent of this gap reflects disparities in trade, investment, savings and commercial lending and in access to global market opportunities, with the bottom 20 per cent of the world population accounting for less than 1 per cent of world trade (UNDP 1994). In fact the share of sub-Saharan Africa in world trade in manufacturing goods fell from 1.2 per cent in 1970 to 0.4 per cent in 1989 (Castells, 1996). These data also reveal a pattern of overconsumption and overproduction in the north and extreme poverty in the south.

Although widespread poverty is to be found throughout the less developed world, the development experience of the south has in fact been highly differenti- ated with the need, perhaps, to consider the older terminology of north/south, First World/Third World as being obsolete (Slater 1997). The blurring of the boundaries between First and Third Worlds is reflected in the growing number of Third World immigrants in the United States, for example, many of whom are making a livelihood based on business links with their home countries (Portes 1997). The increasing differentiation of economic growth, technology capacity and social conditions between countries and within countries throughout the south has also prompted some calls for an end of the Third World to be announced (Castells 1996). The most obvious pattern of differentiation has been between the experience of the NICs of East and Southeast Asia, which until very recently had benefited significantly from the process of globalization, and sub- Saharan Africa and South Asia. This distinction is highlighted by the fact that while private investment flows to developing countries between 1970 and 1992 increased from $5bn to $10bn, sub-Saharan Africa received only 6 per cent of this flow during the 1980s (UNDP 1994). This can be partly explained by the fact that during the 1970s most African countries came under the direct control of policies of international financial institutions that imposed liberalization measures which have had an overall negative effect on trade and investment since then.

Under the dominance of free market conditions both internationally and domestically, it has been suggested that most of Africa, outside South Africa, has 'ceased to exist as an economically viable entity and that the logic of the new global economy has little role for a majority of the African population in the newest international division of labour' (Castells 1996: 135–136). Among the major obstacles hindering the integration of much of Africa within the global economy is the poor level of infrastructure, including that associated with the new information and communication technology.

Despite the gloomy prognosis for much of the African continent, recent shifts in employment, investment and trade from the north to other parts of the south were sufficiently impressive in the early 1990s for some to suggest that developing countries were helping to pull the more developed world out of recession (*The Economist* 1 October 1994). Between 1990 and 1993, for example, Third World countries increased their imports by 37 per cent and their exports also rose by 37 per cent. By 1994 multinational companies employed 12 million workers in developing countries compared with 61 million in the developed world, but the developing countries accounted for almost two-thirds of the total increase in MNC employment since 1985. While these recent trends may look impressive for those parts of the south which are benefiting, it has been estimated that the net outflow of investment since 1990 has reduced the rich world's capital stocks by a mere 0.5 per cent from what would otherwise have been the case (Krugman 1994). Nevertheless this relocation of what represents only a small fraction of global investment has been sufficiently large to cause considerable concern within some political circles both in Europe and in the United States.

The shift towards more flexible and knowledge-intensive specialization, while offering new possibilities to less developed countries, may also make it more difficult for them to attract export-oriented FDI in manufacturing. Apart from some East Asian newly industrializing countries, the 1980s witnessed an 'unprecedented shrinkage' of the main indicators of international technology flows to Third World countries (Cook and Kirkpatrick 1997: 64). An indication of the differential in human resources between the two parts of the world is obtained by the fact that in 1985 the world average of scientific and technical manpower in North America was 126,200 per million population, which was more than 15 times the level in developing countries (Castells 1996: 108).

Economistic models

It is clear from the above data that while some parts of the less developed world have made significant progress in recent decades in integrating their economies within the global economy, the great majority of mankind are still inhabitants of very poor countries and regions which for a variety of reasons have failed to benefit from the enormous expansion in global wealth creation. It could be argued that the dominant model of development which has been promoted throughout most parts of the world has been an economistic model, which has focused primarily on improving economic performance which is frequently measured in terms of competitiveness and productivity.

Many scholars adopt a pragmatic view of development, seeing it as a process which reflects the functioning of free-market enterprise or capitalism within varying levels of state regulation and support. Castells (1996: 113) defined development as 'the simultaneous process of improvement in living standards, structural change in the productive system, and growing competitiveness in the global economy'. Development and restructuring are elements of an economic system which is constantly in a state of flux, involving an on-going search for higher levels of profitability through the exploitation of resources and business opportunities in optimal locations. The geographer has had a traditional preoccupation with the regional dimension of development which reflects the unevenness associated with the spatial pattern of development at different scales, both between and within countries (see for example, Smith (1977, 1994), Lipshitz (1992), Harvey (1973) and Friedmann (1992). Many scholars, particularly those from within the political economy tradition, have tended to view this unevenness as an inevitable outcome of the spatial functioning of capitalism. Harvey (1996: 201) suggests that one of the most striking failures of capitalism has been its inability to produce anything other than 'the uneven geographical development of bland commoditized homogeneity'. Harvey takes his cue from Marx, whom he suggests has provided a thorough explanation of the production of impoverishment, unemployment, misery and disease as 'necessary outcome of how *laissez-faire* free-market capitalism works' (ibid.: 145). Despite his long-held admiration for Marx's contribution, Harvey acknowledges the powerful argument that the market is 'the best mechanism yet devised to realize human desires with a maximum of freedom and the minimum of socio-political constraints' (ibid.: 151). The emphasis in traditional and Marxist theories on the possession of material goods and positions, and their neglect of other forms of oppression, domination and inequality has been criticized by feminist scholars (see Nagel, this volume).

While one of the most significant streams in development thinking can be categorized as economistic, it must be acknowledged that contributions within the political economy tradition have played an important role in providing an incisive critique of market-led development. Booth (1994), however, is particularly critical of the political economy perspective for ignoring complexity, and suggests that the heart of the development theory impasse is 'the reductionist, economistic and epistemologically flawed nature of Marxism itself' (Watts and McCarthy 1997: 73). Corbridge (1998) is also critical of the Left who appeal to the scientific logics of Marxism rather than to ethics and morality in their critique of capitalist development, and who pay little attention to the achievements of development in terms of life expectancy and literacy levels. Booth is also critical of radical development theory for failing to solve the problems of poverty. A major limitation of both approaches, however, is the tendency to view people's lives as secondary to the structural forces of the capitalist economy. In highlighting the limitations of economistic development thinking, it is also necessary to acknowledge the significance of the pragmatic dimension of development in attempting to bring about a real improvement in the lives of communities who remain effectively excluded from the benefits of wealth creation.

Neo-Malthusian perspectives

While tending to subordinate the personal and moral in favour of the material, the political economy tradition has contributed significantly towards a greater awareness of the social justice dimension of development thinking. It is somewhat more difficult to make a favourable assessment of the neo-Malthusian approach towards development, which has been ideologically influenced by a different form of reductionism. Neo-Malthusian thinking about the links between environment, population and resources has had considerable influence on policy formulation, particularly in the Third World context. Harvey (1996: 148) points out that once 'connotations of absolute limits come to surround the concepts of resource, scarcity, and subsistence, then an absolute limit is set on population. And the political implications of terms like overpopulation can be devastating.'

After decades of intensive research, there is no consensus in sight about a relationship between population, development and the environment (Johansson 1995; Furedi 1997). This, however, has not prevented the widespread diffusion of the belief that economic development in the Third World has been severely retarded by high fertility levels. Furedi (1997) has argued that in addition to ideological factors such as eugenics and racism influencing neo-Malthusian thinking, its particular form of reductionism in relation to development issues arises from the methodological error of isolating population as an independent variable, rather than viewing it as one of many factors within a complex of inter-relationships. Rather than isolating population growth as the major factor respon-sible for widespread poverty in the south, Findlay (1995) suggests the need to examine how economic systems are structured, resulting in unequal terms of trade and severe debt-servicing ratios for poor countries. Sen (1994: 117) views the population problem as one of underdevelopment, and he is critical of the approach which suggests that population growth is the 'cause of all calam-ities', while ignoring the political context of major upheavals in regions such as sub-Saharan Africa.

In his critique of World Bank policy documents dealing with sub-Saharan Africa, Williams (1995) notes the simplistic caricatures used to represent Africans as poor, feckless and ignorant, whose reproductive activities have adverse con-sequences, neither intended nor recognized. The highest priority of these policy strategies is the reduction of fertility rates by means of imported contraceptives in order to save Africa from the 'Malthusian trap'. Although the World Bank reports touch on other issues such as civil wars, inappropriate polices (other than World Bank policies) and falling international prices for exports, their primary focus is the nexus of relationships between agriculture, environment and population problems. Little attention is given to the impact of migration, AIDS, tropical diseases and illnesses related to poverty and squalid living conditions.

Duden (1992) traces the emergence of population as a major policy issue to the late 1970s, when the objective of controlling and managing population in policy statements took on images of an explosion of mainly yellow and brown people in countries that could not repay their debts. Once the *underdeveloped* were identi-fied as outbreeding the north, and at the same time frustrating their own pro-gress, controlling population became a newly defined goal. It is clear that fears about racial balance and the demographic marginalization of the West have been important influential factors in policy development. The emergence of what Furedi calls 'demographic consciousness' is not at all unrelated to the realization that there will be a global shift in the demographic centre of gravity towards developing countries during the next 100 years (Findlay 1995).

In outlining what he calls a new 'global apartheid', Richmond (1994) refers to the efforts of the wealthy countries of North America, Europe and Australasia to protect their affluent lifestyles from the imminent threats of mass immigration from poor countries. The term 'global apartheid' was first coined by Gernot Kohler in the 1970s to describe a situation in which the developed 'north' adopts different standards for its own behaviour from those it applies to the 'south' (see Low and Gleeson, this volume). During the past three decades at least 35 million from the south have taken up residence in the north, with around 1 million joining them each year. Another million or so are working overseas on contracts for fixed periods, while the number of illegal migrants is estimated to be around 15 to 30 million (UNDP 1994). It has been suggested that between 20 and 50 million Muslims from the Maghreb countries of North Africa will have migrated to Europe by the year 2025 (Clarke 1986).

It could be argued, therefore, that rather than reflecting a genuine concern for the economic development of poor countries, the population control agenda of affluent countries like the United States has been mainly driven by foreign policy objectives, such as ensuring a guaranteed supply of essential raw materials, minerals and oil, and the forestalling of any potential influx of immigrants and refugees from Third World countries (Grimes 1998). Castells (1996) claims that fear on the part of the north of being invaded by millions of uprooted peasants and workers was the main reason why international aid was channelled to African countries. The predominant consensus within development policy circles has been of the need to diffuse the Western ideology of the small family among the benighted people of the Third World, as a prerequisite for their participation in the development process. Many references are made to the success of population policy in Bangladesh, but there are few indications to date of any significant improvement in the economic welfare of the population. Furedi (1997) refers to this policy approach as 'modernization without development'. It is unlikely that a massive transfer of resources from north to south for broadly defined development will take place, and whatever transfers will be made will be concentrated in activities relating to fertility reduction (Najam 1996).

A radical critique

While most geographers have adopted a pragmatic rather than a narrow neo-Malthusian approach towards development, increasingly scholars with a deeper philosophical background have begun to question more fundamentally the essential nature of development. Some argue that the concept of development has already outlived its usefulness, since it involves nothing more than a new form of colonization with the objective of bringing about a total Westernization of society (Sachs 1992). Sachs traces the interconnections between this view of development and the goal of integrating humanity into one great consumer society. Developmentalists, in his view, have been transferring the Western model of society since the mid-twentieth century to countries of a greater variety of cultures. He describes this development process as 'an ahistorical and delocalized universalism of European origins', which substituted the term 'underdevelopment' for 'savages' and replaced reason with economic performance as the measure of man (Sachs 1992: 104). Watts and McCarthy (1997: 76), however, have been critical of some of these preoccupations, pointing out that since development was essentially about European efforts to deal with capitalism, it was therefore 'necessarily Eurocentric'. Corbridge (1998) is also concerned with what he calls 'an amoral politics of indifference' which he claims characterizes the position adopted by anti-developmentalists, because they pay little attention to the real dilemmas of development and fail to spell out the consequences of the actions or inactions which they propose.

One of the most provocative and useful contributions to this poststructuralist

view of development has come from Escobar (1992, 1995), who has deconstructed the thinking underlying the development of 'backward' regions in the
Third World. In a similar vein, Williams (1995) has been critical of the thinking
portrayed in the reports of the World Bank on Africa, which suggest that because
of the poverty and ignorance of Africans, they are incapable of dealing with their
own problems, which can only be solved by experts from outside. Herman and
Mattingly (this volume) note how those defining the problem are removed from
and more powerful than those who are defined as part of the problem. This
fundamental questioning by radical scholars of the right of experts to evaluate the
worthwhileness of Third World peasant lives, plays the useful and necessary role
of insisting that academics and policy-makers scrutinize the philosophical underpinnings of their development strategies. Roebuck (this volume) argues that we
must move beyond our own frameworks in order to understand 'other' societies
and seek to explain what they do under *their* description.

While acknowledging the need for such a scrutiny, it is also essential to be
aware of the danger of ending up with no clear goals of how to solve the problems
of communities in dire need (Corbridge 1995). This on-going tension between
idealistic questioning and the need for pragmatism is evident in a recent critique
of Escobar's work which points to serious reservations about the idealization of
peasant life in the south (Watts and McCarthy 1997). These authors insist that
there must be a way out of this 'cul-de-sac of postmodern politics', which tends to
view capitalism as a single entity rather than being composed of a complex of
social relationships.

Towards an ethical evaluation

While there is little doubt that a laissez-faire approach to capitalism will continue
to dominate development models throughout the world, bringing about significant increases in wealth for minorities and excluding the majority of mankind
from benefiting in any meaningful way, there are increasing calls from many
sources for the integration of a greater sense of social justice into development
policy. While such calls have frequently been based on a philosophical analysis of
the ethical dimension of development, there has been considerable progress in
specifying in a pragmatic way the implications of such philosophical analysis. Since
the integration of more progressive thinking into policy-making is likely to have
greater success by adopting a gradualist approach, some commentators have been
advising a minimalist method of specifying the integration of social justice (Black
1996; Smith 1997).

Based on his experience in Latin America, Friedmann (1992) has put forward
an incisive critique of the failure of the market-led accumulation model to include
the majority of citizens in benefiting from wealth creation. He calls for an alternative form of development with the concept of 'human flourishing' as a primary
objective, which would involve giving consideration to the fundamental question
of what it means to be human, and would empower individuals and communities
to achieve their full capacity. While Friedmann acknowledges that much of this

thinking has its roots in the Judeo-Christian tradition, other traditions including Confucianism, Ghandiism and Islamic radicalism have also rejected Western development models driven by relentless competition.

In specifying the grounds for an alternative to growth maximization development models, which by their very nature exclude large numbers of people, Friedmann suggests that we must turn to morality rather than simply to facts. Referring to the Universal Declaration of Human Rights (1948), which sets forth a universal code of basic moral conduct for social relationships, he suggests that the basis for seeking a greater degree of social justice through development must include human rights, citizen rights and human flourishing. In expanding on these ideas, Smith (1997) examines the implications for human rights in the more recent UN documents 'Covenant on Civil and Political Rights' and 'Covenant on Economic, Social and Cultural Rights'. While such documents can play a useful role in helping to specify the operationalizing of social justice within the policy arena, the imposition of what is being increasingly defined as 'rights' within a predominantly Western cultural context may prove to be problematic in other cultural settings.

The major challenge facing development scholars and policy-makers in integrating an ethical dimension with the necessary pragmatic approach to development has been highlighted for many years in the work of development ethicist, Denis Goulet. Both Goulet and L. J. Lebret, with whom Goulet studied in France in the 1950s, are regarded as co-founders of the 'new discipline' of development ethics (Crocker 1991). Goulet warns ethicists of the need to deal with the constraints of planners while clarifying the value costs and merits of competing policy proposals. In a similar vein to the concept of human flourishing, Goulet (1991: 10) drawing on the work of L. J. Lebret, introduces the concept of 'transcendence' into development thinking: in this context 'transcendence refers to the ability of all human beings to go beyond their own limitations and reach levels of achievement higher than those previously enjoyed'. Development policy in this sense would need to be based on the important distinction between 'being more' and 'having more'. A key dimension of such an approach is solidarity, by which development is geared towards benefiting all by focusing on the common good. In Lebret's view a development model which results in a small group of nations or privileged groups remaining alienated in an abundance of luxury goods at the expense of the many who are deprived of essential goods, is 'illusory antidevelopment'.

Goulet (1992) sees development as being a two-edged sword which, while bringing many positive elements such as improvements in technology, a wider range of choice, and improvements in welfare, also does away with traditional culture. He is critical of a development model which pays little attention to the values of traditional societies such as religious institutions, local practices, and extended family networks of solidarity. Other authors are also concerned with how market 'developmentalism' diminishes values such as solidarity, generosity, and brotherhood, all of which characterize traditional societies, and which provide social cohesion and meaning in people's lives (Berthoud 1992). Goulet

(1991) suggests that such development thinking is based on the widely held misconception which presumes that traditional societies are inherently obstacles to development.

This critique is very similar to the work of post-structuralists such as Escobar (1995) who have been strongly criticized for idealizing and romanticizing Third World peasant societies. While Friedmann (1992) is acutely aware of the most fundamental ethical principle of reciprocity, which governs social conduct at the household and community levels, he is also aware of the limitations of communities, who tend to see situations from their own perspective, and argues in favour of the state being a major player in promoting greater social justice. Within the European context the problem of social exclusion resulting from predominantly market-driven development models is being increasingly addressed by new forms of policy experiments based on social partnerships between local communities and state agencies (Sabel 1996).

Goulet's ethical evaluation of development thinking has been strongly influenced by the social teaching of the Catholic Church, which is reflected in such writings as *On Social Concern* by Pope John Paul II (1987). The central argument of this document is that the essential nature of development is moral and that the necessary political decisions for overcoming the obstacles to development are moral decisions. By restricting development thinking to the economic realm, humanity becomes reduced to its value in the marketplace as a commodity, and certain sections of the population who, for cultural, technological or other reasons, fail to conform to the market-led model of development, effectively become marginalized.

Smith (1997) agrees that the promotion of development is a moral project, but that the moral content of development has been given little attention to date. His detailed analysis of the moral content of development, however, illustrates the difficulties in connecting some of the relevant concepts from the fields of moral and political philosophy with the practice of development planning. A fundamental issue underlying the many differences in defining the moral aspect of development rests on our understanding of what it means to be human. One of the difficulties he points to is the gap between what might be proposed as basic human needs on the one hand and what particular societies may find acceptable or possible to guarantee. Part of the task identified by Smith, therefore, is to establish some minimalist conception of living standards which would have universal acceptance. While emphasizing this minimalist approach he agrees with Corbridge (1993) who argues that in an increasingly interdependent world, our obligations to other people should not be restricted to the boundaries of everyday communities. More recently Corbridge (1998), while conscious of the difficulty of drawing determinate policy implications from a particular philosophical perspective on development, argues in favour of a 'minimalist universalism' based on the fact that the lives of other people are inextricably linked to our own, and therefore they have a right to call on our resources.

While acknowledging the fact that for most governments and international financial agencies, development still means maximum economic growth and a concerted effort towards industrialization and mass consumption, Goulet (1993: 517) is impressed by 'how thoroughly value-centred and ethical in tone development talk has become'. Noting how value themes such as participation, self-reliance and equity have entered the general lexicon of development planners, he traces the influence of 'dissident strategists' such as Gandhi, Schumacher and Pope John Paul II in helping to bring this about (Goulet 1991).

The influence of an ethical approach to development is evident in policy documents such as the *Human Development Report 1994* of the United Nations Development Programme, which argues for a new global ethic, based on the concept of 'sustainable human development' (UNDP 1994: 19). By putting people at the centre of development, this concept is more inclusive than sustainable development as such, and is based on the view that it is not acceptable to perpetuate the inequities of today for future generations. The underlying philosophy of this concept rejects the exclusive obsession with the pursuit of material well-being, and the classification of people as human capital, arguing instead in favour of a development model based on what it calls 'the universalism of life claims'.

Descending to the more pragmatic level of policy-making, the UNDP argues that developing countries have a strong case for compensation from the rich north because of the dominant flows of labour and trade and also debt payments. Despite the likelihood that industrialized countries with aging populations will increase their dependency on labour flows from the south, these countries are becoming increasingly resistant to immigration, and are contributing to a serious brain drain from developing countries by restricting immigrant permits to a selected number of technical and highly skilled people. In the early 1980s, for example, Ghana lost 60 per cent of the doctors who had been trained locally. Apart, therefore, from an international elite with high levels of skills, there is no evidence of the emergence of a global labour force (Castells 1996). The UNDP suggests that there is a strong argument for compensating developing countries for restrictions on the migration of their unskilled labour. It also suggests that developing countries have a strong case for compensation because of lost trade due to the protectionism being practised by the north. It has been estimated that this loss amounts to at least $50bn a year in the clothing, textile and footwear sector.

While a strong moral case may be made for such compensation, it would appear very unlikely to happen from a political point of view. Black (1996) sees few encouraging signs in the context of European migration policy and suggests that arguing in favour of the individual right to migrate is unlikely to result in positive results. He suggests examining the possibility of pursuing the avenue of special obligations which may result from European countries or the European Union in general having been influential in maintaining or supporting regimes which abuse human rights. He also suggests the need for the EU to examine the role of the Common Agricultural Policy in depressing agricultural prices in poor countries, thereby promoting the necessity to migrate. Such policies may involve a moral obligation on the part of the EU to compensate such countries by means of a more open immigration policy into Europe.

Conclusion

There is widespread awareness in what is increasingly an interdependent world of the significant failure of the dominant market-led model of development to include the great majority of humankind in the benefits of the growing wealth creation which it succeeds in producing. It is clear that the gap between rich and poor is increasing with time and the extent of this gap is of scandalous proportions between the north and the south. Much of the analysis of development by geographers and other social scientists has been within the economistic tradition, and while making a significant contribution to our understanding of the economic processes involved, has failed to suggest a more holistic development policy. Contributions by geographers and others within the political economy school of thought have, despite their tendency to be somewhat deterministic in how they view human agency, nevertheless increased our awareness of the extent of spatial unevenness associated with development and the need to incorporate a greater commitment to social justice.

While some would argue that neo-Malthusian perspectives have also been inspired by a sense of morality in relation to environmental considerations, among others, much of the thinking would appear to be influenced by a very narrow sense of self-preservation resulting from a variety of fears associated with the shifting demographic centre of gravity from north to south. A much more positive contribution has been made by scholars in the poststructuralist tradition who have demanded a fundamental questioning of the objectives of development. Seeing it in terms of a Eurocentric project aiming to integrate all parts of the globe into one large consumer society, this deconstruction of development has been criticized for idealizing peasant society and failing to connect adequately with the pragmatic challenge of reducing mass poverty.

While all of these contributions contain some worthwhile elements for improving our understanding of what is a complex process, the significant failures of development models can only be tackled effectively by incorporating a greater awareness of the ethical dimension. Since development is fundamentally about people, development models need to be based on a more complete understanding of what it means to be human. There is a growing literature which is seeking to highlight the ethical dimension of development, with some studies seeking to specify in policy terms what might be politically feasible in terms of philosophical principles. While we all have the responsibility of presenting our recommendations in a form which can lead to effective operationalization, the need for pragmatism should not discourage geographers and others from promoting development models with the most ambitious targets in terms of social justice.

References

Berthoud, G. (1992) 'Market' in Sachs, W. (ed.) *The Development Dictionary*, London: Zed Books, 70–87.

Black, R. (1996) 'Immigration and social justice: towards a progressive European immigration policy?', *Transactions of the Institute of British Geographers NS*, 21(1): 64–75.

Booth, D. (ed.) (1994) *Rethinking Social Development*, London: Methuen.

Castells, M. (1996) *The Rise of the Network Society*, Oxford: Blackwell.

Clarke, R. (1986) 'Population imbalances: the consequences', *Forum* (Council of Europe), 1: V–VII.

Cook, P. and Kirkpatrick, C. (1997) 'Globalization, regionalization and third world development', *Regional Studies*, 31 (1): 55–66.

Corbridge, S. (1993) 'Marxism, modernities, and moralities: development praxis and the claims of distant strangers', *Environment and Planning D: Society and Space*, 11: 449–72.

—— (1995) 'Editor's introduction' in Corbridge, S. (ed.) *Development Studies – A Reader*, London: Edward Arnold, 1–17.

—— (1998) 'Development ethics: distance, difference, plausibility', *Ethics, Place and Environment*, 1 (1): 35–54.

Crocker, D. (1991) 'Toward development ethics', *World Development*, 19 (5): 457–483.

Duden, B. (1992) 'Population' in Sachs, W. (ed.) *The Development Dictionary*, London: Zed Books, 146–157.

Escobar, A. (1992) 'Planning' in Sachs, W. (ed.) *The Development Dictionary*, London: Zed Books, 132–145.

—— (1995) *Encountering Development*, NJ: Princeton, Princeton University Press.

Findlay, A. (1995) 'Population crises: the Malthusian specter?' in Johnston, R. J., Taylor, P. J. and Watts, M. J. (eds) *Geographies of Global Change*, Oxford: Blackwell.

Friedmann, J. (1992) *Empowerment: The Politics of Alternative Development*, Oxford: Blackwell.

Furedi, F. (1997) *Population & Development*, Cambridge: Polity Press.

Goulet, D. (1991) 'On authentic social development: concepts, content, and criteria' in Williams, O. F. and Houck, J. W. (eds) *The Making of an Economic Vision*, Lanham, MD.: University Press of America, 3–23.

—— (1992) 'Development: creator and destroyer of values', *World Development*, 20 (3): 467–475.

—— (1993) 'Catholic social doctrine and new thinking in economics', *Cross Currents*, Winter 1992–93: 504–520.

Grimes, S. (1998) 'From population control to reproductive rights: ideological influences in population control', *Third World Quarterly*, 19(3): 375–394.

Harvey, D. (1973) *Social Justice and the City*, London: Edward Arnold.

—— (1996) *Justice, Nature and the Geography of Difference*, Oxford: Blackwell.

Johansson, S. Ryan (1995) 'Complexity, morality, and policy at the Population Summit', *Population and Development Review*, 21(2): 361–386.

Krugman, P. (1994) 'Does Third World growth hurt First World prosperity?', *Harvard Business Review*, July–August: 113–121.

Lipshitz, G. (1992) 'Divergence versus convergence in regional development', *Journal of Planning Literature*, 7: 123–138.

Najam, A. (1996) 'A developing countries' perspective on population, environment, and development', *Population Research and Policy Review*, 15 (1): 1–19.

Pope John Paul II (1987) *Solicitudo Rei Socialis* (On Social Concern), London: Catholic Truth Society.

Portes, A. (1997) 'Neoliberalism and the sociology of development', *Population and Development Review*, 23(2): 229–260.

Richmond, A. (1994) *Global Apartheid: Refugees, Racism and the New World Order*, Toronto: Oxford University Press.

Sabel, C. (1996) *Ireland – Local Partnerships and Social Innovation*, Paris: OECD.

Sachs, W. (1992) 'One world' in Sachs, W. (1992) (ed.) *The Development Dictionary – A Guide to Knowledge as Power*, London: Zed Books.

Sauvey, A. (1949) 'The false problem of world population', *Population*, 4(3) (trans. by P. Demeny in *Population and Development Review*, 16 (4): 759–774).

Sen, A. (1994) 'Population: delusion and reality' in Cromartie, M. (ed.) *The Nine Lives of Population Control*, Washington, DC: Ethics and Public Policy Center, 101–128.

Slater, D. (1997) 'Geopolitical imaginations across the North–South divide: issues of difference, development and power', *Political Geography*, 16(8): 631–653.

Smith, D. (1977) *Human Geography: A Welfare Approach*, London: Edward Arnold.

—— (1994) *Geography and Social Justice*, Oxford: Blackwell.

—— (1997) 'Las dimensiones morales del desarrollo', *Economía, Sociedad y Territorio*, 1(1): 1–40.

UNDP (1994) *Human Development Report*, Oxford: Oxford University Press.

Watts, M. J. and McCarthy, J. (1997) 'Nature as artifice, nature as artifact: development, environment and modernity in the late twentieth century' in Lee, R. and Wills, J. (eds) *Geographies of Economies*, London: Edward Arnold, 71–86.

Williams, G. (1995) 'Modernising Malthus – The World Bank, population control and the African environment' in Crush, J. (ed.) *Power of Development*, London: Routledge, 158–175.

6 Virtual geographies
The ethics of the Internet

Jeremy Crampton

Introduction

> Perhaps the most interesting substantive work by geographers on ethics transcends the boundaries between metaphors of space, place, and nature.
>
> (Proctor 1998: 14)

Is geography turning virtual? "From virtual GIS" (Batty 1997) to virtual class-rooms, virtual fieldtrips, and virtual communities, has the discipline long associated with physical experience, with landscape and culture, succumbed to the attractions of the not-quite, the virtual? If so, what does this mean for ethics in geography? In this chapter I wish to briefly address this question in the context of the Internet.

At first it may be difficult to understand why geographers should be concerned about the ethics of the Internet. Unlike spatial technologies such as GIS or digital cartography, geographers did not invent the Internet, nor do they necessarily have a privileged relationship to it. But this is the Age of Information and one of its major means of transmission is the Internet. There is a remarkable geography of the Internet which has three complementary components. First, "where" is the Internet, and more generally, cyberspace? How did it grow? Can the flows of information be mapped to see who is connected and where? What geographic outcomes or practices might it give rise to in finance, politics, or culture? This is here called the *geography of virtuality*. Second, to what degree do these virtual spaces constitute new forms of spatial knowledge? Do they change the way we know the world and the way we think about other people and communities? If so, what are the ethical implications for geographers? This is here called the *virtuality of geography* – the fact that geographic interactions increasingly require or include a virtual component.[1]

But there is also a third area of ethics which intersects with both of the above (although in ways which are not yet clear), and that is professional ethical implications. In some ways these implications – such as the part-timing of the academic labor force, pressures for "corporatizing" the university with the web, digital distance education – are harder and more immediate for us as practitioners of geography. Ironically, they have been comparatively ignored compared to the more speculative attention on the future of human communities in cyberspace, etc. Yet they are far more likely to affect geographers, and geographers are far more likely to be able to affect outcomes in this area.

Because the question of ethics of the Internet is still emerging it is useful to provide a road map of the ethical issues involved – a descriptive ethics. Using the three aspects outlined above, this chapter will focus on trying to capture what I see as the most important descriptive ethical issues facing us, that is, to "provid(e) a rich account of the ways morality interweaves with the geographies of everyday life" (Proctor 1998: 11). It is also necessary to evaluate to what degree the ethical practices we identify are good or bad – a normative ethics.

The ethics of the Internet identified

Does the Internet as a technology have an intrinsic nature or "logic" and if so, what is it? This is perhaps the most basic ethical question because it addresses the moral value of the Internet; that is, whether it has an inherently positive or negative quality. Authors have taken different positions on this question. Critics of the technology, perhaps following the dystopian vision of William Gibson (1984) who coined the term "cyberspace," cite negative aspects such as an inherent "logic of hegemony" associated with all technology because it expresses power relations. In this view any use of technology (such as the Internet) necessarily takes place within, and thus reproduces, power relations. Technology is therefore not a neutral or innocent activity, but an ideology (see Aronowitz 1988). Furthermore, this logic of hegemony presupposes a specific positivist and rationalist framework within which the technology operates. Commodification, or the inevitable transformation of the communications media into a profit-seeking enterprise, is another criticism leveled at the Internet. Critics who adopt this position note that there is a strong trend toward commercial sites on the web (which currently account for more than 60 percent of all web sites), and emphasize moves by an increasing number of educational establishments in the United States to commodify their educational offerings (Noble 1998). A final criticism of the logic of hegemony is that the technology can be used to invade privacy, and provide surveillance measures never before possible.

On the other hand many authors identify cyberspace as enabling and emancipatory, with the heretofore unrealizable potential of building virtual communities and discourses via the web, e-mail, and live chat rooms (Rheingold 1993). These discourses are truly democratic; they necessarily widen the sphere of public communication for consensus-seeking. Some aspects of this "logic of empowerment" include access to government resources such as the Census Bureau, national and international telephone directories for finding people (e.g. missing children), the exchange of opinions and ideas between nations which subvert stereotypes, all sorts of information useful to citizens such as property searches, driving records of potential employees, and so on.

But there is a third position: that the Internet is a topology of competing philosophies without an inherent nature. This non-essentialist claim has been taken by some authors interested in pursuing a progressive agenda in geography (e.g. Grimes and Warf 1997) and it allows recognition of the hyperbole of a Rheingold without adopting a romantic anti-technological stance. Furthermore we can detect an emerging consensus around this position in a related technological field through a reading of the debate about geographic information systems (GIS) and society. In geography, the debate on "logics of technology" has been most visible in the critique and counter-critique of the spatial analytic technology of GIS and its relations to society (see Pickles 1995; Crampton 1993; Lake 1993; NCGIA 1997). The most recent iteration of this debate (see Wright *et al.* 1997a, 1997b; Pickles 1997) saw each side recognizing the other's positions; that GIS *can be* totalizing and positivist, *but* can also be empowering and democratizing, i.e. there is no deterministically inherent logic of GIS. I would like to apply this conclusion to the ethics of the Internet; that the technology gives rise to competing logics. This has an important implication, because it means that which logics become privileged is not so much a factor of the technology itself, but of what we as users do with it (a similar call to activism is made in the final chapter of Rheingold's 1993 book *The Virtual Community*).

What are the ethical factors of these positions for geographers? I have summarized the major points in Figure 6.1. First, what are the implications of new spatial practices and outcomes – what is the geography of virtualization? Is access to the Internet universal, and if not, what are the spatial patterns of access? Within places, who has access and who does not (e.g. males versus females, employees versus management)? Why have these patterns arisen, and will access ever be equitably distributed (Kedzie 1997, Wresch 1996)? Are we producing technological elites (Kaplan 1995)? Furthermore, what are the effects of access, and *should* everybody (e.g. children) have access (Anderson *et al.* 1995; Kedzie 1996). As the Internet continues to grow we might ask whether it is forming a new set of spaces which eclipse national boundaries, and consider what this means for politics, culture and ultimately, the possibility of democracy.

Second, what new forms of spatial knowledge and thinking are produced by the Internet–how is the virtualization of geography taking place? Here again are questions with significant moral content; for example, is the Internet being used to find out about people by invading their privacy? Should this digital information gathering (especially its geographic aspects) be permitted or circumscribed and to what degree (Curry 1997)? Conversely, does the Internet allow us to know more about other cultures which may lead to reduction in stereotyping (e.g. that all Muslims are fundamentalist), perhaps to the extent of democratizing those nations which are more connected (Kedzie 1997)? What are the pros and cons of anonymity, for example in the way we treat strangers (Whittle 1997)?

A significant concern of geographers in the 1990s has been the effect of globalization. Indeed, for some it has already become an organizational theme for introductory geography texts (e.g. Knox and Marston 1998). What role does the Internet play in globalization, and how is spatial knowledge affected (Castells 1996)? Does it extend a network of relations beyond the local (e.g. virtual

New practices and outcomes: a geography of virtualization	New knowledges: the virtualization of geography

Where is cyberspace?
Who is connected?
Spatial and societal differentiation
  e.g. connectivity maps
  e.g. by gender or age
Who should have access?
  e.g. authorities and institutions
  e.g. children
	What and how do we know?
	Invasion of privacy v. democratization?
	  e.g. surveillance
	  e.g. virtual communities
	Knowledge of the other
	Identity/anonymity
	Globalization of space
	The meaning of community?

Professional implications	
Assessment Scholarly value of Internet-based resources? **Employment** Full-time v. part-time **Educational** Effects of technology on student performance? Effects on student research methods? "Corporatization" of education?	

Figure 6.1 The ethics of the Internet*

Note: * Ethical implications arise in three major areas: geographic outcomes or practices, geographic knowledges, and the geographic profession itself. The last intersects the first two in as yet undetermined ways.

communities) or does it lead to a time–space compression due to a hypermobility of capital and finance (Cairncross 1997)? These are important questions concerning the local–global relationship. Additionally, is flexible production encouraged by distributed information, for example by the ability of companies such as the Dell Corporation to configure products individually via the Internet? Conversely, is there still a geography of information which reasserts the local (Clark and O'Connor 1997, Cox 1997)? In sum, is the world coming together or pulling apart (Staple 1997)?

To be sure, these questions are not exclusively ethical ones. For some, the domain of ethics, of deciding what is right and what should be done, has already been answered in practice ("Yes, the Internet has demonstrably increased surveillance of employees") or in principle ("No, children should not have access"). Their positions are well staked out. For other people, while these issues include ethical questions, other issues predominate. While these are valid positions, I argue that the time is right for those of us concerned with ethics to partake in the building of the Internet; to debate and establish good practice; to emphasize its positive aspects and to combat the negative aspects. Part of this agenda is consciousness raising, that is, in documenting the moral dimensions of the Internet in public (e.g. this chapter) and to our students (for an example of how these issues might be introduced in a human geography class, see Crampton and Krygier, 1997).

Whatever one's position on ethics in geography it is likely that few will remain unaffected by developments in the third major area of Figure 6.1: professional implications. Ethics has already played a significant role in academia (especially as "applied ethics," for example, in the treatment of research subjects), but the Internet poses new issues. Among these are questions of how to assess work done outside traditional channels of peer-reviewed publication, and whether that work will be categorized as a component of research, teaching or service. There are also ethical questions about our work as educators: is it right or desirable (not necessarily the same thing) to use educational technology such as the Internet in the classroom? How will it affect student performance? Will students ignore the library now that they have web-based search engines? We can also ask what effects distance education will have on employment of full-time professors. Will it lead to increased "part-timing" of the academic labor force? In fact, distance education may prove to be a defining focus for several competing arguments to do with the future of education. For example, it has been credited both as a means to increase student enrollment in geography (DiBiase 1996) and as a sign of increasing "corporatization" of education (Noble 1998). Further consideration of the ethical implications of distance education would appear to be justified.

These then are broadly the major ethical issues of virtual geographies. In the next section I examine some of these issues in more detail.

Selected Internet ethics in detail

> That we live in a computer age no one seems to doubt. Yet, along with the paeans of praise. . .there is also a growing chorus of criticism and a pervasive mood of doubt about its redemptive features.
>
> (Aronowitz 1988: 3–4)

Competing scenarios

Scenario 1: the Internet as empowerment

Claims for the empowerment of the Internet are most typically found in the hypertext community where hypertext (and its particular implementation in the World Wide Web) democratizes access to information (Bolter 1991; Bush 1945; Landow 1992, 1994). Hypertext is defined as "non-linear writing" which encourages the reader to choose pathways through the text. Hypertext and the Internet are cited for their inherent ability to interconnect people, as well as to challenge established hierarchies. The latter ability, for example, is developed by Jay David Bolter in his influential book *Writing Space* (1991) into a challenge for traditional hierarchical text in the face of non-linear and associational text.

The theoretical articulation of such democratization is best realized in

Habermas' tradition of "communicative democratic action" and "discourse ethics" (Ess 1994). Briefly, this promotes rational, consensus-seeking dialog as the cornerstone of democracy. There is some empirical evidence to support this position, although it is suggestive, rather than definitive. A recent series of studies by the Rand Corporation has found a strong link between the degree of Internet "connectivity" and democracy (Kedzie 1996, 1997) as well as beneficial links between computer networks and human development (Press 1996). Stimulated by maps of "interconnectivity" and "democracy" which looked strikingly alike (high interconnectivity = high democracy ratings), Kedzie discovered a correlation coefficient of 0.73 (significant at the 0.1 level). Indeed, interconnectivity correlated at a higher level than more traditional predictors, such as GDP or schooling (see Figure 6.2).

Kedzie also tested the directionality of this link to see whether interconnectivity leads to democracy or vice versa (for example, because democracies rely on intercommunication and therefore move to interconnect themselves). His findings indicate that interconnectivity is a strong predictor of democracy, while there was no effect of democracy on interconnectivity (that is, democracy does not necessarily lead to interconnectivity).

The Kedzie study proposed a possible third contributing factor that tends to increase both democracy and interconnectivity, namely economic development. It says "to the extent that we, as a nation, aim to influence the development of democracy worldwide, we do so through programs to enhance economic development, education, health, legal reform and so on" (Kedzie 1996: 29). The conclusion indicated is that a strong method of encouraging these programs of health, education and development lies in interconnectivity.

While these are indeed laudable goals, it must be conceded that neither "development" nor "democracy" is an unproblematic term (Bell 1994). The dualism it implies of north–south obscures important spatial differences. And as Yapa has repeatedly pointed out (Yapa 1995, 1996) the links between development, the economy and poverty are highly contested and do not necessarily reveal or encourage "the poor" to achieve power to address problems (a form of denying them agency in that it is assumed that only the "non-poor" can have agency). There is an important point here in that interconnectivity may empower local groups. Interconnectivity, development, and democracy may well be intertwined, but in complex ways (a "virtuous circle" according to Kedzie) with no single determining factor.

Scenario 2: the Internet as hegemony

Perhaps the most widely expressed fear associated with the Internet is the potential for surveillance and other means of privacy invasion. In a 1997 survey carried out by Georgia Tech's Graphics Visualization and Usability Center (GVUC) "censorship" and "invasion of privacy" were the number one and two top ranked answers to the question "What do users feel is the most important issue facing the Internet?" (GVUC 1997). And indeed, the evidence does indicate that the

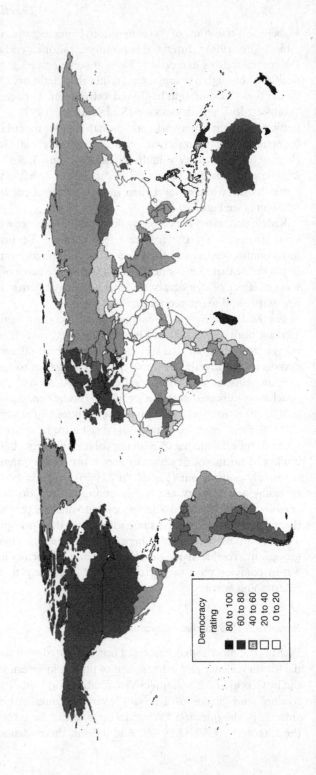

Democracy
rating

80 to 100
60 to 80
40 to 60
20 to 40
0 to 20

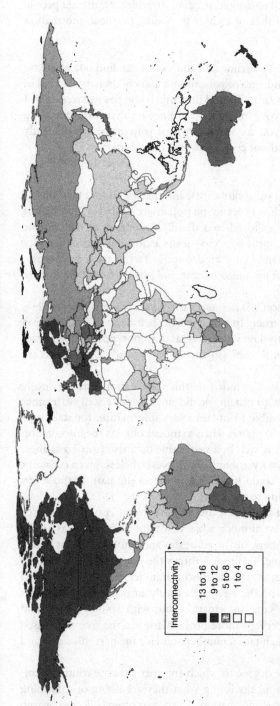

Figure 6.2 Global democracy ratings (top) and global Internet connectivity (bottom)

Source: Redrafted from Kedzie 1997.

Interconnectivity

■ 13 to 16
■ 9 to 12
▥ 5 to 8
□ 1 to 4
□ 0

Internet can be and has been used to exploit, subdue, surveille, and market people and their information. As one author of a guide to finding personal information on the Internet described it:

> Within a few hours – with only a name and address – I can find out what you do for a living, the names and ages of your spouse and children, what kind of car you drive, the value of your home and how much you pay in taxes on it. I can make a good guess at your income. I can uncover that forgotten drug bust in college. In fact, if you are well known or your name is sufficiently unusual, I can do all this without even knowing your address.
>
> (Lane 1997: C3)

Lane's point is twofold: that this capability exists, and that we should take advantage of it for positive reasons such as checking on potential employees, finding lost children or tracking debtors (e.g. "dead-beat dads"). Lane may or may not be right that this capability exists, but it is by no means a closed question whether this is good or bad – it is certainly an ethical one (in fact, Lane's own article was accompanied by one deploring and opposing the privacy breaches of these technologies, see Culnan 1997).

In order to answer this question it is necessary to get an idea of what it really is possible to discover on the Internet. In other words, before an ethical position can be developed, we need to know what is and is not possible. A fuller account can be found in Crampton *et al.* (forthcoming), but some highlights are mentioned here.

The first point to be made is that much of this information was previously available but was either difficult to obtain, or did not integrate well with other datasets. It has always been possible to obtain salary information for state supported schools, but it is a different matter when a student obtains the information in digital form and posts it on the web in a searchable database (top 10 salaries, five-year salary history, etc., see www.roblink.com). Nevertheless, given sufficient effort a lot of this information could have been found in the past by those few (private investigators, etc.) willing to try.

Finding people has become a significant Internet capability. A typical resource can offer the email address, phone number, address, instant map, as well as neighbors' addresses and phone numbers, or the addresses and phone numbers of any street in the USA. As an example, Figure 6.3 shows the names and addresses of residents of N Avenue, Washington, DC (telephone numbers have been excluded although they remain available on the web). Obviously an ordinary white pages phone directory will have the same information, but with far less access, and crucially, not geographically. Perhaps more problematic are the free web-based "reverse-directory" services which take a number and give the person's name and address (even if unlisted).

A final example concerns the degree to which Internet sites are routinely collecting data about visitors without disclosing what they are doing or informing visitors what will be done with the information. The data is often collected by web

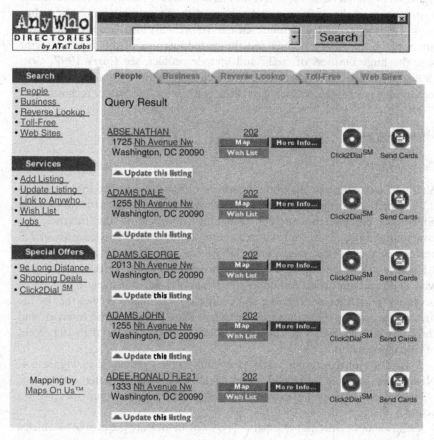

Figure 6.3 A search result of a Washington, DC street reveals telephone listing, address, and geographic neighbors

Source: http://www.anywho.com.

software known as "cookies," and although there are technical methods for turning off cookies or restricting their use most users do not know about, or take advantage of, these methods. And certain services may be denied if cookies are not allowed to work (e.g. online ordering of goods and services). According to a report by a private industry watchdog, OMB Watch, these methods have been most worryingly deployed in federal websites (OMB Watch 1997). The survey looked at 70 federal agency websites and found that 31 of them collected data on names, ages and work history from the public. But only 11 of the sites described their activities, and four of these "probably violated provisions of the Privacy Act of 1974" (O'Harrow 1997: E1).

Often in these cases of data collection without the knowledge or consent of the person (surveillance cameras on high streets, ATM cameras, use of social security numbers on identification cards in the USA) although people are unaware of the high level of the surveillance, or feel they have no choice (they have not been able to participate in the decision-making process) they would in fact object if they knew. What is needed is a much clearer (more informed) debate about the authoritarian use of technology and the vested interests of the financial sector (e.g. the huge business of credit and lifestyle profiles, see Curry 1997, Goss 1995), particularly if it takes place in the context of the pros and cons of the technology, rather than traducing to either extreme of Luddism or technicism.

The ethics of access

Geographically

Although today the vast majority of countries have access to the Internet it has still reached only a tiny proportion of the world's population. It is notoriously difficult to estimate the number of people connected to the Internet, but assuming in 1998 it was somewhere around 100 million, that is only about 1.5 percent of the world's population. Of course, it is likely to be the "right" 1.5 percent; those elites in each country who earn more, are better educated, and who live in urban areas. Although these numbers are not likely to be stable because the Internet is increasing so rapidly (some estimates predict that by the turn of the century there may well be half a billion people on the Internet, and three quarters of a billion by 2001; MIDS 1997), that is still less than 10 percent of global population.

A map of global Internet access as of mid-1997 is presented in Figure 6.4. Notice that the more developed "core" countries are well saturated with access, while the less developed "periphery" and "semi-periphery" countries have less access.

Using a simple "Internet Quotient" (IQ) we can measure the degree to which a country (or community) is connected:

IQ = TH/TP where

    IQ = Internet Quotient
    TH = Percentage of Total Internet Hosts
    TP = Percentage of Total Population

On this measure, any number greater than 1.0 represents a greater proportion of connectivity to share of world population. According to Knox and Marston (1998) the United States has an IQ of 13.2 with 63.7 percent of the world's total Internet hosts and only 4.82 percent of the world's population. However, India, with its huge population and few Internet connections, has a meager 0.0004 rating.

There are geographically uneven distributions at the regional and local level too. Although Figure 6.4 appears to show Africa well connected (43 countries connected by November 1997), this is quite misleading; in fact access is often limited to the capital city and major urban areas, with little or no access rurally.

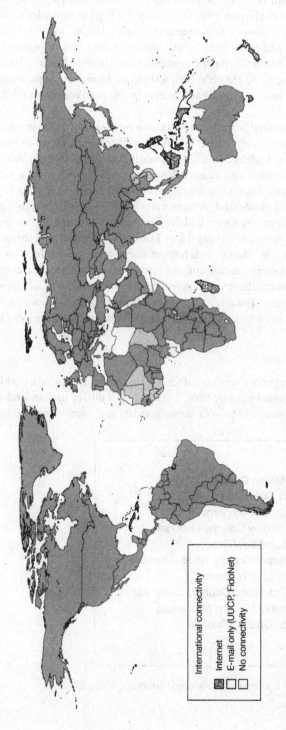

Figure 6.4 International Internet connectivity, June 1997

Source: Redrafted from Internet Society. Copyright © 1997 Larry Landweber and the Internet Society.

International connectivity

■ Internet
☐ E-mail only (UUCP, FidoNet)
☐ No connectivity

This is very significant because about 70–80 percent of Africa's population lives outside the major cities (Jensen 1997a). Among the biggest obstacles to Internet growth across Africa are poor telecommunications facilities, low levels of computerization, scarcity of computers, cost, bandwidth, and lack of regional access points (Jensen 1997b). In Ghana, for example, which is one of the region's better connected countries, there are only 100,000 phone lines for a population of 15 million, and the key issue remains that of access by the rural community (Quaynor 1997).

Even within the United States, which has the bulk of the global Internet connectivity, access varies widely. Places which are well connected tend to be centers of high-tech industry, such as Santa Clara County, California (Silicon Valley), San Mateo County, California and Fairfax, Virginia, or associated with universities, such as Travis County, Texas (University of Texas at Austin) (Figure 6.5). Other places are not so well connected. A recent report (ETS 1997), reveals that high school access to computers varies widely; from about 1 computer for every six students in Florida to one for every 16 in Louisiana. Significantly, the access to computers was lowest in schools with high minority enrollments. In schools with over 90 percent minority enrollment, access was one computer for every 17.4 students, whereas where there was 0 percent minority enrollment, access was 9.7 students per computer. Poverty, too, plays a factor, with the poorest schools obtaining an access of only one computer per 35 students (Sanchez 1997).

Differential societal access

Geographic differences are only one factor in access to the Internet which will interest ethically minded geographers. There are also differential societal degrees of access. These include differences across gender, age, race, income, education,

1 San Mateo County, California
2 Santa Clara County, California
3 Fairfax County, Virginia
4 Washtenaw County, Michigan
5 St Louis City, Missouri
6 Middlesex County, Massachusetts
7 Travis County, Texas
8 San Francisco County, California
9 Hennepin County, Minnesota
10 Fulton County, Georgia

Figure 6.5 Top ten Internet connected counties, USA, 1997
Source: MIDS, Inc.

and increasingly by policy. Gendered differentials are actually decreasing, with about 30 percent of online users identifying themselves as female (GVUC 1997). Other polls, such as one conducted by *Business Week* in April 1997, put the number at over 40 percent (*Business Week* 1997). The same poll found that nearly half of US Internet users were over 40 years old (45 percent) and 37 percent aged 25–40. Minorities are significantly under-represented: 85 percent of Internet users are white. Seventy-three percent of users have high school education or higher (41 percent have college degrees), while 42 percent earn more than $50,000 annually. Many students get online in college (a few technologically advantaged high schools offer student email accounts and webspace as well). In the United States this access is often free, which gets students "hooked" on the Internet, perhaps to the detriment of their studies.

These differentials raise important issues of social justice. Obviously access to the Internet is spatially uneven, a basic condition of resource access which is reflected in studies throughout this book. Addressing these spatial differentials is a difficult problem. In some cases, local organizations and governments have instituted policies which act to restrict access. This is based on the assumption that unfettered access to the web by all people is not desirable, either for users or for providers if it exposes them to legal action. For example, in several counties near Washington, DC, the local library systems have experimented with restricting access to children (Benning 1997). As we saw above, there are perceived dangers of privacy invasion, and of obscene or pornographic material which communities may wish to filter or block. A federal law banning "indecent" or "patently offensive" material was struck down by the Supreme Court as unconstitutional in June 1997, but obscene material remains as illegal on the Internet as it does in the real world (Schwartz and Biskupic 1997). An interesting ethical question raised by these issues is that the Internet fundamentally alters the notion of "community." Can material offend a "community" if it originates thousands of miles away? Can prosecutors file suit in the most conservative communities against material which is not stored there, but available there via the web? In practice, the answer so far seems to be "yes." After the Supreme Court decision, new legislation may be introduced which requires websites to use filters to prevent access to "material harmful to minors." Whether this legislation passes and is constitutional is perhaps less important than the general question of how to make an Internet which is safe, but protects free speech.

Issues of professional ethics

As shown in Figure 6.1 we can identify three areas of professional ethics: assessment, employment and educational. All three areas are as yet lacking in answers and in many cases, discussion, but are very critical questions. First, given the culture of assessment in academia it is unclear how to assess work done with, or on, the Internet, such as websites. In the case of peer-reviewed publications there is an accepted (if not perfect) method for evaluation. But how would you evaluate a website which contained original "content" but had not passed through a peer-evaluation procedure? How would an article in an online geography journal be evaluated even if there was a peer-review procedure? And most problematically, how are websites which require extensive research, but are primarily used in the classroom, to be evaluated? Although the situation changes rapidly, until another generation of geographers become departmental chairs or sit on promotion and tenure committees, it is frankly still advisable to maintain a traditional publishing record. Another recommendation is to amass a series of "peer-website evaluations" in the same manner that peer-teaching evaluations are performed.

A corollary of this question is under what sector of scholarship (teaching, research and service) to classify the work. Although it might be argued that an academic website qualifies as teaching, research and outreach (if the website is available as a service to the global web community), in practice chairs and promotion and tenure committees are still likely to classify the work as teaching. Cases have also arisen where work can be cited negatively if it is used across different educational levels (e.g. high school as well as college). The implication here is that if a work is *too* accessible it does not have sufficient academic rigor. This is obviously a crucial ethical issue for geographers concerned with community outreach or the Geographic Alliances.

A second area of concern lies in the changing structure of the academic profession. As many faculties are aware, the employment trends (at least in the USA) have been towards part-time and adjunct faculty. Would extensive use of the Internet and web-based teaching exacerbate this situation? For example, if one professor can teach the same class at three different universities simultaneously, as a colleague of mine is doing using the web, does that equate to a reduction of two teaching assignments? Or, on the contrary, does the web enable *more* teaching by using scarce teaching expertise (my colleague is an Africanist regional geographer, a fairly specialized interest)? There may never be enough interest for certain specialized classes to be offered unless this kind of distance education is possible.

A third and final question concerns the effect on the student educational experience. Again, this is a multifaceted question with little specific debate. The most mature discussion concerns using educational technology in general, where the consensus seems to be clear that simple instances (e.g. drill and practice) are advantageous, with more complex instances depending on how well they are integrated into the curriculum (Krygier *et al.* 1997). However, other questions abound, such as how student research methods are being affected by the web (a shift from the traditional library or a supplement to it?), how students evaluate material found on the web, student attitudes to educational technology (presumably related to computer exposure and experience at high school, but also perhaps to the quality and support of the technology), and how challenging (how good) the web-based materials are. The outcomes are as yet unclear, but clearly have important implications for how we as geographers wish to see the discipline progress.

Conclusion: toward a participatory ethics of the Internet

What might an ethics of the Internet look like? At this point, it is too early to tell except in broad terms. It is also perhaps too early to develop a code or set of guidelines for adoption in the face of these issues. Rather, this chapter fore-grounds the pervasive issues for geographers in this area as summarized in Figure 6.1. Three main domains were identified: effects on geographic practices and outcomes (the geographies of virtuality); effects on geographic knowledges (vir-tualization of geography); and applied professional ethics. They range from the fundamental (does the Internet have an inherent logic?) to the particular (should children be allowed access in public libraries?). Many of the issues are likely to be unfamiliar to geographers but I suggest that we can take advantage of debates in related areas such as the ethics of GIS to adopt a position which encourages a *participatory* ethics of the Internet.

The goal of a participatory ethics is to address issues through an inclusionary rather than exclusionary process. The debate in GIS was successful to the degree that it included viewpoints from GIS practitioners as well as those critical of GIS practices; but more importantly to the degree with which it erased divisions and increased cooperation between previously competing interests. Naturally, unity of opinion will never be achieved except through some artificial mechanism, legal code or precedent, and interests will presumably always be competing, but they can at least "speak" to each other. At present, there is poor understanding of the geographic ethics of the Internet. For example, most academic institutions have hardly begun to come to terms with the implications on the profession of distance education, differential access to the Internet, peer assessment of academic material on the web, and so on. While many of these issues are still emergent and below the horizon even for ethicists (e.g. professional guidelines on ethics drafted and submitted in 1998 by the Association of American Geographers did not directly address the Internet's role in ethical issues) it remains a fact that decisions about promotion and tenure, hiring, and faculty involvement with the web are made every day. There is a real danger that these decisions and precedents (e.g. a major university declining tenure on grounds of excessive web publishing compared to print publishing) will occur without representation from the stakeholders.

But there is a further sense of participatory ethics to which I wish to appeal. One often hears that the web has little or no quality material, that good material is too hard to find, or that the web is not relevant to scholarship. These are valid viewpoints but too often they are used to cover a lack of familiarity with or dislike of the web and its contents. A more productive approach is to manufacture the content of the web so that it is relevant and of high quality. An analogy can be drawn with the way the Internet (i.e. content providers, Internet providers and companies) adopted self-regulation rather than face government-imposed regulations.

Participation need not entail authoring web pages oneself, although that is an option either individually or with assistance from initiatives such as the Virtual

Geography Department, but may also mean raising the sort of issues discussed here within one's community. Indeed, there are numerous entry points for geographers: spatial differentiation of access providers, implications on "communities," spatial data marketers, impacts on privacy, GIS companies, deans, high school and college classrooms, universities, the Internet Society, the World Wide Web Consortium, commercial educational entities, local, state and federal government. Widely varying as they are, these areas offer our best chance to build an Internet which is sensitive to geographic ethics, is directly and meaningfully relevant to people's lives, and built through participation of stakeholders. Anything less would be unethical.

Note

1 A note on terminology: "cyberspace" and "the virtual" here refer to the conceptual world of networked interactions which are not face to face but physically separated, and which emphasize digital information flows (synchronous or asynchronous). They are also often characterized by servers which can multicast to many recipients, and users or clients who can choose to receive or not receive from multiple sites (i.e. filtering). "The Internet" and the "World Wide Web" are more specific instances of the virtual. For a longer discussion, see Whittle (1997), especially Chapter 1.

References

Anderson, R. H., Bikson, T. K., Law, S. A., and Mitchell, B. M. (eds) (1995) *Universal Access to e-mail, Feasibility and Societal Implications*, Santa Monica: RAND.

Aronowitz, A. (1988) *Science as Power, Discourse and Ideology in Modern Society*, Minneapolis: University of Minnesota Press.

Batty, M. (1997) "Day Trip to the Third Dimension," *GIS Europe*, Vol. 6 (10): 14–15.

Bell, M. (1994) "Images, myths and alternative geographies of the Third World," in D. Gregory, R. Martin, and G. Smith (eds), *Human Geography, Society, Space, and Social Science*, Minneapolis: University of Minnesota Press.

Benning, V. (1997) "Panel says schools slow to halt access to Internet porn," *Washington Post*, November 27: B7.

Bolter, J. D. (1991) *Writing Space, the Computer, Hypertext, and the History of Writing*, New Jersey: Lawrence Erlbaum.

Bush, V. (1945) "As we may think," *Atlantic Monthly*, 176 (July).

Business Week (1997) "Harris poll: census in cyberspace," *Business Week*, May 5.

Cairncross, F. (1997) *The Death of Distance, how the Communications Revolution will Change our Lives*. Boston: Harvard Business School Press.

Castells, M. (1996) *The Rise of the Network Society*, Volume I of *The Information Age: Economy, Society and Culture*, Malden, MA: Blackwell.

Clark, G. L. and O'Connor, K. (1997) "The informational content of financial products and the spatial structure of the global finance industry," in K. R. Cox (ed.), *Spaces of Globalization, Reasserting the Power of the Local*, New York: Guilford Press.

Cox, K. R. (ed.) (1997). *Spaces of Globalization, Reasserting the Power of the Local.* New York: Guilford Press.

Crampton, J. (1993) "The ethics of GIS," *Cartography and GIS*, 22(1): 84–89.

Crampton, J. and Krygier, J. (1997) Companion website for *Human Geography, Places and Regions in Global Context* [Online]. Available: http://www.prenhall.com/knox/. [1997, November 1].

Crampton, J., Krygier, J., and Crum, S. (forthcoming) *A Geographer's Guide to the World Wide Web*, Washington, DC: AAG Publications.

Culnan, M. J. (1997) "What is plain to see, there are ways to protect personal data on the net," *Washington Post*, July 13: C3.

Curry, M. R. (1997) "The digital individual and the private realm," *Annals of the Association of American Geographers*, 87(4): 681–699.

DiBiase, D. (1996) "Rethinking laboratory education for an introductory course on geographic information," *Cartographica*, 33(4): 61–72.

Educational Testing Service (ETS) (1997) *Computers and Classrooms: the Status of Technology in US Schools*, Princeton: ETS Inc.

Ess, C. (1994) "The political computer: hypertext, democracy, and Habermas," in G. P. Landow (ed.), *Hyper/text/theory*, Baltimore: The Johns Hopkins University Press.

Gibson, W. (1984). *Neuromancer*, New York: Ace Books.

Goss, J. (1995) "Marketing the new marketing, the strategic discourse of geodemographic information systems," in J. Pickles (ed.), *Ground truth*, New York: Guilford Press.

Graphics, Visualization and Usability Center (GVUC) (1997) GVU's 7th WWW User Survey [Online]. Available: http://www.cc.gatech.edu/gvu/user-surveys/survey-1997-04/. [1997, November 1].

Grimes, J. and Warf, B. (1997) "Counter-hegemonic discourses and the Internet," paper given at the Association of American Geographers Annual Conference, Fort Worth, Texas.

Jensen, M. (1997a) Internet connectivity in Africa [Online]. Available: http://demiurge.wn.apc.org/africa/afstat.htm [1997, November 25].

—— (1997b) "Policy constraints to electronic information sharing in developing countries," *On the Internet*, November/December: 13–15, 41.

Kaplan, N. (1995) "Politexts, hypertexts, and other cultural formations in the late age of print," *Computer-Mediated Communication Magazine*, 2(3) [Online]. Available: http://sunsite.unc.edu/cmc/mag/1995/mar/kaplan.html. Updated at http://raven.ubalt.edu/staff/kaplan/lit/. [1997, November 1].

Kedzie, C. R. (1996) "Coincident revolutions," *On the Internet*, January/February: 20–29.

—— (1997) *Communication and Democracy. Coincident Revolutions and the Emergent Dictator's Dilemma*, Santa Monica: RAND.

Knox, P., and Marston, S. (1998) *Human Geography, Places and Regions in Global Context*, Saddle River: Prentice-Hall.

Krygier, J. B., Reeves, C., DiBiase, D., and Cupp, J. (1997) "Design, implementation and evaluation of multimedia resources for geography and earth science education," *Journal of Geography in Higher Education*, 21(1): 17–39.

Lake, R. W. (1993) "Planning and applied geography: positivism, ethics, and geographic information systems," *Progress in Human Geography*, 17(3): 404–413.

Landow, G. P. (1992) *Hypertext*, Baltimore: The Johns Hopkins Press.

—— (ed.) (1994) *Hyper/text/theory*, Baltimore: Johns Hopkins University Press.

Lane, C. A. (1997) "My view of your life and why online access is inevitable for us all," *Washington Post*, July 13: C3.

Matrix Information and Directory Services (MIDS) (1997) 1997 users and hosts of the Internet and the matrix. [Online].
Available: http://www.mids.org/press/pr9701.html [1997, November 1].

National Council on Geographic Information Analysis (NCGIA) (1997) Initiative 19: Society and GIS Position Papers [Online].
Available: http://www.geo.wvu.edu/i19/papers/position.html. [1997, November 1].

Noble, D. F. (1998) "Digital diploma mills: the automation of higher education," *Monthly Review*, February: 38–52.

O'Harrow, R. (1997) "Federal sites on web gather personal data," *Washington Post*, August 28: E1, E3.

OMB Watch (1997) A delicate balance: the privacy and access practices of federal government world wide web sites [Online].
Available: http://ombwatch.org/ombw/info/balance.html. [1997, November 1].

Pickles, J. (ed.) (1995) *Ground Truth. The Social Implications of Geographic Information Systems*, New York: Guilford Press.

—— (1997) "Tool or science? GIS, technoscience and the theoretical turn," *Annals, Association of American Geographers*, 87(2): 363–372.

Press, L. (1996) "The role of computer networks in development," *Communications of the ACM*, February: 23–30.

Proctor, J. (1998) "Ethics in geography: giving moral form to the geographical imagination," *Area*, 30(1): 8–18.

Quaynor, N. N. (1997) "Ghana, bringing the Internet to rural areas," *On the Internet*, November/December: 32–33.

Rheingold, H. (1993) *The Virtual Community, Homesteading on the Electronic Frontier*, Reading, MA: Addison-Wesley.

Sanchez, R. (1997) "Poor, minority students lack access to computers," *Washington Post*, May 15: A13.

Schwartz, J. and Biskupic, J. (1997) "1st Amendment applies to Internet, Justices say," *Washington Post*, June 27: A1, A20.

Staple, G. (1997) "Communications beyond frontiers: expansion of national carriers across borders," in E. M. Noam and A. J. Wolfson (eds), *Globalism and Localism in Telecommunications*, Oxford: Elsevier Science Ltd.

Whittle, D. B. (1997) *Cyberspace: the Human Dimension*, New York: W.H. Freeman & Co.

Wresch, W. (1996) *Disconnected, Haves and Have-nots in the Information Age*, New Brunswick: Rutgers University Press.

Wright, D. J., Goodchild, M. F., and Proctor, J. D. (1997a) "GIS: tool or science? Demystifying the persistent ambiguity of GIS as 'tool' versus 'science'," *Annals, Association of American Geographers*, 87(2): 346–362.

—— (1997b) "Reply: still hoping to turn that theoretical corner," *Annals, Association of American Geographers*, 87(2): 373.

Yapa, L. (1995) "Innovation diffusion and paradigms of development," in C. Earle and M. Kenzer (eds), *Concepts in Human Geography*, Lanham, MD: Rowman & Littlefield.
—— (1996) "What causes poverty? A postmodern view," *Annals, Association of American Geographers*, 86(4): 707–728.

Part 2

Ethics and place

The category of place in geography is implicated in ethics quite centrally: not only are all moral schemes emergent from, and tied to, particular places, but the very concept of place involves understanding morality as a persuasive strand in the ways people experience and express their own geographies, lived out in multiple scales of space and place. Michael Curry begins the exploration of links between ethics and place by stressing the normativity of place. While most writers today see places as simply locations or containers within a larger geographical space, he argues that in an important sense place is a more basic category coming prior to space. Because places are basic sites of human activity, a central function which they perform is to define what is possible and allowable within their boundaries. Places are thus fundamentally normative, concerned with what is right and good conduct and where. To say, "That's how we do things here" captures a form of place-specific moral justification which is subject to spatial differentiation. We therefore need to see the study of place, in all its normativity, as the basis for understanding space.

Yi-Fu Tuan then turns the discussion to the neglected subject of evil. He takes up four "dark tendencies" in human nature: identified as destructiveness, cruelty, sadomasochism and compartmentalization. The Moguls exemplified the first two combined, but to those people in that time and place their destructiveness and cruelty were natural, which raises the question of the temporal and spatial specificity of evil. Sadomasochism surfaces the role of domination associated with affection (with reference to pets as well as persons), which touches the moral standing of a culture at a vulnerable point. The concept of compartmentalization is used to suggest that the bounding of space enables particular modes of behavior to find a place (see also Curry's essay), for example drawing lines between identifying indifference rather than care as an appropriate response to others. He raises the fundamental problem of whether it is plausible that we have the potential to develop a version of the good and the evil transcending that of particular times and places.

Gearóid Ó Tuathail considers a specific recent act of evil: the so-called ethnic cleansing of Srebrenica in the former Yugoslavia. Geography made this violence unique in two ways: it took place in what was thought to be civilized Europe, and in a place that the United Nations had declared a "safe area." The author argues that the institutionally proclaimed, bureaucratically supervised and professionally administered ethics of the UN so constricted expression of moral responsibility for vulnerable others that they effectively promoted immoral purposes. The result was acquiescence with genocide on the part of the Bosnian Serb army. Hence his distinction between ethics as a set of institutionalized rules promoting normative behavior (as supposedly laid down by the USA through the UN) and morality as open-ended responsibility towards otherness (with its obvious limits as actual practice which may be manifest in evil).

Caroline Nagel begins by drawing attention to changes in the composition of the population of Western Europe since the Second World War, with the influx of immigrants. This has led to discourses of multiculturalism, and to a politics of identity preoccupied with difference and diversity. However, these movements (sometimes associated with post-structuralism and postmodernism) are themselves now facing increasing criticism. She undertakes an evaluation of multiculturalism through competing conceptions of social justice. The case of Salman Rushdie (a British author threatened with death for supposedly insulting the Prophet Muhammed) is used to exemplify ethical problems posed by the politicization of cultural identities, with particular reference to Britain. Her conclusions point towards a politics of inclusion, of common membership, which recognizes human sameness and affinity rather than promoting exclusionary identity based on irreconcilable difference. Such is the requirement of the heterogeneous populations which inhabit the increasingly porous places of the contemporary world.

7 "Hereness" and the normativity of place

Michael R. Curry

In recent years the concept of place has moved into the mainstream of geographical and other discussions, to the extent that it is now difficult to see that not so long ago it was absent. Before the path-breaking works by Tuan (1971a, 1971b, 1974), Relph (1976), and Buttimer (1974), discussion within geography focused instead on concepts such as space, distance, and especially region.[1] Now only 25 years old, those pioneering works have nonetheless receded into the penumbra cast by the very popularity of the concept. Today we seem constantly to hear of the "death of distance"; at the same time geographers of all stripes proclaim that "place matters," and we are led to imagine that within geographical writing it must always have been so.

Regrettably, as the concept of place has been tamed and naturalized, those pioneering authors have been marginalized, dismissed as too concerned with the discussion of values; they have been termed "humanistic," where humanistic is only a thinly veiled euphemism for "soft," politically suspect, nostalgic, and elitist. Indeed, if the force of those pioneering works has been lost, the evidence of that loss is in the need for this very volume, a volume devoted to ethics and geography.

Of course, it will not do simply to say that for one group there was a moment in geography that focused on places and on the ethical, while for another later group this insight has been forgotten and reinvented. For those now termed humanists were in the 1970s writing against the background of the quantitative revolution, and of a geography that itself proclaimed the irrelevance of issues of ethics in geography, and in science.

In fact, some of those now seen as the new pioneers in the study of space and place were warriors in the quantitative revolution, who viewed the Vietnam War, the civil rights movement, and the beginnings of the women's and gay rights movements from the comforts of the academy, while penning works that claimed to be written without a point of view, and without a need for one.[2]

Ironically, the new critique (or to be honest, rejection) of the humanists of the 1970s – among those who have claimed to have invented the concepts of place and culture or at least to have imported them into a discipline where they had never before been seen – shares with the positivist critique of the 1960s the association with and devotion to a set of practices and institutions that continue this positivist stance (Curry 1996).

But if on this score those who in the 1970s championed the study of places were at odds with the mainstream exemplified in the quantitative revolution (and indeed, in the Hartshornean orthodoxy that preceded it), they seemed often to share with their critics at least one belief. They seemed to believe that one can array places on a continuum, a scale that runs from small to large, from the most intimate place to the grandest, most encompassing space. In a sense, their differences rest on what they make of that fact.

By contrast, I reject this notion of a place–space continuum, and reject its implications, where the difference between place and space may in the end be explained as a matter of "scale" – whatever that is. Too often this way of thinking is the beginning point of research, and not the conclusion; as such, it preordains those conclusions. On the contrary, I would argue that we need as much as possible to recognize as modernist diversions the aphorisms, slogans, and mantras taught us from our earliest exposure to science, and attempt instead to begin from a series of observations. I suggest that if we do so we see the following:

First, places have an existence that is *sui generis*. "Place" is not space viewed through a microscope, any more than space is that same place viewed through the wrong end of a telescope. Rather, the places in which people carry out their lives *are* just that, places. Second, places are human constructions, outgrowths always of an interweaving of human activities. Third, those activities, the creation of the stable places within which people carry on their lives, are themselves practices, routines that are fundamentally normative. The practices by which we create and maintain places can be characterized as right or wrong, good or bad, well done or done abysmally.

Fourth, and as a consequence, those places themselves, the basic sites of human activities, are intrinsically normative. We live in a world in which value is not a post hoc add-on, an after-effect or after-thought, something to be rejected by self-assured academics. Rather, the value that exists in the world is there right from the outset. And finally, because the complex of activities – what Andrew Pickering (1995) has called the "mangle of practice" – through which we construct and maintain places is always constructing and maintaining multiple places, at what some would term different "scales," we are always in more than one place at once. We are at home and in the bedroom and in New Jersey, or at work and in an office and in California, or in an airplane seat and in the United States and in first class. And it is because different activities and objects and ideas "fit" within different places that we are inexorably faced with moral dilemmas. Those moral dilemmas are geographical dilemmas.

And so, the inquiry into the nature of places can shed light on the ways in which the everyday activities by which we make and maintain places involve appeal to ethical and normative concepts usually seen as the provenance of philosophers; at the same time, it can suggest to philosophers what are in the end geographical sources to many of the ethical dilemmas that we face.

On making places

Consider the ways in which the Europeans took over the Western hemisphere. How did a country get the right to a new area, over what was called *terra nullius?* It was not enough just to see a new place and claim it. Something else was required; someone had to carry out a set of actions that created that ownership.

It was in any case required that the person claiming sovereignty have the right to do so. John Cabot, for example, got a patent in 1495 from King Henry VII of England. It gave to him and his sons, assigns, heirs, and deputies the

> full and free authority, faculty and power of navigation to all parts, countries and seas of the east, west and north, under our banners and flags . . . to seek out, discover, and find whatsoever islands, countries, regions or provinces of heathens or infidels, in whatever part of the world they may be, which before this time were unknown to all Christians.
>
> (Keller *et al.* 1938: 50)

Beyond these general features, though, there were differences in the practices used by different countries in claiming land. The Spanish had highly formalized and religious ceremonies. An instruction given to De Solís in 1514 required:

> Cutting trees and boughs, and digging or making, if there be an opportunity, some small building, which should be in a part where there is some marked hill or a large tree, and you shall say how many leagues it is from the sea, a little more or less, and in which part, and what signs it has, and you shall make a gallows (i.e. a judge's bench) there, and have somebody bring a complaint before you, and as our captain and judge you shall pronounce upon and determine it, so that, in all, you shall take said possession.
>
> (Keller, *et al.* 1938: 40).

And Cortez, it is said, "Moved walking on the said land from one part to another, and throwing sand from one part to another, and with his sword he struck certain trees . . . and did other acts of possession" (Keller *et al.* 1938: 41)

Here, almost in a nutshell – and certainly inadvertently on the part of the authors – we have a remarkably complete accounting of the ways in which people create places (see Table 7.1).

There is, actually, missing from this account of the making of places one very important, final element, one that is shown here but not mentioned; people make places by constructing narratives, just as did the narrator of Cortez's actions.

Now it might seem that these sorts of actions are involved in places, but that they are equally involved in the creation and maintenance of community, culture, clan, indeed, of the wide range of human institutions that we think of as *Gemeinschaften*. One might, that is, argue that there is nothing in these activities that is associated directly with places, but rather, that the products of such activities become places simply to the extent that they occur in particular, mappable

Table 7.1 The Europeans' actions, and place-making functions

Particular action	Place-making function
"made a public proclamation"	engage in ritual practices
"unfurled banners"	associate symbols with the place
"cutting trees and boughs"	bring about a change in the "natural" world
"making . . . some small building"	create elements of a built environment
"say how many leagues it is from the sea"	measure and classify
"make a gallows there, and have somebody bring a complaint before you, and as our captain and judge you shall pronounce upon and determine it"	institute laws

locations. Further, even if it were allowed that these actions are involved in the creation of places, one might want to argue that the fact that activities that are normative are involved in the making of places says nothing about the normativity of places themselves. An architect or builder, after all, can do a good job on a hospital or a concentration camp. In the next two sections I shall indicate why I believe both such arguments to be mistaken.

Being here

Central to an understanding of the nature of places are the concepts of "here" and "there." In order to make more clear what I have in mind here, it will be useful to turn to the following question: How do I understand what someone is saying?

One very common answer to this question would go as follows. When someone says to me, "Look at that sparrow," my brain goes into action, searching – quietly, one hopes – for a definition of "sparrow," which might be something like "A small brown bird abundant in the United States, etc." Having come upon the definition, I simply apply it to the object in question. Now, as I say, this is a common enough way of thinking about how people use language. It underlies a great deal of the rhetoric of science, just as for many years it underlay the teaching of second languages.

But philosophers and teachers of language have joined forces, suggesting that this view of language does not and indeed cannot fit the facts (Vygotsky 1962; Wertsch 1985; Wittgenstein 1968). It cannot fit the facts because the user of language is, on this view, placed in an infinite regress. And it does not fit the facts simply because people learn language in a very different way; they learn to use sentences one at a time. And they learn to use them in particular contexts; indeed, to be considered proficient in a language is to be able to say the right thing in the right context (Barnes and Curry 1983).

Wittgenstein characterized the matter in the following way:

To obey a rule, to make a report, to give an order, to play a game of chess, are customs (uses, institutions). To understand a sentence means to understand a language. To understand a language means to be a master of a technique.

(Wittgenstein 1968: Part I: §199)

For Wittgenstein it was important to notice the following. When someone asks me why the line was drawn between Canada and the United States, I can give an explanation, which probably refers to political arrangements. When asked why those arrangements were relevant, I can give an explanation of their source. When asked for an explanation of the source I can give yet another explanation. But in the end, "If I have exhausted the justifications I have reached bedrock, and my spade is turned. Then I am inclined to say: 'This is simply what I do'" (Wittgenstein 1968: Part I: §217).

In the end, "What has to be accepted, the given, is – so one could say – forms of life" (Wittgenstein 1968: Part II: 226).

Over the last several years a number of social theorists have taken this as a starting point. Believing that one needs to begin with the basic elements of human life, and agreeing with Wittgenstein that those elements are human practices, they have attempted to construct an understanding of the world (Bourdieu 1977; Giddens 1979; Winch 1990 [1958]).

It is ironic, as Thrift has pointed out, that those who have appealed to these works have often simply skipped this part of the works of, especially, Bourdieu and Giddens, preferring to begin later in the story (Thrift 1996). And it is unfortunate as well, because in doing so they undercut the possibility of seeing the very basic point at which geography enters into the discussion.

Now one way of adding more detail to Wittgenstein's characterization of the justification for human practices is this: in the end, I throw up my hands and say, "That's just how we do things." Here the "we" might be Americans or Californians or women or the elderly or – some would claim – humans. And it makes a certain amount of sense to say that just to the extent that a group of people are comfortable making this "That's just how we do things" claim, they can be said to belong to a culture or community.

But I would like to suggest that there exist a number – a large number – of cases in which we amend this statement. We say, "That's how we do things *here*." Or there. And just as the initial statement can be seen as constituting or referring to cultures or communities, these statements refer to places. To say "That's how we do things here" is to make a claim about a place, to refer to a place, and to define a place.

And it is *not* at the same time to assert, or even to imply, that essential to that place is some location, the possibility of pointing to the location of that "here" on a map. In some cases we may wish to say that such a location is essential – who would deny that that is true of the North Pole or the Equator – but in a great many others it is simply not. Much of what goes on in a crack house has little to do with what street it is on, or whether it is in New York or Chicago, and a great deal more to do with the fact that it is a certain place, a crack house – and the same can be said of many other places, the fraternity house, the corporate board room, or the casino.

So, far from seeing places as simply small patches of space, or "locations with meaning," I would argue that places need to be defined in terms of a function that they fill in human life, where places are the contexts – or at least one type of context – within which human activities are carried out, and within which they make sense.

On the normativity of places

One might nonetheless wish to argue that having been constructed through this set of admittedly normative practices, places then take on lives of their own, becoming neither good nor bad, demanding nothing of the people there, acquiring values only to the extent that those values are imposed on them. Auschwitz, on this argument, might have become a home for drug dealers, or for unrepentant liberals. To make such an argument is, though, to misconstrue the nature of places – and of the objects that make up the world. It is to imagine that having once been created, an object somehow acquires an essence that "sticks with it."

Yet this fails to understand what it means for an artifact to exist. Consider something as simple as a banana. It may seem an unproblematic item, something that none of us would see as needing much thought. Yet the banana is surely one thing to someone raised in the shadow of the United Fruit Company, where it was long the accompaniment of a brutal corporation and the destruction of a way of life; it means something else to someone who was raised in Central Europe, where, rare indeed, it was a symbol of luxury, of freedom and opulence. It is something else indeed to readers of Thomas Pynchon's *Gravity's Rainbow*, or early fans of the Velvet Underground. And it is something else again to an aspiring political ecologist with his heart set on a plum teaching job.

These examples ought not to suggest that behind these "cultural interpretations" is the "real" banana. For even biologists have questioned the naturalness of botanical categories, of the boundary between the banana and its cousins. And even assuming that those are clear, when is a banana a "real" banana? When fully grown? And when is that? What if it has become fit only for banana bread? And where does one cut it from the stalk? How much is too much?

All of these questions, of course, suggest that a banana becomes and remains a banana only through the *active maintenance* of a range of practices. How one cuts it, how it is graded, when it is acceptable for sale across the counter and not for the commercial food industry, where it is available, who eats it, and so on; all of these practices and their surrounding institutions create the banana, just as a set of practices and institutions are at work in the creation and maintenance of any place. If you remove them, you don't have a natural place, any more than you have a natural banana. What you have is something else entirely.

On the nature of places

If I have suggested that places become places by virtue of the ways in which people appeal to their "hereness" or "thereness," this ought not to suggest that there is nothing systematic to be said about them. Yet that systematically describable element is often obscured by appeals to the wrong sort of metaphor. For example, if as is common we imagine a place necessarily to be something like a location, we are immediately led to a particular way of thinking about it: places can be mapped, their locations specified in numerical terms, their borders found.

This very metaphor, of course, undercuts our ability to see particular places as having to be maintained. Indeed, it would make more sense to compare a place to a conversation. For like places, conversations can be large and small, brief and long-lived. There can be multiple people involved, and the actors can change. Some people are good at keeping a conversation going, and some are not. Over the course of a conversation the nature of the conversation itself can change; who has not been in a conversation that started "normally" and ended in a fight? And finally, when the actors cease to be involved the conversation itself ceases; the social situation is redefined.

This metaphor leads us in a very different direction as we consider places, and in a way that more fruitfully underscores the normativity of places. In particular, it points to three features of places. First, places are structured and hierarchical. Second, for each of us a place exists as a place only in relationship to our body. And third, the elements of places exist in variable relationships one with another.

As it turns out, each of these features of places has been at the heart of a philosophically and scientifically significant theory of space and place. The first of these is the view articulated by Aristotle, in his *Physics* (1984). Central to the understanding of the world propounded there was a belief that things have their proper places. Solid materials – composed of earth – fall to earth because that is where they belong; it rains because water belongs with water. For Aristotle, in this basic principle we can see the structure of the world. As he put it: "all place admits of the distinction of up and down, and each of the bodies is naturally carried to its appropriate place and rests there, and this makes the place either up or down" (Aristotle 1984: 358–59).

If we see this in his physics, we see it as well when we turn to his other works, on biology and politics and ethics (Aristotle, 1941a, 1941b, 1986). And so, Aristotle's view of the world, a view very much at the center of scientific thinking for 2,000 years, included at its core an appeal to hierarchy. This was a world of objects that differed qualitatively one from another; and in this world some things were intrinsically higher and some lower, better and worse.

We see the second feature of places – that for each of us our body is at the center of those places – in Kant's "pre-critical" inaugural essay (1929). Kant based his argument on the problem of incongruent counterparts. Consider, he said, a pair of gloves. If we look at the relationships among the parts, at their angles and lengths, we find that a right-handed glove and a left-handed glove are in fact identical. But just try putting a left-handed glove on your right hand!

For Kant the significance of this seemingly minor matter was profound: we need to see that although space is an absolute, that absolute quality must be seen as extending from the human body. Moved from the center of the world by modernists like Copernicus and Newton, Kant ushered people back in.

Finally, in the early eighteenth century, Leibniz articulated a view that saw space as existing only as the interrelationships among objects (Alexander 1956). For Leibniz it made no sense to talk about a space with nothing in it; space, in fact, only came into existence with the objects that we now think of as filling it.

There are actually two issues here relevant to the issue at hand. First, if we turn to the famous debate held between Leibniz and Newton's stand-in, Samuel Clarke, perhaps the most striking feature of that debate is the overwhelming presence of religion as a motif. In fact, Leibniz attacks Newton repeatedly, for taking a position that lacks humility, that sets limits to what God can do. So more explicitly than many now remember, Leibniz (and Newton) developed scientific arguments that they saw as resting on theological principles; for them, one would not so easily distinguish, as we routinely do today, between facts and values.

But more important here is Leibniz's development in his work on space and in a complementary way in all of his *Monadology* of a view in which the universe consists of elements that are intrinsically connected one with another (Leibniz 1965). Here he focuses on the third central feature of places, as of conversations: to remove an element of a place may very well be at the same time fundamentally to alter the nature of that place. This is clearly true in the case of conversations, as it is in the honeymoon suite.

And so it seems to me that it is useful to reread the history of ideas about space. Here we can see some of the key figures in that history not so much as having been advocates of victorious or vanquished theories of space, but rather as having recognized one or another of the features that we take as a matter of common sense to be fundamental to the places that we create and maintain.

This leads to a final point, or two, actually. First, the idea of scale, so dear to many geographers, is dramatically misleading. It is misleading because it presumes a relationship between space and place. This is not to say that there is no such relationship; there patently is – sometimes. But the creation of space is itself a matter of practices, and practices that take place in places. So too is the discourse about space. To mistake that discourse for something else is to set the stage for greater misunderstandings.

Hence, if those geographers who over the last few years have argued against the appeal to a modernist, "dead" theory of space have been right to develop that critique, insofar as they have focused on space rather than on places, they have been moving in the wrong direction. For it is places that are truly alive. Although places are not, as Pred (1984) claimed, "processes," he was exactly right to the extent that places only exist as long as there *are* processes.

Being in two places at once

Second, paradoxical as it may sound, it is possible to be in two places at once. And this, which may seem a silly aside, is quite the opposite. Indeed, in any conceivable society it is not just possible, but inevitable that at a given moment we are in more than one place. One might go so far as to say that humans are the animals, and the only animals, that *can* be in two places at once. I can be in my office and in California, or at the beach and at a Superfund site.

This is of course because my actions are defined and make sense only within a particular context. While at the beach, I am from one perspective, simply "walking at the beach." If like Richard Nixon I go to the beach while wearing a dark business suit, I am likely to be seen as bizarrely out of place. If I take up a spot directly in front of a person already there, blocking that person's view or access, I am likely to receive an angry stare, and perhaps more. At the same time, if while making my way to the beach I walk across a protected sand dune, the context within which some people would understand my actions would be very different; even if I still think of myself as "at the beach," I am from their perspective somewhere else. And within that place there exist a radically different set of responses and expectations.

Indeed, in a society as complex as ours, we are always in a wide range of places. And as societies become more complex it becomes increasingly difficult to maintain or even develop an awareness of the variety of those places. If even 100 years ago many people could still see themselves as fundamentally – and mostly – residents of one small locality or another, and could at least name the places within which they acted – parts of the home, the neighborhood or locale, the town, the parish, perhaps the county and state – today the range of such places has increased dramatically in scope. We are aware of some of those places – we know, at least, how to respond when asked. Of others we are only dimly aware, or are even unaware.

I would suggest that whatever one takes to be the appropriate form of ethical theory, this fact has important consequences for those creating such theories. This is most clearly the case for someone who wishes to propound a consequentialist theory. For in a world this complex one needs to be able to set limits to the realms within which the consequences of one's actions are to be analyzed and judged; not to do so is to leave oneself in a world that is paralyzingly complex. But if one is to limit the realm within which those consequences are to be judged, then one is faced with a perplexing situation, just because the multiplicity of places within which one acts will almost always lack a well-developed and accepted hierarchical structure, one in terms of which it will be possible to say which is to be accorded primacy.

If, though, we turn to ethical theories that look to the motives that are said to underlie action, then a different problem arises. For just to the extent that my actions are contextually defined, to be in more than one place at once is by that very fact to be doing two things at once. And it is very likely to be doing different things that can be described in terms of different motives.

104 *Ethics and place*

Indeed, here the very underpinning of motive-based theories seems to have come undone. For the attribution that we make of the motives of others is very much based on our "reading" of what "a person like that" in a "situation like that" must have been thinking, of what must have motivated his or her actions. But the proliferation of places makes just that attribution all the more difficult.

I may seem here to be offering a counsel of despair, or as one author (who shall remain nameless) put it, a "council of despair." But although I sometimes feel like at least an honorary member of that council, I would close with two suggestions. First, the view of place that I have outlined here offers in at least provisional form a common ground for discussions between those primarily concerned with the geographical and those primarily concerned with the ethical.

And second, if I seem here to have been denying the possibility of the sort of universalism that underpins so much moral discourse, if I seem to be celebrating the local, I would counter that at the local level, too, there exists the constant draw – and temptation – of the universal. The universal, as Yi-Fu Tuan (1996) has eloquently shown, lives in the local.

Notes

1 See for example Abler *et al.* (1971) or the various essays collected in English and Mayfield (1972).
2 For example, Bunge (1966), Harvey (1969), and Olsson (1975).

References

Abler, Ronald, John S. Adams, and Peter Gould (1971) *Spatial Organization: The Geographer's View of the World*, Englewood Cliffs, NJ: Prentice-Hall.
Alexander, H. G. (1956) *The Leibniz–Clarke Correspondence, Together with Extracts from Newton's Principia and Opticks*, New York: Barnes and Noble Imports.
Aristotle (1941a) "Ethica Nichomachea" in *The Basic Works of Aristotle*, Richard McKeon (ed.), trans. W. D. Ross, New York: Random House, pp. 935–1126.
—— (1941b) "Politica" in *The Basic Works of Aristotle*, Richard McKeon (ed.), trans. Benjamin Jowett, New York: Random House, pp. 1127–1324.
—— (1984) "Physics" in *The Complete Works*, Jonathan Barnes (ed.), trans. R. P. Hardie and R. K. Gaye, Princeton: Princeton University Press, Vol. I, pp. 315–346.
—— (1986) *De anima*, trans. Hugh Lawson-Tancred, Harmondsworth: Penguin.
Barnes, Trevor J. and Michael R. Curry (1983) "Towards a contextualist approach to geographical knowledge", *Transactions, Institute of British Geographers* NS 8: 467–482.
Bourdieu, Pierre (1977) *Outline of a Theory of Practice*, trans. Richard Nice, Cambridge: Cambridge University Press.
Bunge, William (1966) *Theoretical Geography*, Lund: C. W. Gleerup.
Buttimer, Annette (1974) *Values in Geography*, Washington, DC: Association of American Geographers.
Curry, Michael R. (1996) *The Work in the World: Geographical Practice and the Written Word*, Minneapolis: University of Minnesota Press.

English, Paul and Robert C. Mayfield (eds) (1972) *Man, Space, and Environment: Concepts in Contemporary Human Geography*, New York: Oxford University Press.

Giddens, Anthony (1979) *Central Problems in Social Theory*, Berkeley: University of California Press.

Harvey, David (1969) *Explanation in Geography*, London: Edward Arnold.

Kant, Immanuel (1929) "On the first ground of the distinction of regions in space" in *Kant's Inaugural Dissertation and Early Writings on Space*, trans. John Handyside, Chicago: Open Court.

Keller, Arthur S., Oliver J. Lissitzyn and Frederick J. Mann (1938) *Creation of Rights of Sovereignty through Symbolic Acts*, New York: Columbia University Press.

Leibniz, Gottfried Wilhelm (1965) *Monadology, and Other Philosophical Essays*, trans. Paul Schrecker and Anne Martin Schrecker, Indianapolis: Bobbs-Merrill.

Olsson, Gunnar (1975) *Birds in Egg*, Ann Arbor: Department of Geography, University of Michigan.

Pickering, Andrew (1995) *The Mangle of Practice: Time, Agency, and Science*, Chicago: University of Chicago Press.

Pred, Allan (1984) "Place as historically contingent process: Structuration and the time-geography of becoming places," *Annals of the Association of American Geographers* 74: 279–297.

Relph, Edward (1976) *Place and Placelessness*, London: Pion.

Thrift, Nigel (1996) *Spatial Formations*, London: Sage.

Tuan, Yi-Fu (1971a) "Geography, Phenomenology, and the Study of Human Nature," *The Canadian Geographer* 25: 181–192.

—— (1971) *Man and Nature*, Washington, DC: Association of American Geographers.

—— (1974) "Space and place: A humanistic perspective" in *Progress in Geography: International Reviews of Current Research*, Christopher Board, Richard J. Chorley, Peter Haggett and David Stoddart (eds), London: Edward Arnold, Vol. 6, pp. 211–252.

—— (1996) *Cosmos and Hearth: A Cosmopolite's Viewpoint*, Minneapolis: University of Minnesota Press.

Vygotsky, L. S. (1962) *Thought and Language*, trans. Eugenia Hanfmann and Gertrude Vakar, Cambridge, MA: MIT Press.

Wertsch, James V. (1985) *Vygotsky and the Social Formation of Mind*, Cambridge, MA: Harvard University Press.

Winch, Peter (1990 [1958]) *The Idea of a Social Science and its Relation to Philosophy*, second edn, New York: Humanities Press.

Wittgenstein, Ludwig (1968) *Philosophical Investigations*, third edn, trans. G. E. M Anscombe, New York: Macmillan.

8 Geography and evil:
A sketch

Yi-Fu Tuan

Geography has not addressed the question of evil and, with it, the entire realm of morals and ethics that has been and still is a central concern of philosophy both in the East and in the West since ancient times. A reason for this is modern geography's root in a physical science, geology, rather than in a human "science" – history or political philosophy. Morals and ethics are not inherent to a physical geographer's work for the simple reason that his or her subject matter is inanimate. The biogeographer does engage with animate reality, but, interestingly enough, most biogeographers are plant geographers, and plants are not normally seen as possessing feeling, and along with it, the possibility of delight, suffering, and pain that raises moral issues. What if there were more zoogeographers? Would confronting a nature that is "red in tooth and claw" become inevitable? Not necessarily, for human geography is a well established field, and people certainly have feelings – they love, hate, build, destroy, and kill – and yet, until well into the second half of the twentieth century, geographers have managed to avoid morals and morality altogether, or skirt around their edges.

The reasons for this blindness are complex. Among them, I suggest, are the following. The concept called "environmental determinism" treated people as passive. "Victims," to use a currently fashionable term, cannot also be active agents for good and evil. When, by the early twentieth century, geographers started to see people as active agents, moral questions still failed to emerge, for the geographers of that time were mostly converts from geology and so tended to talk about human beings as though they were just another type of geological agent, like wind, water, and ice. When, at last, geographers took people seriously as culture-bearers and powerful cultural agents of change, a triumphalist view of their story rose to dominate geographical thought, one effect of which was to bury once more moral issues, in particular, the ill consequences of human action. "Sequent occupance" is an example of triumphalism. Popular among human geographers from the 1920s to the 1940s, it encourages one to see successive changes as inevitable progress: first forest or brush, then farm and village, then town, finally city and suburb (Mikesell 1976: 149–169). A deeper reason for the neglect of moral questions is the geographer's indifference to events. Events, we seem to feel, are best left to historians. The event of war is prominent in history books. In geography books, it is conspicuously absent. There is of course a geography of the American Civil War, but we have not written it. We map battle-fields – the cool and static aftermath of an event – rather than the clash of beliefs, alliances, and armies, in which courage, cowardice, wisdom, stupidity, good and evil are likely to be displayed.

Geographers have neglected the moral dimension of human reality, but not entirely. There are exceptions. In modern times, an outstanding exception is Carl Sauer, who in a debate with an economist, made his position clear by declaring, "We are moralists" (Leighly 1963: 4). Since at least George Perkins Marsh, geog-raphers have noted how destructive society can be to nature and to humbler folkways when they lie athwart its path to progress. Morality is implicit in much of the Berkeley School of Cultural Geography. Since the 1970s, the more extreme section of the environmental movement has made the moral issue explicit – and, indeed, a clarion call to arms: preservation of wilderness is good, use – even mild use – is somehow bad, an exodus from Eden. From the 1960s onward, geography has finally and seriously come to grips with morality and ethics under the inspir-ation of, first, Marxism, then, feminism and minority rights. Much of the con-temporary geographical literature on ethics addresses imbalances of social power, spatial justice and injustice.

I make these observations so as to prepare the ground for raising a matter that geographers still evade – evil. The word itself is alien to the discipline's lexicon. Even moral philosophers may find it too strong and seek to avoid using it, though theologians cannot. Yet evil is an indubitable fact in human existence and experi-ence (Parkin 1985). Literature would be mere entertainment without its shadows. If the human geographer can avoid confronting it, the humanist geo-grapher cannot do so without depriving the subfield of its seriousness. As a humanist geographer, I wish to inject morality – specifically, the problem of evil – into geographical thought and writing. How to do so is the challenge. No doubt various approaches are possible. One can, for instance, tackle the subject historic-ally, sociologically, or culturally. What I offer here is none of the above, even as outlines. Rather it is a prolegomenon to these approaches, a raising of questions concerning human nature that confronts an unpleasant fact, which is: people generally, and not only those of a particular social class or economic status, can severely – even happily – maltreat one another, as well as plants and animals. The four dark tendencies of human nature that I take up are: destructiveness, cruelty, sadomasochism, and compartmentalization.

Destructiveness

Tennyson's "nature red in tooth and claw" has never quite disappeared from modern consciousness, and this despite the powerful view, promulgated by Romantics, amateur naturalists and environmentalists, that nature is benign and essentially harmonious, and that only humans are destructive, the real snake in the garden as it were, and evil. Nevertheless, even as this benign view reigns, the media and personal experience convincingly show, over and over again, that nature, too, can be wantonly destructive. Wantonly? I use the word not for poetic effect but because it suggests itself – because that's how I and no doubt others, too, feel. Nature can still *seem* evil in its violence, its utter indifference to suffering. And this applies to organic nature as well, for although the Bambi image remains influential, images that stress Tennyson's tooth and claw, torn limbs and bloody carcasses, are increasingly featured, with garish prominence, in photos and films on wild life (Joubert 1994: 35–53; Hammer 1994: 116–130).

But isn't this emphasis on violence and bloodiness in the natural world an attempt to draw on their high visibility and popularity in the human world? Yes, I would say, and that's a surprising turn, because earlier – from the second half of the nineteenth century to the first half of the twentieth century – the influence would seem to have been the other way round: then, at least to those influenced by social Darwinism, it was the ruthless struggle supposedly regnant in the natural world that explained and excused ruthlessness in the human world. Whatever the direction of influence might be in our time, the late twentieth century, one thing is clear: the broad acceptance of the view that humans are exceptionally prone to violence and destructiveness.

Destructiveness as such has an appeal to humans that is not evident in other animals. What is the nature of the appeal? Power and its enlarging effect on selfhood is the beginning of an answer. For many people, destruction is the clearest evidence of their ability to change the world and hence the most convincing proof of their own existence – their own reality and worth. "(Knocking) a thing down . . . is a deep delight to the blood," says the suave and gentle philosopher George Santayana (1980: 81). Parents will agree as they watch an early proud accomplishment of their infant, which is to knock down a pile of wood blocks with great glee. Children old enough to make things retain a fondness for destruction. They may even build for that purpose. At the beach, they make "elaborate reservoirs of sand, fill them with water, and then poke a little hole in one of the walls for the pleasure of watching the water sweeping them away" (Wilson 1963: 167). To Wilhelm von Humboldt, a distinguished humanist and educator, the sight of a force that nothing could resist has always had great appeal. "I don't care," he wrote, "if I myself or my best and dearest joys get drawn into its whirlpool. When I was a child – I remember it clearly – I saw a coach rolling through a crowded street, pedestrians scattering right and left, and the coach unconcerned, not diminishing its speed" (Humboldt 1963: 383–384).

Even the power to build is tied intimately to the power to destroy. The one necessarily precedes the other. Our own body is a wonder of construction, carefully maintained by consuming – that is, destroying other plants and animals. Of course, all animals do that. But, unlike other animals, humans also have culture, that is to say, we make things, even a huge superfluity of things; and everything we make, from a simple bench to a great city, entails prior destruction. The violence of chopping down a tree is forgotten as a bench emerges under our skillful hands. The violence of removing an entire forest, leveling hills and diverting streams, is erased from our memory once a bustling city rises on the cleared site.

Not only nature must make way for the city, so also must prior human occupants and their works. The violence of urban renewal in modern times is as nothing compared with that of antiquity, when the Assyrian king Sennacherib (705–681 BC) could boast:

> The city and its houses from its foundation to its top, I devastated and burned with fire. The wall and the outer wall, temples and gods, as many as there were, I razed and dumped into the Arakhtu Canal. I flooded the city's site with water and made its destruction more complete than that by flood.
>
> (Luckenbill 1924: 17)

Why such destructiveness? The answer would seem to be that ancient builders – Sumerian, Assyrian, and Roman – believed they had to work with a clean slate, uncontaminated by the vengeful spirits of the defeated. All recognized that, in building a civilization, order and lawful conduct might have to be harshly imposed. The Sumerians were exceptional in admitting that evil acts such as deliberate falsehood, violence, and oppression were themselves necessary (Kramer 1963: 125). It is hard to avoid concluding that a core of darkness lies hidden in the best of our cultural accomplishments, including even intellectual accomplishments, for one important route to them – analysis – requires that one develop a habit of taking things apart.

Cruelty

Cruelty is not one of the Seven Deadly Sins of medieval theology, an omission that surprises modern sensibility, for it has come to see the deliberate infliction of pain as possibly the worst evil. "Cruel" and "crude" have the same root: both speak of a rawness that is part of our biological nature which can be removed through acts of cumulative refinement (Partridge 1959: 132; Rosset 1993: 76). Cruelty may thus simply be the effect of an immature mind. Young children are often cruel. As a child, I never hesitated to impale a live worm on the sharp point of a fish hook, something that I would rather not do now. But is cruelty the right word? The child I was then did not intend cruelty: I just never thought of the wriggly thing in my hand as anything other than a bait to be used for a venture that did engage me wholly – fishing. Elizabeth Marshall Thomas provides a more lurid example of thoughtless cruelty. In a much acclaimed book called *The Harmless People*, Thomas presents a charming picture of a Bushman family in Southwest Africa that seems to lend support to the title, but which demonstrates how far primitive life – indeed, any human life – is from harmlessness. Around a fire at camp, Gai the hunter gave his baby son Nhwakwe a tortoise and offered to roast it for him. An old woman, Twikwe, helped. She "held the tortoise on its back, but the tortoise urinated brown urine and Twikwe let it stand up. It stood looking at the flames, blinking its hard black eyes, then started to walk away. But Twikwe caught it again and held it, idly turning it over and over while she talked with Gai about other things." After talking a while, "Gai took the tortoise from Twikwe and laid it on its back." He then applied a burning stick against the tortoise's belly. "The tortoise kicked violently and jerked its head, urinating profuse amounts of the brown urine which ran over Gai's hand, but the heat had its effect, the two hard, central plates on the shell of the belly peeled back, and Gai thrust his hand inside." He pulled out the heart, "which was still beating, and flipped it onto the ground, where it jerked violently for a moment, almost jumping, then relaxed to a more spasmodic beating, all by itself and dusty, now ignored" (Thomas 1965: 51–52).

The picture of a boy sticking an earthworm on the fish hook is innocent enough. So also is the evisceration of a tortoise, although Thomas's vivid account may make some readers feel a little squeamish. How, after all, is human life possible unless we are able to inflict pain unthinkingly? We have to eat, and food preparation – the whole sequence from trapping, rounding up, and killing the animal to its skinning, dismemberment, and cooking (roasting, boiling, frying) – entails violence. A disturbing thought is this: may not the ease with which we can, between bouts of laughter and chatting, cut open a turkey and pluck out its innards, prepare us to commit comparable outrage against human beings we consider less than fully human, when circumstances permit?

At times, children are knowingly, not just thoughtlessly, cruel to creatures weaker than they. What purpose lies behind pulling the legs off a grasshopper? Curiosity? Experimentation? Or the pleasure of absolute power, of knowing that a thing full of life – a grasshopper that is able to leap into the air in one bound – can be reduced at one's whim to a bundle of quivering tissue? Children can be cruel to one another, as adults know from observation and their own childhood experience. Every school has its bully. Overall, children do not show much urge to nurture and protect. Indeed, left to themselves they are inclined to tease and denigrate weaker members and those marked by difference. William Golding's novel *Lord of the Flies* (1954) artfully and persuasively showed how children could quickly turn paradise into a fascist state.

From the late seventeenth to the eighteenth century, at about the time when Europeans turned sentimental in regard to children, they turned sentimental in regard to peoples of simple material culture in distant parts of the world. "*Là-bas on était bien!*" was a slogan of French philosophers, and the "bien" meant wholesomeness and well-being in a state of nature. It is remarkable how this Western attitude persists into the late twentieth century. Not just romantic travelers in search of exotic Edens, but professional ethnographers fall into the trap, as Robert Edgerton has shown in his provocative book *Sick Societies* (1992). In contrast to the West, the Chinese, even the Taoists, were little inclined toward philobarbarism (Levenson and Schurman 1971: 113). I myself am sufficiently Westernized to wish for the existence of uncorrupted humans living in a pristine environment. Until recently, I thought I knew where to find them. My favorite Eden is located in Congo's Ituri forest. The hunter-gatherers there, the Mbuti, are so innocent as to have no concept of "evil." Disasters, when they occur, are attributed not to evil forces, but to spells of absent-mindedness or drowsiness on the part of the all-nurturing forest (Turnbull 1965: 308–309). The Mbuti themselves are a friendly and happy people. Of course they have to work, but it is not hard work and can be done in the mornings. Whole afternoons of leisure remain during which they sing, teach and play with the children. I trusted this picture because it was depicted by reputable anthropologists, outstandingly Colin Turnbull. In their publications, however, they have tended to suppress the dark side, treating it as an inconsequential blemish. Readers, for their part, skip or skim over the shadows to dwell on the sunny scenes.

And what is the dark side, without which the Mbuti will hardly be human? Playing cruel practical jokes – or just plain cruelty – toward the physically handicapped is one. The Mbuti esteem cleverness and despise dumbness, which they associate with animals. A young man who happens to be deaf mute is the camp clown and mercilessly teased for his stuttering speech, "animal noises," as his fellow campers call it (Duffy 1984: 50). Bantu villagers, who share the Ituri forest with them, are also viewed as somewhat beyond the moral pale. Against the Bantu, cheating and stealing are permissible, for, from the Mbuti viewpoint, they are not only human outsiders but also, given their big size and clumsiness, animal-like – a resource to be used (Turnbull 1963: 6). The Mbuti may occasionally show affection for animals, as perhaps all humans do, but they can hardly afford to be sentimental. Killing the larger animals is a triumph of skill and courage. The slaughter of an elephant leaves behind a blood-stained, odorous field of entrails, bones, and skin that is very far from traditional images of paradise. Especially dismaying to Westerners eager to find innocence is that the Mbuti can derive pleasure from watching wounded animals writhe in pain. They may inflict the pain themselves. The domesticated dogs that they keep for hunting, Turnbull notes, have been kicked "mercilessly from the day they are born to the day they die" (1962: 100).

The Mongols: destructive and cruel

When one thinks of massive destruction and unspeakable cruelty, one thinks of large, well-organized societies – empires and nation states. Even if it is granted that these political colossi produce high culture and splendor, and that at their best they elevate the human spirit to great heights, the cost in human suffering and lives and in natural resources is exorbitant. Is it morally acceptable? College-educated young people of our time, many of liberal bent, would probably say no. The word "empire," to them, carries bad odor: because empires depend on power for realization and maintenance, they are necessarily evil. What exactly do young people have in mind when they speak of empire? Very likely European colonial powers of modern times and the American imperium in the American century. Even if the young are aware of older "world states," they may see them as less evil than modern ones, because less darkened by racism and because they presumably lacked the technological power to do great harm.

I would like to offer the thirteenth-century Mongol empire as an example of how far evil can be carried in the absence not only of machine technology, but also of greed driven by capitalism, and of racism, the contempt that one group of people have for another on grounds of certain differences in physical appearance.

The Mongols were pastoral nomads of simple material culture. This, however, is not at all the same as saying that they lacked sophistication – the kind of analytical and abstract thinking that could lead to the amassing of great power. Skill in riding and fighting on horseback made the Mongols, individually, formidable warriors. But more important by far from the standpoint of power is their success in developing what might be called "organizational technology." They produced mega military machines made up of human and animal parts. In the name of efficiency, fighting units broke away from ancient kinfolk allegiances. Their artificiality was emphasized by adopting the decimal system: units came in sizes of 10, 100, 1,000, and 10,000. Once Chinghiz Khan perfected this military machine, it had to be used to expand the empire, or suffer quick disintegration. It was so used, causing the most extraordinary devastation of artifacts and artworks, cities and human lives in vast swathes of the then civilized worlds (Morgan 1990: 90).

Were the Mongols cruel? Objectively, they certainly were, but the Mongols of that period might not even be able to understand the meaning of that word. Treating conquered people as nuisances in the environment, to be eradicated, or as a natural resource to be exploited to the full, was done as a matter of course. On the western edge of their empire, in Transoxiana and eastern Persia, Mongol armies reduced whole cities to rubble, rounded up the surviving inhabitants, and divided them into batches for systematic eradication: that is, each batch constituted a quorum of slaughter for the individual soldier. As the Nazis extracted gold from the teeth of corpses, the Mongols disemboweled their victims to remove swallowed jewels (Saunders 1971: 60). In China, which formed the southern part of the Mongol empire, the conquerors were said to have considered wiping out the entire dense population, as they would brush and forest, to make way for pasture land. They refrained only because they were persuaded that more could be gained by systematically "milking" the population – that is, by taxation. In the first 25 years of their rule, the Mongols made no pretense that at least some of the tax could be used for the public good. To them, a subjugated people had only one reason for existence – as a source of revenue. To the extent that exploitation was restrained, it derived from the knowledge that it made sense to leave peasants with sufficient wherewithal to survive into the next year, so that another year's taxes could be extracted (Morgan 1990: 74, 102). How convenient for the conquerors to have millions of human cattle that dependably yielded without any need to invest in their care!

The evil committed by the Mongol hordes was such that one could justifiably wish a bad end for them. Darker and morally dubious is the wish that the countries they overran should remain desolate as a permanent testimony to human wickedness. Desolation, however, was not to be their enduring fate. By the end of the thirteenth century, although some cities remained ruins, others flourished, thanks in part to the trade routes that the Mongols established and made safe. In Persia, the conquerors left high administration in the hands of talented locals: different populations worked and lived peacefully together to produce a sophisticated, cosmopolitan society. In China, the Mongols were suspicious of the natives and so gave high administrative posts to foreigners like Marco Polo. One unintended result was that the Chinese elite had time on their hands in which to cultivate the arts and literature, which flourished. The Chinese even learned to be grateful to the Mongols for uniting the north with the south, two parts of the country that had been divided for centuries (Morgan 1990: 110–111, 128–129). In short, much good in culture and society emerged in merely two decades after unprecedented destruction. One might be forgiven for thinking this dramatic reversal, even if it was from bad to good, to be the work of the Devil in a deeply cynical mood.

Sadomasochistic dominance and affection

Destructiveness and cruelty are self-evidently bad. What about playful domination? The psychology is far subtler than mindless destructiveness and cruelty. Dominating and being dominated is, to say the least, morally ambiguous, but what if it were done playfully, with affection for the thing dominated and with the dominated thing itself participating more or less willingly? And what if the result of domination were an aesthetic object that gives pleasure? The stickiness of these questions arises from the fact that they touch the moral standing of culture at a vulnerable point. Even if one were willing to concede that economic culture is exploitative, with little to show beyond a superfluity of goods, and that political culture has no other end than power and its use for personal aggrandizement, it is difficult to view harshly aesthetic and affectional cultures that produce such innocent things as gardens and pets. Gardens, we tend to think, are an antidote to the artificiality and greed of cities; pets show how humans properly relate to animals. Yes, maybe so, but only in a relative sense, and sometimes not even that. Play has a dark side that is too often forgotten. To play with nature whimsically, in total freedom, can be even more an expression of uninhibited power than to subjugate it for a limited purpose. Sadism, after all, is a form of play (Tuan 1984).

What are some examples of playing with nature? A pride of great European gardens is the fountain. It is how potentates play with water, making it jump when its nature is to flow down. Such artificiality can be achieved only when different kinds of power work efficiently in harness: the technological power of hydraulic engineering, which Europeans were beginning to acquire in the sixteenth century; the power of organized labor, including the use of disciplined soldiers to dig canals and build aqueducts to bring water from distant sources; and the aesthetic power of great sculptors. Only aesthetic power is ostentatiously on display in a garden, making it seem that fountains are nothing but innocent works of art that give pleasure (Tuan 1984: 41–46).

Water is poetically and metaphorically alive. That's why I choose it as an example of a pet, the pet being something alive that has been taken into the human world, often altered in the process, for the pleasure it gives. But water is only alive in a poetic sense. Plants, by contrast, are truly alive. In great gardens of both Europe and China, they too have been made into pets – that is, domesticated, altered, their limbs twisted, their foliage pruned; and if the twisting of the tree trunk in a Chinese garden can seem grotesque – tied into a knot like the button on a Chinese vest – it is certainly not more so than what can be found in European topiary art: there, shrubbery and foliage are forced into the most unnatural sculptural and geometric shapes. But, to my mind, the supreme example of playful power is the Asian miniature garden, commonly known as bonsai. It is an attempt to reduce the wild glory of nature into something that can be put into a pot and placed in the living room. The technique for stunting plants is torture in its precise and literal sense of twisting. The most common instruments – knife, scalpel, tweezers, shears, screwdriver, weights, and copper wire – evoke images of surgical violence that is obviously not applied for the health of the plants, but for the delectation of their owner (Stein 1990).

Pet, to most people, means an animal that has been taken into the household, where it serves as companion and playmate. In Western civilization, the dog and the cat are the favored pets; in China, the bird and the goldfish. Owner and animal develop a relationship that varies with the animal. A pet bird may be well fed and protected from its natural predators; the cost is a life behind bars. Any confined animal is a picture of curtailed freedom, but this should be especially obvious in a bird, which is born to fly. The goldfish has been genetically altered so that certain strains have become dysfunctional. The so-called Telescope goldfish, for example, has been bred to have grotesquely protuberant eyes that are easily damaged (Anonymous 1909: 37; Hervey 1950: 33). One wonders about an aesthetic culture that could consider beautiful an animal that it has studiously deformed. Cats in Western society are a special type of pet in that, though domesticated and altered through selective breeding, they retain a large degree of independence: they do not "sit" or "roll over" at human command; their dignity is not wholly impaired. Dogs are another matter. They have been bred, historically, for use rather than pleasure. To be useful in hunting, they must be obedient, a trait that is equally desirable in a plaything – a toy. Nearly all the small dogs that we now think of as playthings – terriers, spaniels, and even the poodle – were once bred for hunting. But the Chinese Pekinese may be an exception. As far as we know, it has always been a toy, reduced in size over the generations so that it can be tucked into the sleeve of a woman's coat. The Pekinese is a healthy and frisky animal that can live to the ripe old age of 25 years. More often than not, however, pedigree dogs lose their vigor and intelligence. They become obedient pets, good to look at from their master's point of view, but without the ability to perform the sort of complicated tricks that unpedigreed, circus dogs can (Lorenz 1964: 88–90).

People, too, may be turned into pets; and because they are expensive to acquire and maintain, they tend to carry special prestige. Who are the human pets? In the first place, children – little people who need the affection and training of adults. Nothing wrong with that, for pet status is one that children naturally outgrow. In patriarchal societies, women have been pets to their men – playthings in a doll house. At the same time, women of rank controlled the household and exercised great power over all its underlings, male and female. Women and their menfolk, even when they were kindly by temperament, easily fell into the habit of treating young servants, serfs, or slaves as pets. In an eighteenth-century Chinese novel, one lady said to another:

You can talk to (the young servants) and play with them if you like it, or if you don't, you can simply ignore them. It's the same when they are naughty. Just as, when your puppy-dog bites you or your kitten scratches you, you can either ignore it or have it punished, so with these girls.

(Cao 1980: 157)

In great European households of the eighteenth century, young black domestics, more fancifully dressed than white servants, were prized possessions. A black boy might wear a gold-plated collar with the owner's coat of arms and cipher engraved on it, or some such inscription as, "My Lady Bromfield's black, in Lincoln's Inn Fields" (Shyllon 1974: 9).

Exploited persons can still retain a sense of their dignity: their muscle power is harnessed for another's use, but not their personality – their soul. Human pets, by contrast, find themselves constantly at the beck and call and whim of their master or mistress: their whole being, physical and mental, is at the disposal of someone else. If using another person for one's own selfish end is wrong, playing with another for one's pleasure is, it seems to me, worse – evil. To this dark picture, I add two puzzling psychological components. One is the tendency for the exploited to accept, almost masochistically, their degraded position, provided it is stable. Over time, they learn to find virtue in restful subservience, and even to take a certain pride in it. Confusing necessity with the good is so common-place that it may well be a technique of survival (Moore 1978: 64; Weil 1963: 94–96). The other puzzling component is the psychology of those who hold power: that they should take delight in owning a good-looking animal or human being is understandable, but how is one to account for their pleasure in another's deformity – in ugliness? Great ingenuity and persistence must be exercised to produce a goldfish with monstrously protuberant eyes, or a dog (the *shar-pei*) that looks like an unmade bed, with preposterously folded skin. And why – what fanciful twist of mind – made European potentates of the Renaissance and early modern period want to fill their households with dwarfs, calling them by such mocking names as "king" and "monarch," dressing them up in finery, passing them along to one another as expensive gifts (Tietze-Conrat 1957; Martines 1980: 231)?

Compartmentalization

I have thus far offered relationships – disturbing, evil relationships – between people and nature, people and people. They should at least make us wonder at the depth of destructiveness, cruelty, and contempt that one person or group can have for another when the scales of power are sufficiently out of balance. I now turn to the opposite of relationship – to compartmentalization and its evil con-sequences. A thoroughly compartmentalized world, a world criss-crossed by intransigent boundaries, is a good image of hell. On the other hand, people need physical boundaries to keep out danger, and mental boundaries between self and others so as not to be overwhelmed. The ability to disconnect, separate self from others, separate even the different roles and faces of a single self, may be morally problematic, but it is also a condition of sanity (Cresswell 1996; Sack 1986, 1997: 90–91, 156–160).

Forgetting – that is, disconnecting with the immediate past – is a fact of daily life. We practice it without noticing that anything remarkable has occurred. One moment, for example, I am in the men's room, my face sweaty with the effort to relieve nature, the next I am standing in the classroom, coolly lecturing on Platonic ideals. What are the moral implications of such change? None that I can see. Consider, however, another example. In the morning I fondle a lamb, muttering sweet nothings into its ear; a few hours later in my dining room my jaws chomp appreciatively over lamb chops. I can enjoy the meal only if I have cleanly forgotten the earlier scene. The word "lamb" reminds me of a famous line from the 23rd Psalm, "The Lord is my Shepherd." How wonderfully reassuring these words are! But only on condition that I successfully decouple it from the knowledge that the shepherd nurtures lambs for fleecing or slaughter. In folk and traditional societies, a frequently used technique of compartmentalization is taboo. Suppose there is a conflict between bodily desire and one's sense of right and wrong. Taboo resolves it by creating a temporal barrier. Among cattle-raising Africans, for example, eating meat and drinking milk at the same time is taboo. The proscription helps them to circumvent the untenable moral position of killing and eating the meat of an animal that has nourished them with milk (Saitoti 1988: 73).

Disconnection is necessary from a practical point of view. One must be able to disconnect from one set of demands so as to act competently in the next. On my way to school, I walk by the homeless. If I had seriously tried to help them, I would be too exhausted to be of use to students. So I toss small change and pass on. In a public space, I look at people – if at all – blankly; later, in my classroom, I am attentive and caring. I am acting on the principle of looking after my own first: it is a simple matter of not dissipating limited resources and energy. Drawing the line, to one side of which I am coolly indifferent and to the other side warmly caring, is common sense and a survival technique that is universally acted upon, even if some religious or philosophical thinkers might argue for total impartiality.

Is there anything wrong – that is, morally reprehensible – with these shiftings of scene and role, these acts of dissociation facilitated, indeed promoted, by physical and mental boundaries? Only the most morally sensitive would raise objections. Overwhelmingly, people accept them as a matter of course, too familiar and necessary to justify thought. Indeed, the compartmentalization saves one from thought that is bound to agitate without the promise of resolution. Yet I believe that we need to be aware of, and reflect upon, these disconnections of ordinary life, for, unless we do so, we risk slipping, when circumstances permit, into dissociative monstrosities. In our time, the most extreme example of dissociative monstrosity is the behavior of Nazi SS guards in concentration camps. Even against a background awareness of our own dramatic role-and-personality shifts, it is still a shock to know that the guards who nursed their sick dog in the morning and looked forward to playing a Mozart quartet in the evening, between times had men, women, and children shoved into the gas oven (Todorov 1996: 148–149).

Conclusions

What I have offered here as instances of evil to which all humans are prone, though in varying degree and under the right circumstance, are familiar stuff to humanists – to historians, especially biographers, and literary scholars. Indeed, to storytellers and novelists they are the bread-and-butter of their art. Unless stories show people realistically as mixed bags of good and evil, they can hardly hold the attention of mature audiences and readers. The problem with the humanist or storyteller's approach is that it tends toward relativism and pessimism: it gives the impression that "people are just like that" everywhere, with only minor variations due to custom; that in the moral realm human beings, for lack of transcendental but attainable aspirations, cannot make genuine progress. An alternative and more hopeful view exists within the humanist tradition. It is a philosophy, inspired by Plato and Judeo-Christian religion, that explores seriously and in a sustained manner the nature of the good (Murdoch 1971, 1993).

In distinction to humanists, social scientists (geographers included) are by training inclined to seek causal explanatory factors, such as climate, social class, the economic system, and gender, to account for human oppression and misery. The intent of their work is scientific. The result, however, often has a moralizing flavor, for such accounts tend to ease into a polarization of forces – into good guys and bad guys, as in grade-B movies and literature. Another commonality with such movies and literature is this: the bad guys (enervating climate, ruthless gentry, capitalist exploiters, sexist patriarchs) emerge as vivid agents; by contrast, the good guys and victims seem pale and passive. A final criticism. The contemporary emphasis on exploitation and social injustice in the formerly colonized parts of the world is often written as though, before colonialism, native peoples lived in harmony with one another and with their environment. Romantic exoticism is deeply ingrained in Western thought. Social science papers that seem hard-hitting are often too utopian, for they assume a human nature that, apart from the distortions imposed by class structure, technological and capitalistic hubris, is free of evil.

In a visionary mood, I see a future moral geography that combines the best of both humanist and scientific perspectives. I would go further and say that, unless these two perspectives are combined, what we have is moralistic rather than moral geography. For a moral geography to be broadly and firmly grounded, it seems to me that we must confront our species' moral nature, whether it is true that we always live under the twin imperatives of appropriateness and inappropriateness, right and wrong, good and bad; and whether it is plausible that as intelligent and moral beings we have the potential to develop a vision of good and evil that transcends the insights and practices of particular times and places.

References

Anonymous (1909) *Japanese Goldfish: Their Varieties and Cultivation*, Washington, DC: W. F. Roberts.

Cao, X. (1980) *The Story of the Stone*, vol. 3, Harmondsworth: Penguin.

Cresswell, T. (1996) *In Place/Out of Place: Geography, Ideology, and Transgression*, Minneapolis: University of Minnesota Press.

Duffy, K. (1984) *Children of the Forest*, New York: Dodd, Mead & Company.

Edgerton, R. (1992) *Sick Societies: Challenging the Myth of Primitive Harmony*, New York: The Free Press.

Golding, W. (1954) *Lord of the Flies*, New York: Putnam.

Hammer, W. M. (1994) "Lions of darkness," *National Geographic* 186, 2: 35–53.

Hervey, G. (1950) *The Goldfish of China in the Eighteenth Century*, London: The China Society.

Humboldt, W. (1963) *Humanist Without Portfolio*, Detroit: Wayne State University Press.

Joubert, D. (1994) "Deadly jellyfish of Australia," *National Geographic* 186, 2: 116–130.

Kramer, S. N. (1963) *The Sumerians: Their History, Culture, and Character*, Chicago: University of Chicago Press.

Leighly, J. (ed.) (1963) *Land and Life: A Selection from the Writings of Carl Ortwin Sauer*, Berkeley: University of California Press.

Levenson, J. R. and Schurman, F. (1971) *China: An Interpretive History*, Berkeley: University of California Press.

Lorenz, K. (1964) *Man Meets Dog*, Harmondsworth: Penguin Books.

Luckenbill, D. D. (1924) *The Annals of Sennacherib*, Chicago: University of Chicago Press.

Martines, L. (1980) *Power and Imagination: City-States in Renaissance Italy*, New York: Vintage Books.

Mikesell, W. M. (1976) "The rise and decline of 'sequent occupance'" in D. Lowenthal and M. Bowden (eds) *Geographies of the Mind*, New York: Oxford University Press.

Moore, Jr. B. (1978) *Injustice: The Social Basis of Injustice and Revolt*, White Plains: M. E. Sharpe.

Morgan, D. (1990) *The Mongols*, Oxford: Blackwell Publishers.

Murdoch, I. (1971) *The Sovereignty of Good*, New York: Schocken Books.

—— (1993) *Metaphysics and Morals*, New York: Allen Lane Penguin Press.

Parkin, D. (ed.) (1985) *The Anthropology of Evil*, Oxford: Blackwell Publishers.

Partridge, E. (1959) *Origins*, New York: Macmillan.

Rosset, C. (1993) *Joyful Cruelty: Towards a Philosophy of the Real*, New York: Oxford University Press.

Sack, R. D. (1986) *Human Territoriality: Its Theory and History*, Cambridge: Cambridge University Press.

—— (1997) *Homo Geographicus: A Framework for Action, Awareness, and Moral Concern*, Baltimore: The Johns Hopkins University Press.

Saitoti, T. O. (1988) *The Worlds of a Maasai Warrior*, Berkeley: University of California Press.

Santayana, G. (1980) *Reason in Society*, New York: Dover Publications.

Saunders, J. J. (1971) *The History of the Mongol Conquests*, London: Routledge and Kegan Paul.

Shyllon, F. O. (1974) *Black Slaves in Britain*, London: Oxford University Press.

Stein, R. A. (1990) *The World in Miniature: Container Gardens and Dwellings in Far Eastern Religious Thought*, Stanford: Stanford University Press.

Thomas, E. M. (1965) *The Harmless People*, New York: Vintage Books.

Tietze-Conrat, E. (1957) *Dwarfs and Jesters in Art*, London: Phaidon Press.

Todorov, T. (1996) *Facing the Extreme: Moral Life in the Concentration Camps*, New York: Henry Holt.

Tuan, Y.-F. (1984) *Dominance and Affection: The Making of Pets*, New Haven: Yale University Press.

Turnbull, C. (1962) *The Forest People*, Garden City: Doubleday Anchor Books.

—— (1963) "The lesson of the Pygmies," *Scientific American*, January 1963, reprint.

—— (1965) "The Mbuti Pygmies of the Congo" in J. L. Gibbs (ed.) *Peoples of Africa*, New York: Holt, Rinehart & Winston.

Weil, S. (1963) *Gravity and Grace*, London: Routledge and Kegan Paul.

Wilson, C. (1963) *Origins of the Sexual Impulse*, London: Arthur Baker.

9 The ethnic cleansing of a "safe area"

The fall of Srebrenica and the ethics of UN-governmentality

Gearóid Ó Tuathail

On Tuesday July 11 1995 a Bosnian Serb army led by General Ratko Mladic triumphantly entered the eastern Bosnian town of Srebrenica. The town, named after its long historic association with silver mining, was at the center of a multi-ethnic region of 36,666 people in 1990, 75.2 percent of whom identified themselves in the 1990 Yugoslav census as "Muslims" while 22.7 identified themselves as "Serb" (Honig and Both 1996: xviii). Swelled by refugees from years of civil war, the population of Srebrenica had reached 50,000 in 1993. Certain groups had been able to flee the town and as the Bosnian Serb army encroached others tried to escape their clutches. However, an estimated 30,000 people were eventually surrounded by Mladic's army near the United Nations compound at Potocari, on the outskirts of the town.

Over the next four days, the Bosnian Serb army, with the reluctant acquiescence of the Dutch UN troops, expelled an estimated 23,000 women and children, permanently evicting them from their lands and homes. The men were treated differently. Separated from their families, they were driven off in buses to various locations, to an abandoned gymnasium, an athletic field, and clear patches in forested areas. There almost all were murdered, either by being enclosed and shot from a height or by mass executions at close quarters with bullets to the back of the head. A few miraculously survived, left for dead by those charged with liquidating them. An estimated 3,000 men were killed. A further 4,000 people were murdered as they tried to outrun the Bosnian Serb army which had organized "hunting expeditions" to track, stalk and kill them. In total, over 7,000 people, the vast majority of them men, are missing and presumed dead as a consequence of the fall of Srebrenica.

The ethnic cleansing of Srebrenica was not an unusual act of violence in the post-Cold War world. In Afghanistan, Algeria, Azerbaijan, Cambodia, Chechnya, Croatia, Rwanda, Sri Lanka and many many other places political, ethnic and religious conflicts have degenerated into bloody wars of often shocking brutality. The New World Order promised by President Bush and the United Nations immediately after the Gulf War has become a new world disorder where anarchy, chaos and brutal violence are widespread. Yet geography made the violence of Srebrenica unique in two ways.

The first was its location in Europe. In the scorching July of 1995 I was visiting

The ethnic cleansing of a "safe area" 121

Italy for the first time and watched with horror what was unfolding only a few hundred miles south-east of Trieste. Bosnian Serb television footage of Mladic addressing those captured was broadcast around the world. What was happening in Srebrenica was close both geographically and visually to "us," to the safe and civilized world of the European Union. Subsequently, the violence of that July was represented as "Europe's worst massacre since World War II" (Rohde 1997), the European location granting the violence unusual significance.

Srebrenica was also special because it had been declared by the United Nations a "safe area" in April 1993. Designating the town as a "safe area" represented an effort by the international community to legislate a special zone of order and security amidst the generalized disorder and warfare in Bosnia. The United Nations demonstrated its commitment to the town by placing a battalion of troops there. Airplanes from the NATO base of Aviano in northern Italy and from ships in the Adriatic were charged with protecting these troops and making sure that the safe area remained safe. Srebrenica was part of a United Nations ethical order imposed upon the new world disorder. In driving past the Aviano air base that July, it was clear that the high tech warplanes based there were on full alert. Yet with the exception of one minor attack, no serious effort was made by the UN to repulse the Bosnian Serb army as it began transforming the "safe area" into an expulsion and killing zone. In violently rearranging the human geography of Srebrenica, the Bosnian Serb army revealed the limits of the United Nations' commitment to the people of Srebrenica, exposing its ethics as a self- rather than other-directed code of bureaucratic procedures and professionally delimited response-ability (Campbell 1994).

This chapter seeks to explain how the United Nations' governmental and ethical system pronounced Srebrenica a "safe area" yet nevertheless allowed this "safe area" to become the site of the worst massacre in Europe since the Second World War. The chapter addresses three themes; first, the strategic and ethical re-spacing of world order by the Western alliance system after the Cold War; second, the establishment of a United Nations governmentality in Bosnia as a particular strategic and ethical order; and finally, the contradictions and failure of this ethical order to take moral responsibility for Srebrenica. While many of the essays in this volume understand "ethics" as positive normative reflection upon codes of behavior, this chapter considers how codes of behavior are already implicitly ethical orders sustaining certain forms of normative behavior that may not necessarily be moral or reflective. I wish to suggest that, in this case at least, there is an important difference between ethics as *a set of socially institutionalized rules promoting normative behavior* and morality as *a primordial and open-ended responsibility towards otherness.* Bauman (1993, 1995), Herzfeld (1992) and others have argued that ethical orders routinely produce moral indifference and suppress open-ended moral responsibility towards otherness. I wish to argue that the institutionally proclaimed, bureaucratically supervised and professionally administered ethics of the United Nations in the former Yugoslavia so constricted expressions of moral responsibility that they effectively promoted immoral purposes and ends (Barnett 1996). In the case of Srebrenica at least, professional ethical orders tried to capture and control morality, to strait-jacket it within a code of conduct and bound it with rules of engagement and delimited responsibility. The result was an evasion and occlusion of morality, and eventual acquiescence with the genocidal practices of the Bosnian Serb army.

The new world (dis)order and UN-governmentality

The Cold War between an Eastern bloc led by the Soviet Union and a Western bloc led by the United States organized the geopolitics of the world order for over 40 years after the Second World War. Its alliance systems and economic institutions organized international space into two distinct zones of allegiance, while a third non-aligned movement attempted, with a variable record of success, to distinguish itself from both of these zones. In giving international affairs a geopolitical intelligibility the Cold War also helped establish a geography of strategic responsibility and obligation through its systems of alliances, treaties and international organizations. These specified, often in legal and contractual detail, certain structures of authority, spheres of influence and systems of obligation and military security. As a consequence, politics and diplomacy during the Cold War was conducted in a world marked by reasonably distinct maps of proximity and difference in international affairs. Certain countries were recognizably close to the Western alliance system and its way of life while others were perceived as distant from the imagined ideals of "the West." Proximity and distance in international affairs encompassed but was not reducible to territorial proximity and distance. A state like Cuba, for example, was territorially close to the United States but beyond its self-constructed civilization of values, a satellite orbiting in the foreign ethical universe of the Soviet Union.

Revolutions in communication and transportation together with economic globalization were already forcing a re-spatialization of international affairs before the collapse of the Communist dictatorships in Eastern Europe and the break-up of the Soviet Union in 1991. The end of the Cold War made such a re-spatialization an imperative and when Iraq invaded Kuwait in 1991 President Bush proclaim a New World Order with the United States and its allies working through the United Nations and re-directed Cold War alliance systems like NATO to thwart aggression against lawful sovereign states and maintain peace and security in the international system. For a moment it seemed that the United Nations could become something it was not during the Cold War, an institutional expression of a universal will and the organizational center of a more inclusive and proactive international community.

As the rhetorical hubris of Bush's New World Order faded, however, the daunting challenges of the actually existing new world disorder became more apparent. In many places, like Yugoslavia, Somalia, Rwanda and Haiti, states were breaking apart in violent struggles and ceasing to function. When states failed and became ungovernable, the United Nations was frequently called upon to act as an international rescue service, an emergency paragovernmental service of last resort for states that had descended into chaos. Governmentality, which Foucault (1991) defines as "the right disposition of things so as to lead to a convenient end," had first emerged as a modern administrative and ordering project focused on state building in eighteenth-century Europe. In the new world disorder of the late twentieth century, with state governmentalities failing across the globe, the United Nations became the institutional locus of a world governmentality. Its various administrative agencies and peacekeeping missions to particular failing states represented forms of UN-governmentality in places that had lost their own local governmentality and become ungovernable (Luke and Ó Tuathail 1997). The United Nations represented the hope and promise of a universal modernity, the possibility of a minimal form of governmentality in regions that appeared to have become engulfed by anarchy and to have reverted to tribal and feudal systems of governance.

One region of chaos where the United Nations was soon deployed was in the former Yugoslavia. A nominally non-aligned location during the Cold War, Yugoslavia was of marginal strategic interest to the United States and NATO after the end of the Cold War (Zimmermann 1996; Ó Tuathail 1999). After the wars triggered by the push to break up Yugoslavia started, it was quickly agreed in the United States that Yugoslavia was a regional problem that was best left to the European Union. While the European states were geographically closer to Yugoslavia, they also initially had little strategic stake in the conflict. The dominant European interest was represented as "humanitarian," which in practice meant, first, concern with population displacements and refugee flows into the states of Western Europe and second, concern with human rights abuses and genocide. However, this last humanitarian interest was to prove the most troubling of all and over time made the wars in the former Yugoslavia a more strategic challenge to the West than first anticipated. The proliferation of new media technologies like direct satellite broadcasting and new media programming like 24-hour news broadcasting made wars in distant locations appear much closer and more visible than ever before. This put the bloody brutality of war in Europe's own backyard into the homes of Europeans and Americans on a nightly basis. It created moral pressure and imperatives for the Western powers to do something about this brutality while also conditioning the possible nature of their response (whether they could use national troops and what they could or could not do).

The European and United Nations' response to the Yugoslavian wars was shaped by all these factors. There developed an ethic of engagement with Yugoslavia that comprised, on the one hand, an exhaustive search for a diplomatic solution with, on the other hand, the deployment of a "peacekeeping" force with a purely humanitarian mandate to secure the delivery of food and medical supplies to civilians that most needed them. This ethic of engagement was a clearly circumscribed and limited one. The European Union and the United Nations would facilitate efforts to find a diplomatic or an undiplomatic solution (i.e. surrender by the Bosnian Muslims) to the conflict but would not impose any solution. Progress towards such a solution was ultimately dependent upon the warring parties themselves. The United Nations Protection Force

(UNPROFOR) sent into Bosnia in 1992 had a mandate that required it to adhere to a policy of strict neutrality towards the varying warring parties. UNPROFOR could defend itself if attacked but it was under strict instructions not to aid one side or the other in the conflict. The response of the European Union, the Western alliance and the United Nations towards the crisis in Yugoslavia was thus enframed by a strong ethics of professionalism, the professionalism of the diplomat who must always be neutral and willing to negotiate with any of the warring parties, and the professionalism of military commanders charged with carrying out their mandate and strictly following their chain of command, rules of engagement, and standard operational procedures.

The problem with these ethics of engagement is that they became substitutes for morality as responsibility for and to otherness. They became, in effect and practice, substitute moral choices that propelled diplomats for the Western powers and the United Nations to negotiate with war criminals and UNPROFOR to remain neutral in a country where acts of indiscriminate mass murder and crimes against humanity were being committed. Ethics substituted for and attempted to contain the "unreasonable" demands generated by Bosnia as a challenge to morality.

Ethical engagement without moral responsibility: the creation of "safe areas"

Until 1995 the disposition of the West and the United Nations towards the Bosnian war is best summarized as one of ethical engagement without moral responsibility. Bosnia was represented as part of a general universal of obligation on the part of the West and the UN but this obligation was a circumscribed "humanitarian" one that defined and described itself in terms of diplomatic and peacekeeping professionalism. Bosnia was also consistently represented by the leading powers in NATO as a place beyond its domain of strategic obligation and responsibility. While many European NATO members were willing to provide the troops necessary for UNPROFOR to establish and carry out its "humanitarian" mandate, no NATO state was willing to have its troops take a side in the Bosnian war and fight in the region. The United States was not even willing to have any of its troops deployed on the ground. Military neutrality was the best course. Bosnia, in short, was not worth dying for.

Sustaining this attitude of military neutrality were a number of strategic calculations by NATO. Yugoslavia as a whole and Bosnia in particular, as we already noted, was generally represented as having marginal strategic value and interest to the West. The region contained no major resources vital to the West's way of life and had no weapons of mass destruction that could potentially threaten members of NATO. The war in the region was also frequently depicted as a centuries-old conflict between competing tribal identities. It was not the postmodern or even the modern war that NATO was trained and equipped to fight but a particularly brutal "premodern war" between combatants locked in history. Those who represented the war as one between the values of multiculturalism and fascistic nationalism or between Western democracy and Communism – and certain influential figures in the West like Margaret Thatcher and George Schultz publicly called for greater Western military involvement in the conflict – were generally represented as naive and simple-minded, figures that did not understand the supposed complexity of the conflict and tended to idealize one side over the other. Many Western military leaders, including some of those with leadership positions in UNPROFOR, repeatedly noted that all parties were at fault in the Bosnian war and that the Bosnian Muslim side in particular were not the victims they were often portrayed as in the Western media. A number of military commanders held the Bosnian Muslim army in contempt and tended to bond more easily with Serbian military leaders. The strategic thinking of NATO was also shaped by the danger of an emergent Islamic fundamentalism amongst Muslims in Bosnia (Rieff 1995).

As a consequence, many Western military and diplomatic leaders tended to reason in a *realpolitik* manner about the conflict in Bosnia. *Realpolitik* reasoning on Bosnia represented ethical engagement without moral responsibility *par excellence*; it was the product of a masculinist culture of professionalism and expertise that defined itself by its ability to suspend moral questions and judgments, often recognized and coded in feminine terms as "passion" and "emotional" arguments, in order to "see things in a realistic and hardheaded way" and to eventually "get the job done" (Ó Tuathail 1996a). The military diplomacy of General Rose and the civilian diplomacy of David Owen, in particular, are examples of this masculinist culture of *realpolitik* reasoning (Owen 1995). For them and many others, the most realistic solution to what was represented as the "Balkan quagmire" was for the Bosnian Muslim army to face up to the fact that it was militarily weaker and effectively defeated on a number of fronts. The Bosnians, as a consequence, needed to think seriously about surrender.

One place where *realpolitik* reasoning dictated that the Bosnian Muslims should cut their losses and surrender was Srebrenica. Initially seized and raided by paramilitary Bosnian Serb militias (Arkan's Tigers) in April 1992, the town had been recovered by the Bosnian Muslims under the leadership of Nasar Oric, only to be surrounded and in dire need of food by February 1993. In March the United States began airdrops of food and medical supplies to it and other cities in eastern Bosnia, the detached aerial action a manifestation of the circumscribed US involvement in the conflict. The United States was willing to treat the "humanitarian" consequences of warfare in Bosnia from a distance but refused, along with the other NATO countries, to do anything substantive about the cause of this "humanitarian" suffering.

The Serb advance on Srebrenica and other Bosnian Muslim-held towns in eastern Bosnia imperiled even the "humanitarian" mandate of UNPROFOR and the other key UN agency, the United Nations High Commissioner for Refugees (UNHCR). It was within the international humanitarian aid community that the idea of creating "safe areas" in Bosnia was first broached. Honig and Both (1996: 99) credit the idea to the president of the International Committee of the Red Cross in Geneva but the general concept of secure zones had first been

implemented at the end of the Gulf War when the allied coalition declared northern Iraq a "safe haven" for Kurds fleeing an Iraqi army counter-attacking after a
Kurdish uprising in the region. Forged at the birth of what was supposed to be a
New World Order, the concept of a "safe haven" was a new type of space in
international affairs, a zone within the territory of a sovereign state that was
militarily protected from the air by an outside coalition of states. The concept
helped cover up the embarrassment to the Western-led UN coalition caused by
the Kurdish refugee crisis at the end of the Gulf War. It represented the desperate
desire to maintain the pretense of a New World Order in the face of an actually
existing new world disorder.

The concept of a "safe area," as it came to be delimited by the international
diplomatic community, was different from a "safe haven." Whereas the "safe
haven" in northern Iraq was a relatively large territorial area declared by a victorious coalition against a state they had defeated in war, "safe areas" or "protected
zones" were envisaged as demilitarized areas which required the prior consent of
the combatants in order to be established. Safe areas were conceived as humanitarian islands of relative peace and security in a sea of warfare. Unlike "safe
havens," they required that the international community be politically and militarily neutral in their establishing and administration of these zones. They represented territories of UN-governmentality that sought to maintain the conceit that
the United Nations and the international community could avoid taking a side in
the Bosnian war. As such, they represented an intensification of the principle of
ethical engagement without moral responsibility. The international community
wanted to strengthen its ability to carry out its "humanitarian mission" yet without imperiling its neutrality and the supposed moral authority that derived from
this.

Most members of the Western military community and many Western diplomats opposed the concept of "safe havens." The Pentagon considered them
unrealistic and unsustainable. If implemented they held the risk of dragging the
Western alliance further into the "Balkan quagmire," contributing to what Joint
Chiefs Chair Colin Powell saw as "mission creep," the type of expansion in
mission that the United Nations undertook in Mogadishu and eventually led to a
humiliating withdrawal of US forces from Somalia (Ó Tuathail 1996b). David
Owen and Cyrus Vance felt they would further encourage ethnic cleansing by
implicitly designating other areas as unsafe and abandoned by the international
community, and therefore fair game for ethnic cleansing. The permanent
members of the Security Council were also reportedly skeptical while the
UNHCR produced a study arguing that they should be used only as a last option.

Despite all of these reservations and concerns, the UN Security Council nevertheless passed Resolution 819 on April 16 1993 declaring Srebrenica a United
Nations' "safe area." Hastily adopted in the face of the imminent collapse of
Srebrenica and the justifiable fear that the Bosnian Serbs would subsequently
brutalize the population, the resolution offered the UN Security Council the
pretense that it was "doing something" but did little to clarify or resolve the
problems already identified with the concept. It did not define the extent of

the "safe area" and absurdly placed the onus upon the Serbs and the Muslims to make Srebrenica safe. UNPROFOR's role would be to observe and monitor the humanitarian situation in Srebrenica (Honig and Both 1996: 104). While no doubt well-intentioned, the resolution continued to cling to the notion that the international community could remain neutral in the conflict and that it could designate and work within a sphere of "humanitarian" interests without making moral choices about the larger military, political and moral context of the war. In responding as it did, however, the UN Security Council was making larger moral choices by default, choosing to care but only in a way delimited and circumscribed by an ethics of neutrality and professionalism.

Over the following weeks, the possible meaning of the concept of a "safe area" was debated within the international community and worked out on the ground by the competing warring parties and UNPROFOR leaders. Srebrenica's defenders were to be demilitarized by UNPROFOR soldiers and the geographical extent of the "safe area" eventually defined as the frontlines between the combatants. In New York, the non-aligned countries on the UN Security Council with the support of US Ambassador Madeleine Albright extended the concept of "safe areas" to other embattled towns in Bosnia. Two competing visions of "safe areas" were outlined in a French draft paper, a "light option" which would spread 9,600 UNPROFOR troops throughout six enclaves with a mandate to "deter aggression" or a "heavy option" with 35,000 to 40,000 troops to "oppose any aggression" (Honig and Both 1996: 111).

Eventually, on June 4 1993, the UN Security Council passed Resolution 836 which extended the mandate of UNPROFOR "to deter attacks against the safe areas, to monitor the ceasefire (and) to promote the withdrawal of military or paramilitary units other than those of the Government of the Republic of Bosnia and Herzegovina" (cited in Honig and Both 1996: 114). Rather than oppose aggression, UNPROFOR was only allowed to "deter" and, as further specified by the resolution:

> acting in self-defense, to take the necessary measures, including the use of force, in reply to bombardments against the safe areas by any parties or to armed incursion into them or in the event of any deliberate obstruction in or around those areas to the freedom of movement of UNPROFOR or of protected humanitarian convoys.
>
> (cited in Honig and Both 1996: 114)

The key phrase was "acting in self-defense" for it specified UNPROFOR's responsibility as ultimately a self-centric one. For UN troops in "safe areas," the rules of engagement were to respond to use of force by the Bosnian Serbs only if they themselves came under direct threat. UNPROFOR's core universe of obligation was to itself and not to the refugees, civilians and soldiers fighting for their lives in the besieged enclaves. UNPROFOR's mandate, its professional ethics and code of behavior, ruled out any independent moral response-ability to the vulnerable others – the victims of ethnic cleansing and rape, those turned into refugees

and defenders by injustice, criminality and brutality, those fighting to live –
encountered by UNPROFOR troops in the "safe areas." In what became a deadly
irony, UNPROFOR codified an ethics of UN-governmentality organized around
its "humanitarian mandate" and the protection of "safe areas" which stifled moral
response-ability in Bosnia and produced anti-humanitarian consequences including the ethnic cleansing of "safe areas."

The ethics of UN-governmentality in practice

To understand how this happened we only need examine the self-imposed limitations, the bureaucratic chain of command, the rules of engagement and the professional codes of conduct governing the UN establishment and administration of
the "safe area" in Srebrenica. First, despite the passage of the "safe areas" resolutions in the UN Security Council, nearly all states refused to provide troops to
help create and run the "safe areas." Of all the Western states, only the Dutch
were willing to send national troops under UN command to the eastern enclaves
(Honig and Both 1996: 126). The United States had been active in supporting
passage of the "safe areas" resolutions yet it, like other members of the UN
Security Council, refused to embody its commitment to these areas by placing the
bodies of its troops on the line.

Second, the commitment made to those troops who did end up going to the
"safe areas" was minimal. On March 3 1993, 570 Dutch troops officially relieved
a force of 140 Canadian soldiers in Srebrenica, yet their weaponry was light
and logistical supplies heavily dependent upon Serb cooperation. While some
relief convoys got through others were turned back or plundered. Ammunition
was particularly scarce with troops reduced to 16 percent of their operational
ammunition requirements by July 1995 (Honig and Both 1996: 128).

Third, the chain of command governing the ability of UNPROFOR to "deter
attacks against safe areas" was extremely cumbersome and bureaucratic. Control
over the use of force by UNPROFOR was shared by the UN Secretary General,
charged with carrying out the Security Council's resolutions, UNPROFOR's
military commanders, and NATO, the alliance whose military forces would be
used to carry out any response. This system proved to be ineffective for a number
of reasons. For a start, any use of force potentially compromised the "humanitarian mandate" and the supposed neutrality of UNPROFOR in Bosnia. Clinging to
the hope of progress in diplomatic negotiations, Boutros Boutros Gali and his
special representative in Yugoslavia, Yasushi Akashi, were extremely reluctant to
approve any use of force for fear of its political impact. This fear was compounded
by the fact that the only practical way of responding to attacks was with NATO air
power which, as an instrument of force, provided policy-makers with a limited
gradation of force (the two choices were widespread strategic bombing and close
area support; calculating, as the UN wished to do, "proportionate response" was
as a consequence difficult) and was subject to numerous conditions and qualifica-
tions (flight time to the region, cloud cover, weather conditions, surrounding
terrain, etc.). UNPROFOR commanders tended to be reluctant to approve force

for fear it might further endanger vulnerable UNPROFOR troops. The commander of all UN forces in the former Yugoslavia by 1995, Lieutenant General Bernard Janvier, was a particularly cautious and "by-the-book" commander (Rohde 1997: 368). In addition, the UN and NATO often had different interests, with both organizations driven by bureaucratic imperatives to protect their own image and shore up their increasingly tarnished credibility.

Fourth, UNPROFOR's rules of engagement were circumscribed by its mandate to protect the safe areas but only to respond to force by acting in self-defense. Interpreting and operationalizing this mandate in a practical way as orders, procedures and codes of encounter to the foot soldiers stationed in the "safe areas" actually left an important degree of latitude with commanders and officers on the ground. Over the course of two years, this interpretative latitude became codified as increasingly conservative and restrictive rules of engagement by General Janvier as he conducted a campaign to re-consolidate UNPROFOR troops in central Bosnia and effectively abandon the "safe areas" in eastern Bosnia. Janvier's proposal was rebuffed by the UN Security Council in late May 1995 after which his office issued new guidelines governing UNPROFOR troops in the eastern "safe areas." Seeing no real political or military will to defend these areas, Janvier ordered that outlying observation posts in the "safe areas" were to be abandoned, instead of defended by troops and NATO planes, if attacked. This is precisely what happened in Srebrenica in July as the Bosnian Serbian army cautiously sought to capture the "safe area." Over the course of the intense week of July 6 to 13, chronicled in detail by Rohde (1997), Janvier, Akashi and others managed to block numerous calls by the Dutch commander in Srebrenica for close air support to defend the "safe area", thus fatally undermining the credibility of UNPROFOR and the international community. General Janvier's May 29 1995 directive to Rupert Smith, the British commander of UNPROFOR in Bosnia-Herzegovina and an advocate of a more forceful response to the aggression by the Bosnian Serbs, is perhaps the starkest statement of the ethics of UN-governmentality. "The execution of the mandate," Janvier wrote, "is secondary to the security of UN personnel. The intention being to avoid loss of life defending positions for their own sake and unnecessary vulnerability to hostage-taking" (sic, cited in Honig and Both 1996: 156). The ungrammatical last sentence underscores the bureaucratic and institutionally self-centric sentiment of the memo. UNPROFOR's personnel were more important than its mandate. Some of Srebrenica's defenders had already grasped this. The Muslim officer who told a Dutch first lieutenant in Srebrenica that 30 Dutch were more important than the lives of 30,000 Muslims was correct (Rohde 1997: 68–69). In not executing their mandate Janvier and UNPROFOR were making it easier for the Bosnian Serbs to execute Muslim men.

Conclusion: ethics versus morality

While the Bosnian Serb army is ultimately responsible for the mass murder of Muslim men after the fall of Srebrenica, the international community bears considerable responsibility, given what was already known about the behavior of this army and its leaders, for allowing this to happen, especially in a place it had designated a United Nations' "safe area." The Bosnian Serb capture and ethnic cleansing of Srebrenica revealed the limitations and conditionality of the United Nations' ethics of engagement in the former Yugoslavia. It revealed, as many commentators have noted, a systematic failure of leadership on the part of the United Nations and the dominant powers on its Security Council to respond to the Bosnian war by making explicit strategic and moral choices (Mendlovitz and Fousek 1996; Gow 1997). Rather, the United Nations and most of the dominant powers within the international community suspended moral judgments and occluded moral choice between the conflicting groups. Responding only to a de-contextualized "humanitarian nightmare" (in George Bush's words) and not to the origins, nature and immorality producing this nightmare, the international community through the United Nations rendered the war itself a matter of moral indifference. The response to the war was one of adaiphorization, the rendering of it as an object on which ethical authorities do not feel it necessary to take a stand (Bauman 1995: 152–158). In fact, what came to be constituted as the "ethical approach" to the war by the United Nations in Bosnia was precisely and scrupulously to maintain adaiphorization by assiduously avoiding making a moral choice between the conflicting parties.

This logic of adaiphorization structured all the activities of UNPROFOR in Bosnia. It expressed itself not only in the conceptual fiction of the "humanitarian mission" but also in the ethical codes of conduct governing negotiations with the conflicting parties, the administration of aid, the response to threats, the use of force, and an even-handedly limited responsibility towards the victims of war. Enforcing this UN-governmental system of ethical order required structured divisions of authority and expertise, precise calculations of means and ends, constant evaluations of the organizational consequences of actions and the assertion of professional rationalities amidst the chaos of the new world disorder. UNPROFOR invented Bosnia using an ethical map that defined it within a narrowly delimited "humanitarian" universe of ethical obligation but not within an open-ended universe of moral responsibility. Bosnia may have been somewhat close to "us" but it was nevertheless represented as sufficiently far away from being considered "us."

The UN's ethical map of Bosnia could not, however, contain the war in that country unfolding as an insistent moral challenge to the international community. The horrific violence produced by the Bosnian Serb army in randomly shelling the cities it surrounded and in ethnically cleansing territory it coveted incessantly deconstructed the ethical map UNPROFOR used to situate itself in the region. In a land of ethnic cleansing and genocide, UNPROFOR was proclaiming its neutrality. The moral order represented by its governmental ethics was being exposed as immoral, its "humanitarian mandate" revealed as a cover for a lack of humanitarianism.

The fall of Srebrenica is a parable of geography, ethics and morality. It is yet another reminder that ethics and morality are not necessarily the same thing, and that ethical orders often produce and institutionalize moral indifference. Morality is not a foundational state but an insistent challenge to our identity. It cannot be contained by institutions and delimited by ethical codes. It does not have a calculus of rationality and a geography of limits. It is the challenge to be with and for the other no matter what the distances involved, to embrace this responsibility knowing that moral situations in the new world disorder are ambivalent and open-ended, and that moral choices are often "irrational" according to governing standards of rationality. Morality exceeds the borders and boundaries of ethical orders and selves. It is the call to transgress geographies of ethics in the name of a responsibility without limits.

References

Barnett, M. (1996) "The politics of indifference at the United Nations and genocide in Rwanda and Bosnia," in T. Cushman and S. Mestrovic (eds.) *This Time We Knew: Western Responses to Genocide in Bosnia*, New York: New York University Press, 128–162.

Bauman, Z. (1993) *Postmodern Ethics*, Oxford: Blackwell Publishers.

—— (1995) *Life in Fragments: Essays in Postmodern Morality*, Oxford: Blackwell Publishers.

Campbell, D. (1994) "The deterritorialization of responsibility: Levinia, Derrida, and ethics after the end of philosophy," *Alternatives*, 19: 455–484.

Foucault, M. (1991) *The Foucault Effect: Studies in Governmentality*, G. Bruchell, C. Gordon and P. Miller (eds), Chicago: University of Chicago Press.

Gow, J. (1997) *Triumph of a Lack of Will: International Diplomacy and the Yugoslav War*, New York: Columbia University Press.

Herzfeld, M. (1992) *The Social Production of Indifference*, Chicago: University of Chicago Press.

Honig, J. W. and Both, N. (1996) *Srebrenica: Record of a War Crime*, New York: Penguin.

Luke, T. and Ó Tuathail, G. (1997) "On videocameralistics: the geopolitics of failed states, the CNN International and (UN)governmentality," *Review of International Political Economy*, 4 (4): 709–733.

Mendlovitz, S. and Fousek, J. (1996) "Enforcing the law on genocide," *Alternatives*, 21: 237–258.

Ó Tuathail, G. (1996a) "An anti-geopolitical eye? Maggie O'Kane in Bosnia, 1992–94," *Gender, Place and Culture*, 3: 171–185.

—— (1996b) *Critical Geopolitics*, London: Routledge.

—— (1999) "A strategic sign: the geopolitical significance of 'Bosnia' in US foreign policy," *Society and Space* 17.

Owen, D. (1995) *Balkan Odyssey*, New York: Harcourt Brace and Company.

Rieff, D. (1995) *Slaughterhouse: Bosnia and the West*, New York: Simon and Schuster.

Rohde, D. (1997) *Endgame: The Betrayal and Fall of Srebrenica, Europe's Worst Massacre Since World War II*, New York: Farrah, Strauss and Giroux.

Zimmermann, W. (1996) *Origins of a Catastrophe: Yugoslavia and Its Destroyers – America's Last Ambassador Tells What Happened and Why*, New York: Times Books.

10 Social justice, self-interest and Salman Rushdie

Reassessing identity politics in multicultural Britain

Caroline Rose Nagel

Introduction

The post-Second World War era has witnessed massive changes in the social composition of Western Europe with the influx of millions of migrants from Africa, the Arab world, and South Asia (Castles and Miller 1993; S. Smith 1993; Kofman 1995). Newcomers have encountered host societies reluctant to embrace the heterogeneity that increasingly characterized their societies – reluctance manifesting itself in both violent attacks and in more subtle, everyday practices of exclusion and discrimination. Marginalized socially, politically, and economically in their host societies, "foreigners" – many invited in to re-build Europe's post-war economy – have mobilized to contest such notions of nationhood and the impediments to full participation that circumstances presented to them.

In the 1970s, multiculturalism emerged as a major discourse of struggle and resistance of minority groups. Multiculturalism has developed along several ideological trajectories, ranging from efforts to facilitate integration by recognizing the cultures of new immigrants, to recent anti-racist struggles espousing more radical and transformative goals. Here, multiculturalism is used broadly to refer to the entire spectrum of movements and ideologies that have sought to de-link nation, race, and ethnicity from membership and participation in society (Joppke 1996). In fighting discrimination, these movements have also called for a celebration of social identities and for recognition of cultural difference by the state. As such, they have created a particular brand of identity politics in which the affirmation of difference and diversity is paramount.

Successes have been notable in many regards, with the politicization of identity and difference leading to state support for mother-tongue and culture classes in schools, the provision of facilities for immigrant and minority communities, and legislation to combat discrimination in housing and employment. In recent years, however, the politics of identity and the relentless pursuit of diversity in society have met with increasing criticism and resistance. Claims are made among pundits, politicians, academics, and ordinary people, that the preoccupation with identity and difference has gone too far and has outlived its original purpose and intent, becoming less a movement of resistance and more a self-perpetuating form of self-interest. Such criticisms come not solely from anti-progressive forces and right-wing elements, but from groups that have sympathized with resistance politics and liberation movements of various forms. Such debates appear to have created a crisis, evident in recent bouts of navel-gazing, among those who have embraced the politicization of identity.

It seem pertinent, then, that those who have supported the agenda of diversification evaluate the legacies of identity politics. The task of evaluating multiculturalism, identity politics, and diversity raises normative, ethical questions concerned not just with what is, but also with what should be, and how we ought to go about enacting ideals and moral standards. In appraising the legacy of multiculturalism, we also make judgments regarding the legitimate bases of contemporary social movements, the relative validity of conceptions of power and inequality, and the role of the academy in promoting social change. The issue, then, is one of *social justice* – that is, conceptions of what is fair and unfair, and of the social arrangements necessary to ensure that members of society are treated justly.

In this chapter, I will be evaluating multiculturalism through competing conceptions of social justice. I will focus in particular on three bodies of thought which have been influential among geographers, broadly classified as liberal/ distributive, radical/class-based, and poststructuralist/identity-based viewpoints. While distinct and often formulated in opposition to each other, these theories have several important commonalities that provide guideposts for social justice concerns. Most importantly, they highlight the contexts in which identity politics have taken place while refocusing attention on the societal institutions and political arrangements through which inequalities and power disparities can be mediated and negotiated. Using the case of Muslim political activism in Britain, and especially the Salman Rushdie affair, I will illustrate the practical implications of these theoretical approaches for understanding and evaluating identity politics. I will begin by introducing the Salman Rushdie case, which exemplifies the controversy and theoretical and ethical confusion posed by the politicization of cultural identities.

Identity and politics among Britain's Muslim communities

In 1989 the Rushdie affair burst into the public consciousness. Encouraged by a death warrant issued by the aging Ayatollah Khomeini against author Salman Rushdie, hundreds of British Muslims called for the banning of Rushdie's controversial book, *The Satanic Verses*. Protests involved street demonstrations, condemnation of the author by community leaders, and in a few cases, the burning of copies of the book. The book, protesters claimed, vilified and degraded the Prophet Muhammed, showing him, for instance, to be cavorting with prostitutes.

While considered one of the defining moments of immigrant–host society relations in Britain, the Rushdie affair was just one of many events in Britain and Europe in which a Muslim minority had asserted themselves in a vocal manner. The "Muslim identity" employed during the Rushdie affair has developed through decades of struggle between Asian minorities and the white majority, and within Asian communities themselves. It is also a product of multiculturalism, which, having outweighed earlier emphasis on immigrant "assimilation," has fostered a sense of Muslim difference by providing an infrastructure of community networks and services.

The ideology of multiculturalism and diversity is intended to promote social harmony by emphasizing that cultural differences are valid and, indeed, important to the society as a whole (Anthias and Yuval-Davis 1992). Multiculturalism, in a sense, has turned around the notion "separate but equal" used to justify racial discrimination, to "all different, all equal" (see, for instance, European Youth Campaign 1996). Explicitly challenging notions of a British identity as essentially English – white, Anglo-Saxon, and middle-class – multiculturalism argues that discrimination is alleviated by positive assertions of difference. Hence, state schools allow for religious, language, and cultural instruction, sometimes regularizing the instruction of minority languages by offering General Certificate of Secondary Education (GCSE) courses in Bengali, Urdu, Arabic, and the like (Steward 1992). Multiculturalism has led to a growing representation of Asians, Muslims, and other minorities in cities such as Bradford with large "black" communities, and has, in general, raised awareness of the fact that Islam is rapidly becoming the country's second major religion (Nielsen 1995).

Thus, identity-conscious movements have brought discrimination to the forefront of societal consciousness, leading to an awareness of cultural differences, promoting cosmopolitanism rather than a narrow sense of nationhood, and compelling British society to adhere to the "Western" values of tolerance and democracy that it claims to promote. But multiculturalism has also created a number of dilemmas for minority groups and their sympathizers. The recognition of group difference, first of all, has often entailed assumptions or even enforcement of a unity and homogeneity that in reality does not exist in minority communities. As Anthias and Yuval-Davis (1992) assert, multiculturalist practice places individuals into categories, seeing them as belonging first and foremost to their "ethnicity" or "race," and assuming that interests, goals, and values are non-conflicting. This categorization, when incorporated into the structure of the welfare state, has generated a great deal of divisiveness in Muslim and other minority communities, as individual power brokers struggle to maintain a monopoly over the representation of certain groups (Lewis 1994; Werbner 1991).

Assumptions of homogeneity as manifested in cultural diversity policies, have, according to some activists, become particularly detrimental to anti-racist struggles in racially-mixed neighborhoods of London. A recent report sponsored by anti-discrimination committees in the Borough of Greenwich asserts that contemporary racist attitudes are shaped by a sense of unfairness toward whites, whereby the problems and grievances of whites appear to be ignored by the press and local authorities (Hewitt 1996). Equally significant is that white working-class students have been disaffected by the "celebration of cultures" engendered by multiculturalism. If for ethnic minority students, representations of their "cultures" seem removed from their lived experience, they seem even more alienating to white students. White students, Hewitt argues, "experience themselves as having an invisible culture, even of being *cultureless*. . .To some extent, (they) seem like cultural ghosts, haunting as mere absences the richly decorated corridors of multicultural society" (Hewitt 1996: 40).

The legacies of multiculturalism have been debated and scrutinized in other regards. Some argue, for instance, that the politics of identity promoted by multi-cultural movements has done little to address many substantive problems that affected minorities and other marginalized groups (Harvey 1996). It has also not put an end to discrimination, evident in tightening restrictions on New Commonwealth immigrants (Collinson 1993). My own ethnographic research with Muslim immigrants in London reveals an opinion among some that overt assertions of difference have simply exacerbated racism by making groups visible targets and by alienating the "mainstream society" in which minorities have a *duty* to "integrate." Relatedly, some assert that identity politics have created a number of predicaments for young Muslims, who must negotiate their membership in what are conceived to be conflicting social entities (such as "Islam" and "the West") (A. Ahmed 1992).

Multiculturalism, then, has come under intense scrutiny as its inconsistencies and contradictions have become more visible. The politics of identity, diversity, and multiculturalism have consequently become vulnerable to a strong backlash in British society since the early 1980s. In London, this backlash culminated with the ousting of the "Loony Left" and the dismantling of the Greater London Council in 1986 (Jones 1996). Reactions against multiculturalism are likewise evident in claims of the erosion of British society by tribalism and factionalism (an argument paralleling American reactions against political correctness), of which the Rushdie affair is seen as the prime example. And a backlash is evident in the subtle, discrete re-assertions of "English" identity, as in John Major's "imagin-ing" of England as warm beer, cricket and spinsters cycling to evensong (Lunn 1996: 87).

Finally, the backlash against multiculturalism, interestingly enough, has also emerged from those purporting to uphold its basic tenets. In recent attempts by some Muslim parents to secure state funding for Islamic schools, local authorities have responded with the argument that such schools compromise the goals of multiculturalism by stifling contact between groups (Dwyer and Meyer 1996). Such arguments raise questions regarding the goals of multiculturalism and, in particular, the place of Muslims in a "multicultural society."

Identity politics and academia's dilemma

Academia has been put on the defensive in this backlash against identity politics. Many supporters of the broad goals of multiculturalism and diversity have come from the academy and, indeed, scholarly institutions have been an important ground on which the battles of multiculturalism have been fought. Groups previ-ously marginalized in academic disciplines, including women and minorities, have increasingly challenged established ways of knowing and speaking for "others," and have successfully argued that different voices must be heard and different approaches, methodologies, and concerns considered to reflect the interests and needs of under-represented groups (Geiger 1990; Gilbert 1994; Rocheleau 1995). The introduction of "multiculturalism" into academic settings has corresponded with postmodernist and poststructuralist projects, involving a dismantling of grand narratives which ignore social difference in explaining societal relationships and a new concern with identities of researchers and of research subjects (Cloke *et al.* 1991). These challenges to the production of knowledge are part and parcel of changing political commitments by academics. Rejecting notions of dispassionate and "objective" research, scholars actively involve themselves in the social movements which they observe and analyze – often those revolving around the political, social, and economic rights of minority groups (Kobayashi 1994; Laws 1994).

But the contradictions in multiculturalism are problematic. On the one hand, multiculturalism articulates emancipatory goals, using group identity to challenge discrimination and cultural imperialism. But on the other hand, as evident above, it assumes a homogeneity of interests and, like nationalism, it harbors exclusionary ideologies that essentialize group characteristics and suppress differences within (Bondi 1993; Joppke 1996). Identity politics, in other words, can be profoundly reactionary. These tensions within multiculturalist thought have troubled academia. First, with regard to epistemological and methodological concerns, the preoccupation with identity has raised questions about the nature of objective knowledge and the "truth," about whose voices count as "authentic," and about the representation of marginalized voices in academia (Nagar 1997). Second, the questioning of discourses of difference has manifested itself in ambivalence toward identity-based social movements outside of academia (and toward the participation of academics in them) (Kobayashi 1994; Gilbert 1994). In academic literature, the goal among activist-scholars has been to embrace the voice of difference, to provide a forum for subaltern voices to contest dominant paradigms and Western ways of viewing reality and to bring the "other" from the margins to the center of new academic discourses. Yet particular cases, such as Muslim politics in Europe and in the Middle East, do not always elicit wholehearted sympathy. In demanding separate schools, separate facilities for boys and girls, and greater state support for religious observance, for instance, Muslim activists, as noted above, often evoke dichotomies reminiscent of Orientalism, claiming natural and irreconcilable differences between "Islam" and "the West" (L. Ahmed 1991). More troubling still to academics, some Muslim activists have turned to groups typically labeled "reactionary" – i.e. those supporting segregation because of beliefs in innate racial differences and goals of keeping groups separate (Dwyer 1993).

Under such circumstances, academics appear reluctant to address difficult normative questions relating to identity politics. At times this reluctance resembles hypocrisy, whereby academics seem hesitant to encourage "difference" that is not in line with their accepted notions of "emancipatory." It has been quite easy to participate in campaigns against discrimination and to call for diversity within society and the academy. It becomes less comfortable to support "difference" that seems to approach religious "fundamentalism."

But in fairness, many have been caught in the ideological confusion and the moral relativism that surrounds identity politics. On the one hand, the contradictions inherent in identity politics and multiculturalism confound notions of emancipatory and reactionary, progressive and regressive, left and right that have usually guided normative evaluations. On the other hand, poststructuralist thought, which has guided scholarship on identity, has challenged hegemonic moral codes to such an extent that it is difficult to establish standards of normative evaluation (Harvey 1996). In focusing on the fluidity and hybridity of multiple identities, academics have hedged the more volatile issue of whether identity-based claims have value or validity. And in challenging established modes of revealing truth and reality, they have rendered measures of value and validity quite ambiguous. Academics, to be sure, are no less committed to moral concerns: revealing injustices, fighting inequality, and making inquiries into unfair circumstances. But while the commitment to social justice is clear, it is stymied by the conflicting impulses present in the politicization of identity and difference.

There is, I believe, a great deal of danger in wallowing in confusion and indecision. Academics seem to be relinquishing their role in raising issues of ethics and morals in society and questioning dominant mores and established attitudes. While the concern is clearly there, it is muddled in the contradictions of identity politics and poststructuralist thought, and presented in a way that does not make sense to a non-academic audience. Meanwhile, decidedly non-progressive forces have staked out the moral high ground, appealing to a wide audience by speaking in a language that people can understand and making normative judgments in a way that is logical, rational, and definitive.

Academic responses to the Rushdie affair illustrate these dilemmas. Many academics rightfully came forward to criticize the press for unfair coverage of the event and for making Muslim protesters against Rushdie's book appear uniformly as fanatics and fundamentalists (Cottle 1990). Academics clearly articulated the problems associated with coverage of Muslim activism in Britain and with government responses to the burning of Rushdie's book (Asad 1990). They attempted to show why representations of the Prophet Muhammed in Rushdie's novel were so deeply offensive to many Muslims. Yet academic commentary, in emphasizing the social construction of the incident and its significance for later articulations of Muslim identity, effectively avoided much more difficult questions. Were all the claims being made by demonstrators valid? How do we measure rights of minorities to protest and be protected from offensive material against the rights of others to express themselves freely? Are certain "rights" unequivocally legitimate in a multicultural setting? And how do we create political forums in which these issues can be deliberated in a way that results neither in the burning of books, racist remarks, or questionable claims of representativeness of a group interest? Clearly, it is easier to deconstruct and analyze identity than it is to make such normative appraisals of identity politics or to imagine what lies beyond politicized difference.

In the following section, I will outline how existing theories of social justice provide some guideposts with which to answer such questions and, more generally, to evaluate multiculturalism and the politics of difference. In the section following that, I will show how three major sets of theories employed in geographical analysis may help us to understand, evaluate, and promote social justice. Finally, I will return to the Rushdie affair to explore these theoretical arguments.

Social justice in geographical inquiry

Claims of social justice are inherent in most forms of geographical inquiry. That is, modes of describing, modeling, and theorizing geographical phenomena contain some sense òf what is right or wrong, equal or unequal, fair or unfair. Some geographers in recent decades have drawn more explicitly on theories of social justice as a way of asserting and clarifying normative goals and political commitments. In the following section, I will describe the interplay of three broad categories of theoretical literature – liberal/distributive, radical/class-based, and poststructuralist/identity-based – that permeate normative analysis in the discipline.

One of the major themes found in theories of justice has been the distribution of resources in society. Injustice, in this sense, is conceptualized in terms of the inequality of socially defined goods and/or the mechanisms by which inequality is created. One of the most influential expositions of distributive justice comes from John Rawls, whose 1971 work, *A Theory of Justice*, attempts to establish a universal sense of justice and "the right" independent of conceptions of the "greater good" or "public good" (such as those found in utilitarianism). Justice, he argues, does not reflect some intrinsic ultimate "good," but a fair distribution of resources as defined in a particular societal context. According to Rawls, the need for some standard of justice stems from the scarcity of resources and the self-interest of individuals, who pursue their own versions of the "good life" often without considering the consequences for others. To achieve social justice, or a fair distribution of resources, he suggests that all social primary goods be distributed equally unless an unequal distribution is to the advantage of the least favored. To do so requires that priorities be made between equality, liberty, and efficiency, with equality taking precedence unless it serves to further disadvantage marginalized groups. (Rawls 1971; D. Smith 1994).

Rawls' theory of distributive justice has been extensively critiqued. In *Social Justice and the City* (1973), for instance, David Harvey, while inspired by Rawls' vision of equal distribution, takes issue with Rawls' faith in liberal, democratic institutions to achieve a fair distribution of goods and resources in society. Injustice cannot be regulated by the liberal state, he argues, because injustice is rooted not in individual self-interest, but in the dynamics of capitalist production which underpin the entire society – including the state itself. From this perspective, urban space and the inequalities within it are actively produced by a mode of production which is premised on the exploitation of labor and the extraction of profit from urban space. Inequality, in other words, is built into the capitalist system and cannot be eradicated unless an alternative to capitalist production is forged.

While Harvey's work inspired countless geographers and brought Marxism into the mainstream of geographic inquiry, Marxist views on social justice have come under fire by feminists and others since the 1980s. Feminists argue that Marxists have ignored issues outside of the workplace and have wrongly assumed capitalism's blindness to the social identities of workers (England 1994a). While certainly not dismissing capitalist production and class relations, feminist geographers assert the primacy of gender relations and ideologies in structuring cities and "politics" (defined broadly), directly challenging Marxist tendencies to "collapse cultural processes into economic ones, and to marginalize gender oppression in a tale of class oppression" (Pratt 1990: 595).

The emergence of feminism in geography has been accompanied by a theoretical engagement with poststructuralism, which shifts attention from material production and class politics to cultural production/reproduction and the politics of "difference" – that is, how culture, society, and social relations are social relations created and mediated not just by forces of material production but by social constructions of "self" and "other." This poststructuralist approach has influenced theorists of social justice, including Iris Marion Young (1990, 1996), whose work has been influential in geography in the 1990s (see, for instance, Laws 1994; Staeheli 1996; Harvey 1992, 1996). Young criticizes the preoccupation in theories of social justice with distribution, asserting that these ignore "issues of decision-making, divisions of labor, and culture" (Young 1990: 16). Traditional and Marxist theories, she claims, focus on the possession (or lack thereof) of material goods and positions, and seek to ensure the even distribution of goods across what is usually conceptualized as a homogenous polity. This obscures different forms of oppression, domination, and inequality that are embedded in institutions, social practices, and cultural meanings which cannot be accurately conceptualized as forming "bundles" of goods. Furthermore, these theories tend to ignore the group-based nature of social organization and oppression, leading to the fallacious notion that justice can be achieved through uniform action or by disregarding or suppressing group difference.

For Young, then, social justice requires not the achievement of an ideal distribution of goods, but the undermining of social structures and institutional contexts that lead to group-based oppression. Young proposes an egalitarian but non-essentializing "politics of difference" that does not attempt to eliminate differences, but seeks instead to assert positive difference and to address group-based oppressions in democratic forums.

Intersections

While quite distinct from each other, these three theoretical perspectives intersect at several points, providing a basis for a more general understanding of what constitutes social justice, and how justice can be evaluated and implemented in "real life." First, all three perspectives reject the utopianism that characterized many nineteenth-century theories of just societies. While all suggest an "ideal" whereby elements of unfairness are remedied, this ideal serves more as a way to define what is indeed unfair than as a representation of a timeless static condition devoid of context. Second, while emphasizing the contexts and contingencies of justice, and rejecting universal or metaphysical definitions of justice, all suggest that social justice appeals to some widely-held – even if societally-specific – notions of fairness and equality. There is, then, a sense that even if groups differ over how fairness and equality should be defined and achieved, they generally share an understanding of what constitutes fairness and equality and can make normative judgments regarding the outcomes of certain policies, actions, and social arrangements. And finally, all of these theories indicate that defining and achieving social justice is a contentious matter, and that justice emerges from debate and conflict. Justice, then, is a *political* concept, and the forums in which politics take place are of critical importance.

It is this third point, I believe, which generates difficulties, but also many possibilities, for academics in evaluating social justice movements and in promoting normative goals. In each of these approaches, justice involves the active redress of substantive forms of oppression – be it framed in terms of class exploitation, unequal distribution, cultural denigration, or social marginalization. Such redress necessitates forums – movements, institutions, and social arrangements – in which power can be negotiated and contested. Each perspective offers a distinct vision of the ideal political forum, reflecting very different notions of power and social relationships. Rawls (1993), for instance, speaks of liberal, democratic institutions in which individuals enter into public debate on equal terms to present equally compelling and reasonable conceptions of the public good. But Rawls neglects to address relations of power by which citizenship, "the public," and democratic institutions have traditionally been made less accessible to certain groups, such as women, minorities, and the working classes (Walby 1994; Staeheli 1996; Mitchell 1996). Recognizing this, Harvey moves political conflict outside state institutions and into civil society, where he emphasizes the importance of class-based social movements in contesting inequalities in capitalist society. But this vision is also exclusionary in that it neglects non-class-based forms of oppression. Young, in contrast, sees multiple power relations being addressed both in civil society, through social movements, and in political institutions providing representation for oppressed groups. But the difficulty encountered in Young's vision is how to promote group-based representation and a politics of difference without creating divisiveness and fragmentation in the state and civil society, and without, in the words of Harvey, condemning all universal appeals for justice to "the abyss of formless relativism and infinitely variable discourses and interest grouping" (Harvey 1992: 594).

Power relations and political forums

The core problem, then, centers on how political relationships are conceptualized and how political institutions are to be formulated. Faced with the problematic outcomes of multiculturalism, many activist-scholars are attempting to theorize a new form of identity politics – one which addresses multiple levels and relationships of power but which does not involve the assertion of "essentialized difference." Bondi (1993), for instance, advocates the strategic deployment of identity, while Said (1993) calls for "post-nationalist" movements to liberate groups oppressed by social, political and economic imperialism without oppressing those excluded from the boundaries of nationhood. And Mouffe (1995) advocates a "radical democracy" in lieu of a group-based political system that promotes debate and deliberation and that allows for constant realignment and reformulation of group boundaries and identities.

Somewhat ironically, such ideas tend toward Rawls' notion of liberal democracy in which individuals, forming non-permanent unions with other individuals, deliberate and negotiate their equally-valid beliefs, values, and interests. The institutions and mores of liberal democracy have long been criticized for embodying patriarchal, racist, and classist interests under the guise of "individualism" and "equality." Theories of liberal democracy has also been critiqued for locating politics in the state and the "public sector," which peripheralizes political relationships and political actions taking place in households, workplaces, and localities. But liberal democracy is being re-examined by poststructuralist scholars for the potential it offers to the development of a "post-identity" politics and the development of democratic ideals that permeate all spheres of decision-making. Rawls' vision of liberal democracy and poststructuralist viewpoints share the belief that the purpose of democratic institutions is neither the advancement of rational self-interest nor the enforcement of collective good and universal consensus. On the contrary, both contain a very realistic sense of democratic institutions as a site of political conflict for individuals not bound by membership in any group. Thus, Young (1996) and Benhabib (1996) champion the reformulation of democracy, arguing that democratic institutions and practices allow for deliberation and conflict to take place and for power to be negotiated between social groups in the state and civil society. The expansion of democratic decision-making allows for power relationships to be addressed directly and tangibly not only in the state, but also in workplaces and social organizations (see also Gould 1988).

The theoretical difference between traditional and poststructuralist views comes with the latter's more nuanced understanding of power and recognition that not all groups have equal access to decision-making channels in order to press their valid claims. What is needed, then, is to make democratic institutions more accessible to different groups rather than to institutionalize special group representation – to focus on the potential of individual-oriented democratic thought rather than to dwell on past injustices embodied in democratic systems. In this way, political struggle generates pragmatic and substantive outcomes. Rather than chasing down elusive identities, the goal becomes, in part, to find ways to negotiate competing claims and resolve issues of unfairness and inequality – whether framed in terms of identity or not – in work places, local government, schools, electoral systems, and so on.

Affinity and common bases of membership

In making such arguments, scholars have switched emphasis from "difference" to sameness and from identity to affinity. Recent methodological discussions in geography, for instance, have reconsidered the focus on situatedness and positionality, arguing that how one is situated should not be construed as a complete barrier to understanding others or to addressing inequalities and injustice (England 1994b; Kobayashi 1994; Rocheleau 1995).

If social difference has so frequently compromised democracy, then it seems that some level of commonality or affinity must be forged between people if democracy is to work. To be sure, the quest for commonality has as checkered a past as the assertion of difference, and they are, in many respects, two sides of the same coin. As Young (1990) remarks, notions of individualism found in liberal thought have involved the rejection and marginalization of social differences – a forced assimilation by which members of a society must conform to a norm specified by more powerful groups. The challenge is to replace exclusive nationalisms with "a 'civic' nationalism expressing allegiance to the values specific to the democratic tradition and the form of life that are constitutive of it" (Mouffe 1995: 264). It is the common membership, borne of participation and interaction in many spheres, sectors and institutions, which must be the basis of political action and political institutions. The crux of the matter is to ensure that participation and interaction is truly free and equal such that people may choose, rather than be compelled, to enter into particular social, political, cultural, and ideological groupings.

While rejecting his class-based arguments, this argument parallels Harvey's call to identify "common values" in the midst of striking heterogeneity of lives, beliefs, and identities. To find common values and to mobilize a progressive politics is a geographical venture, requiring a sense of place as embedded in multiple social processes that extend far beyond a particular locality. The ability to abstract social relations – to understand how one's life is tied to peoples and geographies elsewhere and to wider social processes – allows local struggles to move beyond the parochial fragmenting politics of group identity. Without a sense of common membership, moral arguments become relativistic and fragmented, with various groups asserting different systems of beliefs and values over which they claim full authority. The rigid boundaries upon which oppression is based are solidified and placed within a hierarchy of difference rather than dismantled to facilitate egalitarianism.

In sum then, the dilemmas posed by multiculturalism have forced us to re-evaluate the goals of identity politics and the means by which we attempt to achieve social justice. Poststructuralist analysis, I have argued, does not require an uncompromising commitment to identity politics. In identifying the group-based nature of social oppression, whether in class, gender, racial, or caste terms, we should not insist that the hardening of such categories is the only means by which such oppression is remedied. If the goal is to undermine power and oppression and to come to some consensus on equality and fairness and justice, then we must recognize our common identities and our mutual stake in the institutions of our societies; we must emphasize our very real and substantive relationships and interactions rather than our shifting and imagined differences.

Conclusions

By way of a conclusion, I return to the civil unrest provoked by *The Satanic Verses*. The Rushdie affair reveals the immense complexity of social justice claims in a multicultural society. In doing so, it provides a real-life situation to evaluate how the ideals of social justice and democracy play out in everyday affairs.

The social activism unleashed during this incident, as described earlier, reflects many of the dilemmas of multiculturalism. The incident was portrayed as evidence of a seemingly irreconcilable clash between group rights and freedoms of speech and expression, and was decried as evidence of tribalism and societal fragmentation. But in the media hubbub surrounding the Rushdie affair, efforts by protesters to broaden (rather than to limit) the terms of membership in the British polity have been lost. More than a clamor for separatism, the claims of many protesters reflected a desire for equal application of Britain's blasphemy laws to Muslim offenders. Beyond the Rushdie affair, Muslim activism has focused on attaining access to state funds for religious-based schools equal to that of Christian and Jewish groups (Dwyer and Meyer 1996). And more generally, Muslim activists have challenged a cultural hierarchy – evident in media reactions to the Rushdie affair – which positions Muslims as a threat to cultural unity in Britain, and Islam as somehow antithetical to "Western" values of free expression (Cottle 1990; A. Ahmed 1992). Muslim activists have directly and insistently addressed their marginalization, demanding that they be not simply tolerated, but fully accepted as social, economic, and political participants in British national life.

Protesters, in other words, while asserting themselves as Muslims, also engage in a struggle with themselves and with society-at-large over meanings, ideologies, and policies in a much larger sphere of political activity than their "parochialism" would suggest (Layton-Henry 1990; Phillips 1996). In mobilizing and protesting, they are often requesting not that they be given "special treatment," but rather that they be treated the same – that is, to be given equal treatment commensurate with their membership in the British polity.

Political mobilization along the lines of identity, to be sure, is an ambivalent undertaking. But clearly, the utilization of group identity is not entirely detrimental and parochial. More importantly, it seems unrealistic to suggest that politics does not appeal to identities – particularly in challenging group-based oppression. As poststructuralist analysis emphasizes, we draw upon social identities every day, and our political consciousness and actions are guided, in part, by a sense of our membership in particular groups. Perhaps, then, the main goal for activists is to move the debates generated by identity-based social movements into civic forums in which all members of society can participate – in which seemingly group-specific concerns become pertinent to others.

To move beyond a politics of difference is contingent on promoting an overall

144   *Ethics and place*

political system that does not preserve or institutionalize difference or privilege, either for majorities or minorities. The notion of de-institutionalizing group difference inevitably raises difficult questions about the future of affirmative action, and group-conscious race relations policies, race- and gender-based quotas for political representatives, separate schools, blasphemy laws, and the like. But such questions became more manageable and palatable in an institutional context that is inclusive, that allows and encourages individuals to enter into discussion on equal terms, treats its members on equal standing, and does not enforce conformity, but rather provides opportunities for debates to take place.

What is being argued here, therefore, is neither multiculturalism as we know it nor a reversion to life before multiculturalism; it is, rather, a practical and realistic approach to politics which recognizes that while harmony is not attainable, the management of power and conflict and debate is. And more often than not, the search for "possible worlds," as Harvey (1996) terms it, requires breaking down barriers wherever possible rather than building them up, and the fostering of common membership in the polity in substantive terms rather than promoting illusory differences.

References

Ahmed, A. (1992) *Post-Modernism and Islam*, London and New York: Routledge.

Ahmed, L. (1991) *Women and Gender in Islam: Historical Roots of a Modern Debate*, Princeton, NJ: Princeton University Press.

Anthias, F. and Yuval-Davis, N. (1992) *Racialized Boundaries: Race, Nation, Gender, and the Anti-Racist Struggle*, London and New York: Routledge.

Asad, T. (1990) "Multiculturalism and British Identity in the Wake of the Rushdie Affair," *Politics and Society* 18, 4: 428–480.

Benhabib, S. (1996) "Toward a deliberative model of democratic legitimacy," in S. Benhabib (ed.) *Democracy and Difference*, Princeton, NJ: Princeton University Press.

Bondi, L. (1993) "Locating identity politics," in J. Penrose and P. Jackson (eds) *Constructions of Race, Place, and Nation*, Minneapolis: University of Minnesota Press.

Castles, S. and Miller, M. (1993) *The Age of Migration*, New York: Guilford Press.

Cloke, P., Philo, C. and Sadler, D. (1991) *Approaching Human Geography: An Introduction to Contemporary Theoretical Debates*, New York: Guilford Press.

Collinson, S. (1993) *Europe and International Migration*, London: Pinter for the Royal Institute of International Affairs.

Cottle, S. (1990) "Reporting the Rushdie affair: a case study in the orchestration of public opinion," *Race and Class* 32, 4: 45–66.

Dwyer, C. (1993) "Constructions of Muslim identity and the contesting of power: the debate over Muslim schools in the United Kingdom" in J. Penrose and P. Jackson (eds.) *Constructions of Race, Place, and Nation*, Minneapolis: University of Minnesota Press.

Dwyer, C. and Meyer, A. (1996) "The establishment of Islamic schools: a controversial phenomenon in three European countries," in W. A. R. Shadid and P. S. Van Koningsveld (eds), *Muslims in the Margins: Political Responses to the Presence of Islam in Western Europe*, The Hague, The Netherlands: Pharos.

Dwyer, C. and Meyer, A. (1999) "Veiled meanings: young British Muslim women and the negotiation of differences," in *Gender, Place and Culture*, 6, 1.

England, K. (1994a) "From *Social Justice and the City* to women friendly cities? Feminist theory and politics," *Urban Geography* 15, 7: 628–643.

—— (1994b) "Getting personal: reflexivity, positionality, and feminist research," *Professional Geographer* 46, 1: 80–89.

European Youth Campaign (1996) Commission for Racial Equality website, http://www.open.gov.uk/cre/crehome.htm.

Geiger, S. (1990) "What's so feminist about feminist research?" in *Journal of Women's History* 2, 1: 169–182.

Gilbert, M. (1994) "The politics of location: doing feminist research at 'home,'" *Professional Geographer* 46, 1: 90–96.

Gould, C. (1988) *Rethinking Democracy: Freedom and Social Cooperation in Politics, Economy, and Society*, Cambridge, UK and New York: Cambridge University Press.

Harvey, D. (1973) *Social Justice and the City*, London: Edward Arnold.

—— (1992) "Social justice, postmodernism, and the city," *International Journal of Urban and Regional Research* 16, 4: 588–601.

—— (1996) *Justice, Nature, and the Geography of Difference*, Cambridge, MA: Blackwell Publishers.

Hewitt, R. (1996) *Routes of Racism: The Social Basis of Racist Action*, London: Trentham Books Ltd. for the International Centre for Intercultural Studies, University of London.

Jones, E. (1996) "Social polarization in post-industrial London" in J. O'Loughlin and J. Friedrichs (eds) *Social Polarization in Post-Industrial Metropolises*, Berlin and New York: Walter De Gruyter.

Joppke, C. (1996) "Multiculturalism and immigration: a comparison of the United States, Germany, and Britain," *Theory and Society* 25: 449–500.

Kobayashi, A. (1994) "Coloring the field: gender, race, and the politics of fieldwork," *Professional Geographer* 46, 1: 73–80.

Kofman, E. (1995) "Citizenship for some but not for others: spaces of citizenship in contemporary Europe," *Political Geography* 14, 2: 121–137.

Laws, G. (1994) "Oppression, knowledge, and the built environment," *Political Geography* 13: 7–32.

Layton-Henry, Z. (ed.) (1990) *The Political Rights of Migrant Workers in Western Europe*, London and Newbury Park, CA: Sage Publications.

Lewis, P. (1994) *Islamic Britain: Religion, Politics, and Identity Among British Muslims*, London: IB Tauris.

Lunn, K. (1996) "Reconsidering 'Britishness': the construction and significance of national identity in twentieth century Britain" in B. Jenkins and S. Sofos (eds) *Nation and Identity in Contemporary Europe*, New York and London: Routledge.

Mitchell, D. (1996) "Political violence, order, and the legal construction of public space: power and the public forum doctrine," *Urban Geography* 17, 2: 152–178.

Mouffe, C. (1995) "Post-Marxism: democracy and identity," *Environment and Planning D: Society and Space* 13: 259–265.

Nagar, R. (1997) "Exploring methodological borderlands through oral narratives" in J. P. Jones, H. J. Nast, and S. M. Roberts (eds) *Thresholds in Feminist Geography: Difference, Methodology, and Representation*, Lanham, MD: Rowman and Littlefield.

Nielsen, J. (1995) *Muslims in Western Europe* (second edn), Edinburgh: Edinburgh University Press.

Phillips, A. (1996) "Dealing with difference: a politics of ideas or a politics of presence?" in S. Benhabib (ed.) *Democracy and Difference*, Princeton, NJ: Princeton University Press.

Pratt, G. (1990) "Feminist analyses of the restructuring of urban life," *Urban Geography* 11, 6: 594–605.

Rawls, J. (1971) *A Theory of Justice*, Cambridge, MA: Belknap Press of Harvard University Press.

—— (1993) *Political Liberalism*, New York: Columbia University Press.

Rocheleau, D. (1995) "Maps, numbers, text, and context: mixing methods in feminist political ecology," *Professional Geographer* 47, 4: 458–466.

Said, E. (1993) *Culture and Imperialism*, New York: Knopf.

Smith, D. (1994) *Geography and Social Justice*, Oxford, UK and Cambridge, MA: Blackwell Publishers.

Smith, S. (1993) "Immigration and nation-building in Canada and the UK" in J. Penrose and P. Jackson (eds) *Constructions of Race, Place, and Nation*, Minneapolis: University of Minnesota Press.

Staeheli, L. (1996) "Publicity, privacy, and women's political action," *Environment and Planning D: Society and Space* 14: 601–619.

Steward, S. (1992) "Introducing Arabic teaching into state schools in Britain," paper presented at second Arab Communities Conference, 19–20 November 1992, London.

Walby, S. (1994) "Is citizenship gendered?" *Sociology* 28, 2: 379–395.

Werbner, P. (1991) "The fiction of unity in ethnic politics: aspects of representation and the state among British Pakistanis" in P. Werbner and M. Anwar (eds) *Black and Ethnic Leaderships: The Cultural Dimension of Political Action*, London and New York: Routledge.

Young, I. M. (1990) *Justice and the Politics of Difference*, Princeton, NJ: Princeton University Press.

—— (1996). "Communication and the other: beyond deliberative democracy" in S. Benhabib (ed.) *Democracy and Difference*, Princeton, NJ: Princeton University Press.

Part 3
Ethics and nature

Nature and the natural enjoy a long tradition of intellectual interest among geographers, and have served as powerful, though often unexamined, grounds for personal values and moral codes. James Proctor begins with a challenge to philosophical convention, asserting that facts and values are not separate but mutually interdependent features of human understanding of the natural world. Proctor argues that prevailing moral perspectives on global environmental change are built on naïve resolutions of the fact/value distinction. Where one major perspective champions the moral authority of facts, its alternative, informed by recognition of an environmental crisis and by such sentiments as the need to heal the earth, celebrates values in an unreflective way. A third perspective argues that moral claims about global environmental change come largely from the political, economic and cultural interests of groups, who line up both facts and values to their causes. In their place, Proctor proposes another perspective: that we understand ourselves as inhabitants of a "moral earth," which is to understand the inescapable moral threads running through our practices and their traces on the earth. This perspective, which erases hard distinctions between facts and values, does not provide a ready answer to the many moral issues connected with global environmental change, but does offer an intellectually richer point of departure.

Sheila Hones also links the natural and the human, in discussing the effects of associating history with a view of nature as a process tending towards stability and geographical order. This is elaborated through the concept of the "natural war," which plays a role in the narrative of national identity as an organic unity. Her argument is based on a reading of the American periodical *The Atlantic Monthly* for part of the 1880s, which reveals the connection made between a particular version of "nature" and "the natural" and a particular ethical interpretation of the history of the United States. The War of Independence was portrayed in terms of natural and inevitable, while the Civil War was described in images of the un-natural, disorientation and disruption. Both rest on a conception of nature as a stable, harmonious system directed by some form of moral intelligence. A critique of these writings, in the spirit of the deconstruction of text popular in contemporary methodology, helps to subvert the association of national history with a natural geo-spatial inevitability, thus opening up opportunities for alternative understandings.

A different angle on the environment as well as a different spatial setting is adopted by Jeremy Tasch. He takes the Russian Far East to illustrate how places, as contextualizations of socio-ecological relations, can serve as sites in which to locate self and community in a "moral space." The background is the formation of post-Soviet society, with its changing power relations, spatial circulations and discourses on nature. Under the Soviet Union an exploitative attitude towards nature prevailed, promoted from the center and with no room for alternative or local understandings. Post-Soviet times have seen expressions of local unease as to the utilization of the region's resources, now open to pressures arising from integration into global economic structures. This global/local interface is crucial to the dynamics of how nature will be assessed and controlled, and by whom. The author argues that an effective place-focused environmental ethics is required, if development is not to ignore local nature–society relations.

Alice Dawson extends the discussion of environment/nature into the treatment of animals. Our attitude to animals is complex (as shown in part of Tuan's essay earlier in this volume): both exploitative and protective. Dawson takes the pig to exemplify the issues involved, reviewing the treatment of the pig as an agricultural commodity, as a medical resource and as a pet. The use of the pig (and other animals) as a source of spare parts for humans, in the context of the kind of science which (re)produced or cloned Dolly the Sheep in 1997, dissolves something of the distinction between at least some humans and animals – bits of which become "part of us" as living beings and not merely as food, if never "one of us" as members of some human community. Animals raise important ethical issues at the intersection of human culture and environment, which geography occupies.

11 A moral earth

Facts and values in global environmental change

James D. Proctor

Joy and Woe are woven fine
A Clothing for the soul divine
Under every grief and pine
Runs a joy with silken twine.
William Blake,
Auguries of Innocence

Departure

A transformed earth

Above my childhood home in the US Pacific Northwest is a peak that somehow got the oxymoronic name of Canyon Mountain. I've ascended this peak quite often, passing clearcuts covered with snow and forests cleared by fire. And when I'd get to the top I'd climb an old navigational beacon, look out at the forests of southern Oregon, and witness firsthand the magnitude of the human transformation of nature. As far as I could see, they had been altered by human hands, plundered of their original timber wealth and then – with varying degrees of success – replanted to produce more. Aside from the effects of fires that roam the hills of southern Oregon in the dry summers, these forests have undergone a magnitude of environmental change in the last half-century unmatched by non-anthropogenic forces over the last several millennia.

Each one of us has, from our youth, encountered a transformed earth. The dynamic biophysical processes that shape the world around us have been joined in the last several millennia by the transformative power of humans, who have altered the earth's landscapes, utilized its vast resources, profoundly modified its biota, and to a certain extent influenced its major biogeochemical cycles. The human transformation of nature is no monolithic process: it has been intentional as well as accidental, ephemeral to long-lasting, both local and global. Yet, overall, it has vastly accelerated in magnitude and spatial extent over the last several hundred years. The earth – at least the critical, thin life-supporting biosphere enveloping the earth – is now in many ways a product of humankind.

Global environmental change is, one could reasonably say, a fact. But it is not just a fact: the tremendous recent surge of scholarly and lay interest in global environmental change is driven as much by concern as curiosity. Indeed, facts and values are found together in many accounts of global environmental change. Consider the following two summaries, for example. Even though the first was written from the perspective of geography, which (this volume aside) resides primarily in the realm of facts, and the second from the perspective of environmental ethics, which concerns itself more with values, they are remarkably similar:

> Its ingredients have become well-known. Massive burning of Amazonian forests, indiscriminate logging in South-east Asia and food and fuel needs of Africa's fast-growing population are destroying tropical rain forests, the Earth's richest repositories of biodiversity. Soil erosion, desertification, improper irrigation, inadequate recycling of organic matter and excessive use of farm chemicals are reducing the extent and the quality of arable lands. The Antarctic ozone hole and a possibility of its Arctic duplicate are causing fears of extensive damage to crops, animals and human health. And the anticipation of rapid climatic change is moving nations toward a formulation of global co-operative policies designed to forestall the burdens of reduced harvests, declining economies and masses of environmental refugees.
>
> (Smil 1994: xii)

> In the late 1980s, the second wave of the twentieth century's environmental crisis began to crest. Word reached the public that a "hole" in the planet's protective membrane of ozone had been discovered over the Antarctic . . . Each spring the hole has grown larger. Because of the continuing increase in carbon dioxide and other "greenhouse" gases in the earth's atmosphere, most scientists now agree that the planet will warm up, with potentially disastrous environmental consequences. The assault on Earth's girdle of moist tropical forests, home to half the planet's complement of species, has intensified. Our generation may preside over a rare episode of abrupt, mass species extinction . . . The environmental crisis – discovered in the industrial West in the 1960s, plastered over with regulative legislation in the 1970s, then forgotten only to return with a vengeance in the 1980s – is now global in scope and focus.
>
> (Callicott 1994: xii–xiii)

There are no valueless facts, no factless values here; Smil, a geographer, and Callicott, an environmental ethicist, both speak fact-values, value-facts in their assessments of global environmental change.

Well-worn paths, constraining perspectives

This essay asks a question it will not answer: what are the ethical implications of global environmental change? I ask this question because, in contrast to the reams of literature that have recently emerged on the biophysical, political, and economic dimensions of the phenomenon, relatively little inquiry has been explicitly devoted to the ethics of global environmental change (Callicott and a few others aside; see for instance Jamieson 1996). Yet, ironically, the moral terrain of global environmental change has already been too well traversed, the principal routes too well demarcated, to allow us passage without finding ourselves in a pre-established rut. I thus will not answer this question because I do not believe we are ready to do so.

This non-answer first requires a bit of terminological distinction between ethics and morality. I use the term "moral" to refer generally to existing, often taken-for-granted schemes of good/bad, right/wrong, and so forth. I use the term "ethics" to refer to intellectual reflection on morality. When I argue that we cannot yet consider the ethics of global environmental change because its moral terrain is too deeply rutted, what I intend to warn against is the sort of ethical journey that mindlessly follows existing, highly partial perspectives on coming to moral terms with global environmental change.

In their place, I will suggest an alternative manner of departure, one that recognizes, as suggested in the above narratives, that analyzing the ethics of global environmental change is not so much a matter of adding values onto a primarily factual discourse as of carefully exploring the values contexts that already enframe the ways we make sense of global environmental change. This approach is informed by my identity as a geographer. I consider ethics as a geographer because I believe that ethical questions are too important to leave only for philosophers (whose intellectual rigor I do not question) to clarify – let alone answer – for the rest of us. Geography offers a very important perspective for ethical reflection, one that has only begun to be elaborated in our discipline.

Indeed, geography and ethics run deep in the meaningful fabric of our lives, whether or not we identify ourselves as geographers and/or ethicists (and, of course, most people don't). Geography (literally, "earth writing") and ethics (moral reflection) matter at a very fundamental level, because we inhabit a moral earth. It is moral precisely because we inhabit it. The values we have woven into our existence on earth are not necessarily the best ones possible, nor certainly are they self-evident, but there is never some value vacuum we must fill; the earth is already a moral place.

Understanding ourselves geographically as inhabitants of a moral earth may not lead to tidy resolution of the complex ethical questions surrounding global environmental change, but it will at least remind us that they are already there – indeed, the worth of ethical reflection often lies in the reflective attitude itself as much as the elusive answers we seek. From the perspective of a moral earth, values issues are not beyond the pale of science, restricted to policy implications, or some human add-on to a biophysical phenomenon. Global environmental change is about facts and values: it concerns facts because there are indeed more and less factually robust ways of understanding it, yet it also concerns values because there is literally no way that we can utter a sensible statement about the biophysical process of global change and its implications without bringing our (moral) humanity into the mix. Rather than follow some purification ritual of fact–value separation, then, I suggest that we proceed from the axiomatic point of departure that facts and values are – as joy and woe – woven fine.

Reflection

Facts

Global environmental change is as much a product of science as an emerging reality (Buttel *et al.* 1990; Wescoat 1993; Wynne 1994). It has roots in a number of scientific disciplines (*The Economist* 1995), and has benefited from post-Second World War international research efforts running from the International Geophysical Year of 1957–58 to the ongoing International Geosphere-Biosphere Program (IGBP), launched in 1986 (International Council of Scientific Unions 1986). But atmospheric science has played a special and leading role (Fleagle 1994), due in no small part to space technology competition between the once-Soviet Union and the United States and the related post-Cold War desire to keep space budgets aloft: the National Aeronautics and Space Administration (NASA), for instance, is slated to receive fully three-quarters of projected 1998 US Global Change Research Program funding (Subcommittee on Global Change Research 1997: 79).

Though scientists themselves do not manifest a settled position on the facts of global environmental change – after all, science is not simply an accumulation of facts, and at any rate research budgets require questions more than answers as justification – it is not surprising that the dominant moral perspective on global environmental change today privileges facts over values, arguing that the imperative is to get the facts straight and design appropriate corrective policy measures where necessary (Herrick and Jamieson 1995). The values decisions inherent in policy-making should be primarily informed by good science; in this way, the fuzzy realm of values is offered some secure footing in the less inherently contestable realm of facts. Consider this summary to the first Intergovernmental Panel on Climate Change (IPCC) report, a statement of crucial significance to the ongoing debate over climate change policy:

> This . . . report considers the scientific assessment of climate change. Several hundred working scientists from 25 countries have participated in the preparation and review of the scientific data. The result is the most authoritative and strongly supported statement on climate change that has ever been made by the international scientific community . . . It will inform the necessary scientific, political and economic debates and negotiations that can be expected in the immediate future. Appropriate strategies in response to the issue of climate change can now be firmly based on the scientific foundation that the report provides.
>
> (Houghton *et al.* 1990)

The notion that science-as-facts is fundamental to evaluating environmental change can, however, be taken in the opposite direction; indeed, one of the major criticisms of concern over global environmental change is that it is based more on hype than science. Gregg Easterbrook, for instance, argues that anthropogenic greenhouse gas contributions are insignificant in the biophysical scheme of things:

> The present rate of increase in human-causes greenhouse forcing . . . works out to about 0.002 percent per annum of the total effect . . . People assume that the twentieth-century increase in artificial carbon dioxide emissions has an overwhelming impact on nature. From nature's way of thinking the impact may still be so minor it is difficult to detect.
>
> (Easterbrook 1995: 23)

Easterbrook calls for "ecorealism," based on the principles that "logic, not senti-ment, is the best tool for safeguarding nature; that accurate understanding of the actual state of the environment will serve the earth better than expressions of panic" (Easterbrook 1995: xvii).

Whether or not these sorts of counter-assertions are correct, they do suggest the fragility of this perspective. As supporting facts are called into question, the whole moral house of cards collapses. Ethical inquiry becomes silenced in cases where their very complexity leads to disagreement over factual matters. And thus emerges a particular problem in adopting this perspective on facts and values: as spatial scale increases, so often does complexity, such that environmental change at the global scale becomes in many ways a much more difficult factual matter than the environmental changes that have taken place in the forests I wandered as a child. This echoes the argument of Anthony Weston:

> This earth eludes us . . . It eludes the computer modelers, who still, appar-ently, even now, have no idea where a billion tons of carbon dioxide – a seventh or more of the total dumped into the atmosphere from human sources – goes every year, though some of them confidently go on to predict, or deny, global warming anyway. And this earth eludes our fatalism. Predic-tion is dangerous, as E. F. Schumacher once said, especially about the future.
>
> (Weston 1994: 176)

The problem with this perspective is not, however, simply that facts are contest-able or elusive. It also resides within the very term "fact," which derives from the Latin *factum*, meaning a deed or something done. The most common under-standing is that a fact is an item of *knowledge* that is true by virtue of its cor-respondence to *reality*. To assert, for example, that it is a fact that anthropogenic emissions have boosted greenhouse gas concentrations in the atmosphere is to claim that this knowledge-statement is true because it corresponds to actual occurrences. Facts are true knowledge-claims about reality. Fictions, in contrast, are demonstrably false knowledge-claims about reality, hence the apparent need to distinguish between the two – between "eco-facts and eco-fiction," as the recent title goes (Baarschers 1996).

Thus the realms of ontology (derived from the Greek *einai*, to be) and epistemology (derived from *episteme*, the Greek word for knowledge), are conflated in this sense of "fact" – my Random House Dictionary, for instance, defines a fact as "Something that actually exists; reality; truth." In addition to its fairly naïve correspondence theory of truth, a major problem with this conflation of ontology and epistemology lies in what Roy Bhaskar (1975) has termed the "ontic fallacy." According to the ontic fallacy, knowledge is reduced to reality, truth-assertions point immediately to the conditions they assert to be true, our ways of understanding appear as faithful mirrors of the things we strive to understand. This sense of "fact" then masks the very human qualities of facts about global environmental change. The perspective that privileges facts therefore privileges far more than what most people think of as facts.

Values

Clearly there is a need to look at global environmental change in a manner that takes values more seriously than the previous perspective; and indeed a values-based perspective on global environmental change has thrived in the recent past. Yet there are, predictably, pitfalls with championing values as well. Consider the following observation made by the President of the Royal Society of Canada in the preface to a book on global environmental change:

> The human being is an animal that has moved out of ecological balance with its environment. Humankind is a wasteful killer and a despoiler of other life on the planet. This normal and apparently acceptable behaviour has been licensed by a belief that our use of the Earth's resources is God-given, and encouraged by an economic system that emphasizes short-term profit as a benefit ... Humankind is now dominant in effecting perhaps irreversible change on the Earth's surface, and I suggest that we do not know enough to decide how to run this planet.
>
> (McLaren 1991: xiv)

McLaren's account is not at all squeamish about values. But is this ethics? It feels more like an unfettered proclamation of right and wrong, founded on a fairly settled (and generally apocalyptic) reading of the facts of global environmental change, than a critical reflection on morality. It thus resembles in important ways the perspective that privileges facts – though of course to the dogmatic factist it must feel like the tail wagging the dog.

Outside of the strictly scientific literature on global environmental change, values run rampant. They influence many of the overriding themes in recent popular books about global environmental change: witness the earth poised in the balance between destruction and preservation (Gore 1993), the imperative to heal the planet (Ehrlich and Ehrlich 1991), or the need to adopt new metaphors of the earth such as Gaia (Myers 1993) and to achieve a steady-state economics that values the earth (Daly and Townsend 1993). Even the undergraduate environmental science curriculum has moralistic moments; for instance, students can take a telecourse on environmental science with a study guide entitled *Race to Save the Planet* (Wolf 1996).

I do not wish to be overly critical; as with the factist perspective, a genuine impulse underlies efforts in this vein. Take, for instance, the abundant literature on crafting a new "earth ethic." Environmental philosopher Holmes Rolston states:

> The home planet is in crisis . . . Our modern cultures threaten the integrity, stability, and beauty of Earth and thereby of the culture superposed on Earth. Beyond the vision of one world is the shadow of none. We are searching for an ethics adequate to respect life on this home planet.
>
> (Rolston 1996: 162)

As Rolston's passage suggests, a strong sense of looming and existent crisis leads authors in this genre to argue for the need to craft a new normative ethics, to fashion a moral imperative to "heal the earth" (Harrington 1990). A number of metaphors are invoked, from thinking of the earth as a garden (Allsopp 1972) to conceiving of the earth inclusively as a scene of plenishment, where good and pain alike must be embraced (Ross 1995). Many are attempts to bring religion to bear in crafting an earth ethic (Stone 1971; Murphy 1989; Rasmussen 1996); one recent work in this genre explicitly intends to build a global environmental ethic based on the spiritual and other traditions of diverse cultural groups of the world (Callicott 1994).

Though the attempt to craft an earth ethic is laudable, it is more of an ethics of "being ethical" or "doing good" than ethics as critical reflection on being ethical or doing good. As such, it resembles McLaren's preface, Gore's *Earth in the Balance*, and the other works cited above in promulgating a fairly clear sense of right and wrong. Yet values are rarely evaluated as simply as this. Where the factist perspective discounts values, its alternative celebrates values in an excessively unreflective manner.

Interests

A third prevalent perspective on facts and values in global environmental change offers an important corrective to the two other perspectives discussed above. Consider the following statement from a recent text on global environmental change:

> We must therefore ask two . . . crucial questions. How far is global environ-mental change really about the advanced countries of the North keeping their control (hegemony) over the poorer countries of the South? And, how far is global environmental change about scientists persuading concerned

governments to continue to fund their expensive research, despite the fact that few people really want it, or actually benefit from it?

(Moore *et al.* 1996: 198).

From this perspective, the predominant factors at work in making claims about global environmental change are the political, economic, cultural, or other interests that motivate an individual or group; facts and values are then lined up to support a particular platform of interests.

Consider, for a moment, Earth Summit 1992, formally known as the United Nations Conference on Environment and Development (UNCED), held in Rio de Janiero. Earth Summit 1992 gathered the heads of state from over 100 countries of the world, with 178 states participating altogether, to discuss global environmental problems and find ways to solve them that embraced the need for what has become known as sustainable development. Maurice Strong, Secretary-General for UNCED, summarized the motivation for the Earth Summit in a manner suggestive of the moralistic perspective above:

> We, the world community, now face together greater risks to our common security through our impacts on the environment than from traditional military conflicts with one another. We must now forge a new "Earth ethic" which will inspire all peoples and nations to join in a new global partnership of North, South, East, and West . . . Earth is the only home we have; its fate is literally in our hands.
>
> (Strong 1992: 115)

Yet development scholar and critic Wolfgang Sachs was not so enamored of Earth Summit, stressing its political motivations:

> As ecological issues have moved to the top of the agenda of international politics, environmentalism appears in many cases to have lost the spirit of contention, limiting itself to the provision of survival strategies for the powers that be. As a result, in recent years a discourse on global ecology has developed that is largely devoid of any consideration of power relations, cultural authenticity and moral choice; instead, it rather promotes the aspirations of a rising eco-cracy to manage nature and regulate people worldwide.
>
> (Sachs 1993: xv)

And from a very different political viewpoint, Walter Kaufman similarly questioned the scientific basis of UNCED:

> As environmentalists began their massive public relations campaign for the Rio summit, many scientists recognized that propaganda was again to be passed off as science. They saw that committees meeting to prepare the summit's agenda had few scientists, and those who were included were

seldom specialists in the area being discussed . . . the Rio summit would have
noble motives, eloquent speeches, and a distinct unscientific bias.

(Kaufman 1994: 84–85)

Privileging interests over values and facts is, however, rarely thoroughgoing, as to
be entirely consistent one would have to cynically dismiss one's own perspective
as little more than interests-based. Indeed, though Sachs questions the dominant
values underlying Earth Summit, he still points to the possibility of "cultural
authenticity and moral choice"; and while Kaufman dismisses the factual basis
underlying UNCED, his very critique suggests that good science was ignored.
Thus, the cynical eye informed by this perspective seems generally to gaze
outward rather than inward; it is a perspective of the other.

The error with this perspective lies not in its truistic assertion that interests play
a major role in the ways people come to moral terms with global environmental
change; indeed, we all legitimately speak out of certain interests, and this must be
recognized in designing more participatory approaches to crafting policy
responses. Rather, the problem lies in the extent to which the interests-based
perspective suggests a reductionistic attitude toward facts and values while retain-
ing its own moral voice. From this cynical extreme, the politics of the other
washes out any hope of epistemological or ethical clarification, while the politics
of the self are somehow more genuine.

A moral earth

Rather than attempt to purify our understandings of values as from the factist
perspective, or celebrate some unfettered normativity as from the moralistic perspective, or smother values entirely (save, perhaps, our own) as from the cynical
side of the interests perspective, I offer another way to think of values in the
context of global environmental change. Some geographers have argued that our
identities and our ideas of the world around us are linked. Anne Buttimer has
observed that "*humanus* literally means 'earth dweller'" (Buttimer 1993: 3) –
that, whether or not one enjoys looking at maps or celebrates Earth Day, our lives
are fundamentally geographical. Robert Sack argues that our lives are fundamentally geographical because our identities are constituted in relation to the real
and imagined places we inhabit – hence the title of his most recent work, *Homo
Geographicus* (Sack 1997). And Clarence Glacken has argued that conceptions of
nature through Western history abound with references to the fitness of the earth
as a home for humans, an "orderly harmonious whole, fashioned either for man
himself or, less anthropocentrically, for the sake of all life" (Glacken 1967: 36).

The geographical tradition of conceptions of self-in-relation-to-the-world pre-
dates the last few decades, and certainly has been expressed in far stronger terms
than the above. The French geographer Elisée Reclus began his first volume of
L'Homme et la Terre with the statement, "L'Homme est la nature prenant con-
science d'elle-même" ("Man is nature becoming self-conscious") (Reclus 1905),
which one commentator summarizes as the argument that "Humanity must come

to understand its identity as the self-consciousness of the earth, and that it must in
its own historical development realize the profound implications of this identity
(Clark 1997: 119).

Even more grandiose notions have emerged from the geographical tradition. In
1834, a man well into his sixties wrote a letter to a close friend, tracing the
contours of an unfulfilled dream:

> I have the crazy notion to depict in a single work the entire material universe,
> all that we know of the phenomena of heaven and earth, from the nebulae of
> stars to the geography of mosses and granite rocks . . . It should portray an
> epoch in the spiritual genesis of mankind – in the knowledge of nature. But it
> is not to be taken as a physical description of the earth: it comprises heaven
> and earth, the whole of creation . . . My title is *Cosmos.*
>
> (quoted in Botting 1973: 257)

This was no idle dreamer. The man was Alexander von Humboldt (1769–1859),
an explorer and naturalist, famous in his time, a friend of Jefferson and Goethe,
author of, among other publications, a 30-volume chronicle of his research travels
to Central and South America, and, by many accounts, a founder of modern
geography. Humboldt's vision was "impressed . . . with the analytical potential of
Enlightenment science, and equally convinced of the values proclaimed by its
romantic critics" (Buttimer 1993: 170). As his biographer, Douglas Botting,
stated, "Humboldt saw nature as a whole and man as part of that whole" (Botting
1973: 259).

When I suggest that we understand ourselves as inhabitants of a moral earth I
am simply turning on its head the longstanding geographical tradition of viewing
human identity in relation to the earth. It is the realization of a world already
laden with moral meanings, and not anthropomorphic excess, that leads me to
suggest this as an alternative point of departure for reflecting on the ethics of
global environmental change. The earth is a moral place by virtue of being
inhabited by people who have acted in certain morally-relevant ways, and justified
their actions and condemned others with reference to existing moral notions.

What are the implications of this perspective? The most fundamental theoretical implications are twofold. First, facts and values are not as separate as the two
distinct terms imply (Proctor 1998). Given the dual, ontological/epistemological
contribution to a "fact" as suggested above, this means that reality, knowledge,
and ethics are intertwined. This very important point is one a geographer would
not miss – and indeed, the connections between ethics and ontology are the focus
of Nicholas Low and Brendan Gleeson's essay elsewhere in this volume (Chapter
3), as are the connections between ethics and epistemology in Tim Unwin's essay
(Chapter 19). The second major implication is the positional and relational notion
that values connected to our outer worlds (the earth) and inner worlds (our
identities) are joined. As Paul Roebuck's essay in this volume (Chapter 2) argues,
the Enlightenment-derived objectivist notion of nature as manipulable other has
driven a wedge of incomprehension between our senses of self and the world.

One brief example should illustrate the values-embeddedness of global environmental change, and its relations to senses of self. We need look no further, in fact, than the value-added modifier "global." Many commentators on global environmental change seem to feel as if this were the one true moral scale of things, where all subglobal others pale in comparison:

> Immersed in the world ecosystem, we have not grasped the meaning of our true environment. We have fragmented our surroundings, and constructed fields of knowledge, disciplines, educational systems, departments – an entire culture of arts and sciences – around the fragments. But revelations from outer space of the environmental whole, interpreted by ecological understanding, are challenging age-old ideas of human preeminence and purpose that have brought the world to the brink. The unity is the ecosphere – literally the home-sphere, the global "being" whose inseparable physical/biological parts have evolved together for 4.6 billion years.
>
> (Rowe 1991: 331)

Critics have argued, however, that the very sense of "global" implied in global environmental change is a logical conclusion to the longstanding process in which modernity has conferred a separated view of the earth, a view of people as not so much inhabitants as onlookers (Ross 1991: 221; Ingold 1993; Cosgrove 1994). Indeed, even von Humboldt has been accused of contributing to this tradition of "world-as-exhibition" (Gregory 1994: 40). Separating our identities from the world, separating facts from values, our concern over global environmental change thus reproduces the flawed perspectives we invoke to make moral sense of it. The perspective of a moral earth aims to make explicit the moral threads that weave through our existing webs of significance, a project which must necessarily precede asking how these webs could be woven differently.

Return

How would the forests of southern Oregon now appear to me from the perspective of a moral earth? They probably would not look markedly different, at least at first glance. I would still see the evidence of transformation – by fires, insect infestations, and certainly logging – and variable regrowth, the interwoven natural and human history of the landscape. But I believe I would see more than the facts of history: I would also see in the landscape historical traces of the moral imaginations of people who lived there, as well as people who lived far away. Those who lived close by worked the forests, lived in community with those who did, or otherwise built and justified their identities around the rapidly-transforming landscape. Those who did not, those whose only connection to southern Oregon was as a source of resource-based capital or building material, nonetheless had in their moral imaginations of progress or well-being or stability a material thread tying them to the region. I would, from this perspective, see myself in a different light as well, as a person whose identity and moral sensibilities have been shaped by growing up in, then moving far away from, this small forest community. I would thus become more aware that the good and bad I see in human-induced environmental change reflects in complex ways the good and bad I see in myself.

One possible objection to the perspective of a moral earth arises from the consideration that notions such as good/bad and right/wrong sound like platitudes to many of us; why, then, need we over-moralize everything? My response is that, along with less ideal-typical polarities – justified/unjustified, honorable/ reprehensible, understandable/inadmissible, sly/devious, and so forth – these are the moral tensions that accompany our practical engagement with the earth and with each other. Our moral imagination is inescapably a part of our earthly lives. To conceive, then, of a moral earth is not so much to look at everything with colored glasses as to notice the moral threads running through our practices and their traces on the earth.

I, for one, find a great deal of rich content for ethical analysis in the existing moralities I observe around me. This point of departure for doing ethics is not perhaps as intellectually glamorous as others – indeed, I often feel as if I remain on the ground long after others have done loops and spirals around me (perhaps this is why I am a geographer and not a philosopher). To readers who are still looking for the final answer – even a tentative answer – on whether, for whom and in what ways global environmental change is a good or bad thing, I can only hope that they will take the next steps in this direction. My humble contribution is to observe that this question can never be posed, much less answered, in a moral vacuum: the fact-values, value-facts of global environmental change are simply too compelling for us to ignore this question or, more generally, to ignore our presence on a moral earth.

Acknowledgments

James Birchler, Helen Couclelis, David M. Smith, Billie Lee Turner, and James Wescoat kindly commented on an earlier draft of this essay; any remaining obscurities, whether intentional or otherwise, are fully my responsibility.

References

Allsopp, B. (1972) *The Garden Earth: The Case for Ecological Morality*, New York: William Morrow and Company, Inc.
Baarschers, W. H. (1996) *Eco-facts and Eco-fiction*, London: Routledge.
Bhaskar, R. (1975) *A Realist Theory of Science*, Leeds: Leeds Books.
Botting, D. (1973) *Humboldt and the Cosmos*, New York: Harper and Row Publishers.
Buttel, F. H., Hawkins, A. P. and Power, A. G. (1990) "From limits to growth to global change: constraints and contradictions in the evolution of environmental science and ideology," *Global Environmental Change* 1: 57–66.
Buttimer, A. (1993) *Geography and the Human Spirit*, Baltimore: The Johns Hopkins University Press.

Callicott, J. B. (1994) *Earth's Insights: A Survey of Ecological Ethics from the Mediterranean Basin to the Australian Outback*, Berkeley: University of California Press.

Clark, J. (1997) "The dialectical social geography of Elisée Reclus," in A. Light and J. M. Smith (eds) *Philosophy and Geography I: Space, Place, and Environmental Ethics*, Lanham, MD: Rowman and Littlefield Publishers Inc.

Cosgrove, D. (1994) "Contested global visions: One-World, Whole-Earth, and the Apollo space photographs,' *Annals of the Association of American Geographers* 84: 270–294.

Daly, H. E. and Townsend, K. N. (eds) (1993) *Valuing the earth: Economics, ecology, ethics*, Cambridge, Mass.: The MIT Press.

Easterbrook, G. (1995) *A Moment on the Earth: The Coming Age of Environmental Optimism*, New York: Penguin Books.

Ehrlich, P. R. and Ehrlich, A. H. (1991) *Healing the Planet: Strategies for Resolving the Environmental Crisis*, Reading, Mass.: Addison-Wesley Publishing Co., Inc.

Fleagle, R. C. (1994) *Global Environmental Change: Interactions of Science, Policy, and Politics in the United States*, Westport, Conn.: Praeger Publishers.

Glacken, C. J. (1967) *Traces on the Rhodian Shore: Nature and Culture in Western Thought from Ancient Times to the End of the Eighteenth Century*, Berkeley: University of California Press.

Gore, A. (1993) *Earth in the Balance: Ecology and the Human Spirit*, New York: Penguin Books.

Gregory, D. (1994) *Geographical Imaginations*, Oxford: Blackwell Publishers.

Harrington, R. F. (1990) *To Heal the Earth: The Case for an Earth Ethic*, Surrey, BC: Hancock House Publishers.

Herrick, C. and Jamieson, D. (1995) 'The social construction of acid rain: Some implications for science/policy assessment,' *Global Environmental Change: Human and Policy Dimensions* 5: 105–112.

Houghton, J. T., Jenkins, G. J. and Ephraums, J. J. (eds) (1990) *Climate Change: The IPCC Scientific Assessment*, Cambridge: Cambridge University Press.

Ingold, T. (1993) "Globes and spheres: The topology of environmentalism" in K. Milton (ed.) *Environmentalism: The View from Anthropology*, London: Routledge.

International Council of Scientific Unions (1986) "The international geosphere–biosphere program: A study of global change" Ad Hoc Planning Group on Global Change Report prepared for the ICSU Twenty-First General Assembly, Berne, Switzerland, 14–19 September

Jamieson, D. (1996) "Ethics and intentional climate change", *Climatic Change* 33: 323–336.

Kaufman, W. (1994) *No Turning Back: Dismantling the Fantasies of Environmental Thinking*, New York: Basic Books.

McLaren, D. J. (1991) "Preface", in C. Mungall and D. J. McLaren (eds) *Planet Under Stress: The Challenge of Global Change*, Toronto: Oxford University Press.

Moore, P. D., Chaloner, B. and Stott, P. (1996) *Global Environmental Change*, Oxford: Blackwell Science.

Murphy, C. M. (1989) *At Home on Earth: Foundations of a Catholic Ethic of the Environment*, New York: The Crossroad Publishing Company.

Myers, N. (ed.) (1993) *Gaia Atlas of Planet Management* London: Gaia Books Ltd.

162   *Ethics and nature*

Proctor, J. D. (1998) "Geography, paradox, and environmental ethics," *Progress in Human Geography* 22: 234–255.
Rasmussen, L. L. (1996) *Earth Community, Earth Ethics*, Maryknoll, NY: Orbis Books.
Reclus, E. (1905) *L'Homme et la Terre*, Paris: Librairie Universelle.
Rolston, H., III (1996) "Earth ethics: A challenge to liberal education" in J. B. Callicott and F. J. R. da Rocha (eds) *Earth Summit Ethics: Toward a Reconstructive Postmodern Philosophy of Environmental Education*, Albany, NY: State University of New York Press.
Ross, A. (1991) *Strange Weather: Culture, Science and Technology in the Age of Limits*, London: Verso.
Ross, S. D. (1995) *Plenishment in the Earth: An Ethic of Inclusion*, Albany, NY: State University of New York Press.
Rowe, J. S. (1991) "Summing it up" in C. Mungall and D. J. McLaren (eds) *Planet Under Stress: The Challenge of Global Change*, Toronto and New York: Oxford University Press.
Sachs, W. (1993) "Introduction" in W. Sachs (ed.) *Global Ecology: A New Arena of Political Conflict*, London: Zed Books.
Sack, R. D. (1997) *Homo Geographicus: A Framework for Action, Awareness, and Moral Concern*, Baltimore: The Johns Hopkins University Press.
Smil, V. (1994) *Global Ecology: Environmental Change and Social Flexibility*, London: Routledge.
Stone, G. C. (ed.) (1971) *A New Ethic for a New Earth*, Andover, Conn.: Friendship Press.
Strong, M. F. (1992) "Environment and development: The United Nations road from Stockholm to Rio," *Interdisciplinary Science Reviews* 17: 112–115.
Subcommittee on Global Change Research (1997) *Our Changing Planet: The FY 1998 U.S. Global Change Research Program*, Washington, DC: SGCR, Committee on Environmental and Natural Resources Research of the National Science and Technology Council.
The Economist (1995) "A problem as big as a planet," *The Economist* 333: 83–85.
Wescoat, J. L., Jr (1993) "Resource management: UNCED, GATT, and global change," *Progress in Human Geography* 17: 232–240.
Weston, A. (1994) *Back to Earth: Tomorrow's Environmentalism*, Philadelphia: Temple University Press.
Wolf, E. C. (1996) *Race to Save the Planet: Study Guide*, Belmont, Cal.: Wadsworth Publishing Company.
Wynne, B. (1994) "Scientific knowledge and the global environment," in M. Redclift and T. Benton (eds) *Social Theory and the Global Environment*, London: Routledge.

12 Natural and unnatural wars

Sheila Hones

The good, the bad, and the natural

Two useful books about American history and shared memories – *The Good War* and *The Bad War* – sit next to each other on my bookshelf. The titles worry me. I swivel round in my chair and look at them, and I wonder about other American wars and the way we label them. Was it a revolution, or a war of independence? A civil war, or a rebellion? Looking at my bookshelf, I think about how we live in the wake of past wars and participate in present wars, calling them "good" or "bad," "civil" or "rebellious," crazy, unavoidable, or "natural." I wonder about the ways we have of writing and remembering the past, and the effects they have on our ways of living in the present. In this chapter, I'd like to think about the effects of associating national history with a view of nature as an intelligently-directed process tending towards stability and the establishment of geographical order. I'd like to think about how this leads to the concept of "the natural war," a concept that may not only close down opportunities for reconsideration of the national past but may also inhibit reform-oriented questioning of the national present.

Tim Unwin, in his contribution to this collection (Chapter 19), argues that we need to contribute to social reform through the study and criticism of aspects of our world "with which, for whatever reason, we as individuals feel ill at ease." Popular versions of national history which rely on metaphors of natural development, are, for me, one such aspect, and so this chapter is intended as a constructive critique of the idea of "natural" and "unnatural" wars in national history. The chapter is closely focused on a case study of a particular version of national history shared by one group of Americans, but its wider relevance comes from the suggestion that a currently inhibiting sense of the inevitability of national history and of particular forms of national identity may rest in part on ways of thinking which we have inherited from the past.

An influential group of Americans writing in the 1880s had, in their version of national history, a "good" war that was part of a natural process and a "bad" war that was the result of an unnatural disruption. The "good" war was remembered in images of stable nature; the "bad" war was remembered in images of disruption and disorientation. For this group of people, ethical judgment about war and history was linked to assumptions about the meaning of "nature," and about the relationship between "nature" and national identity. In his contribution to this collection (Chapter 13), Jeremy Tasch illustrates the enormous power of the normative appeal to nature by reference to the way in which the Soviet government was able to dictate to its citizens "who they were, what they wanted, and what they should become," in part through its control of what could be considered "natural." Many geographers would agree with Neil Smith's point that " 'Nature' is an established, trenchant and powerful weapon in 'western' discourse" because of the way in which the identification of some events as "natural" and others as "unnatural" appeals to the "full authority of an inevitable, suprahuman nature" (Smith 1996: 41). As Smith explains, the power of this appeal relies

> on the slippage from the externality to the universality of nature. The authority of "nature" as a source of social norms derives from its assumed externality to human interference, the givenness and unalterability of natural events and processes that are not susceptible to social manipulation. Yet when this criterion of "naturalness" is reapplied to social events, processes and behaviours, it necessarily invokes the assumption of a universal nature, a sufficient homology between human and nonhuman natures.
>
> (ibid.)

The "natural war" of national development is thus a dangerously powerful concept, first because, in its appeal to nature, it sidesteps the ethical engagement involved in the idea of the "just war" (Walzer 1992; Graham 1997). While the "just war" can be defined as a war fought in a just cause and carried out within a set of ethical boundaries, the "natural war" is defined not in relation to pre-war purposes or to war-time conduct but retrospectively, in relation to a post-war vision of the nation as an organic whole. Such a war is "natural" not by reference to any concept of natural rights or natural justice, but because of the role it plays in a narrative of national identity articulated through the metaphor of organic unity. In this way, the identification of "natural" and "unnatural" wars rests on a view of the nation as a form of ecosystem, erasing the distinction between a country and its people critical to just-war theory (Graham 1997: 103). Where "just war" theory problematizes and demands moral choice, the discourse of the "natural war" simply naturalizes.

This case study of "natural" and "unnatural" American wars is based on a reading of the American periodical *The Atlantic Monthly*, 1880–1884, a period when the magazine still occupied a position of considerable cultural authority (Sedgwick 1994). Following a convention in the discussion of periodical literature, the study treats ten volumes of the *Atlantic* as a single text with a "corporate author," and also reads the text as the characteristic voice of its implied audience (James 1982). As Marc Brosseau has suggested, geographical work based on literary evidence can be most productively focused on the analysis of details of textual expression, "its singular use of language," in an investigation of the way in which the text "generates norms, particular models of readability, that produce a particular type of geography" (Brosseau 1994). In line with this approach, my description of the connection that exists in the magazine between a particular version of "nature" and "the natural" and a particular ethical interpretation of American history relies on an analysis of the figurative language and the narrative structures which hold the two (nature and history) together within the text. This study of figurative comparisons and conventional narrative forms provides us with access to connections the text assumes between geographical description and ethical judgment.

As I have suggested above, it seems to me that the way in which the text creates and then uses its particular versions of two culturally iconic wars is significant not only because it clarifies a particular historiographical moment or a culture-specific view of society–nature relations, but also because it uncovers the roots of some still current assumptions about US history and national identity. Writing about geographical studies of disability, elsewhere in this collection (Chapter 16), Rob Kitchin suggests that one of our concerns in practicing critical geography should be the extent to which our work promotes people's control over processes that affect their daily lives. If we shift and widen the focus here to include all people affected by normative versions of national history and identity, I'd like to suggest that one way to participate constructively in the work of critical geography might be to subvert the association of national history with natural or geo-spatial inevitability, and thus promote opportunities for as wide a range of people as possible to take control over their understanding of history, and to participate in the creation of their shared pasts, presents, and futures.

Metaphors and history

Suggesting even in the title of his 1884 essay in the *Atlantic*, "The Embryo of a Commonwealth," that the revolutionary beginnings of the United States were in fact organic and natural, Brooks Adams considers the "general law" that permanent governments require long gestatory periods.[1] The American government, he concludes, is destined to be permanent because it is "not the ephemeral growth of a moment of revolution" but rather "the reincarnation of a longstanding tradition." The "germ" of the US Constitution can be found "at the dawn of English history."[2] This image of "germination," the figurative incarnation of the independent states as natural growths or offshoots of the parent plant, is the key to the *Atlantic* version of the revolutionary war as anything but revolutionary.

A writer discussing in 1883 the potential of American history for providing plot material for the stage dismisses the revolution as "singularly lacking in dramatic properties." We are misled, it seems, by its popular name: "the American Development would be a truer phrase." While the French Revolution was a "real" revolution, that "shook to the center an old order of things," the American Revolution merely "set the seal to a foregone conclusion." Existing political relations were "disturbed," but "new ones of a higher order" were already "germinant." Thus the "American Development" was simply part of a natural progress: the new order had already been seeded by the old.[3]

George E. Ellis's 1884 article on Thomas Hutchinson, the last colonial governor of Massachusetts, relies on this organic image in its description of the inevitability of American independence.[4] Even Governor Hutchinson, Ellis argues, was aware that the colonies were moving naturally towards independence, and for that very reason he was wrong in accepting the governorship, for he must have known "that civil as well as religious independence of the mother country germinated in the first field-planting of the colony, and had been bearing and resowing its own crops, strengthening on their stalks throughout the generations."

This view of American independence as a natural growth has been confirmed, Ellis says, by "the sagacious judgment" of opinion "on both sides of the water during the last two score of years." These "consenting minds" have come to agree:

> that at the period of our revolutionary strife the fitting time had *nearly* come for the colonies to drop away from the mother country by a natural, unaided, unimpeded ripening, as mature fruit drops from the tree. Some idealists have ventured to assert that this process of severance might have been peaceful and propitious. We have emphasized the adverb *nearly*, which in its place is significant. For the question left now is whether the process, a little premature, was violently hurried by one party, by pounding and shaking the tree, to anticipate the fruit before it was ripe; or whether the process was blindly and perversely, and also violently, resisted by the other party, in an obstinate refusal to allow the natural and the inevitable.
>
> (ibid.)

Ellis admits that the fact "that the patriot party did throw stones at the tree to anticipate and hurry the severance of the fruit would seem proved by several of the incidental accompaniments of our rupture with the mother country." But he argues that:

> when we, from our side, look across the water to judge if the king and his ministry were not stupidly and obstinately setting themselves to retard and baffle the natural dropping off of the colonies that had come to full age, we seem to see evidence of extreme obstinacy and folly in their course.
>
> (ibid.)

Thus, the American Development: stone-throwing in an orchard. The *Atlantic's* secession, on the other hand, could never be read as a "natural dropping off," no matter how much the image was stretched to allow for pounding and shaking. The magazine's organic metaphor argues in quite the opposite direction, for in 1880 it takes the "friendly feeling" that Confederate soldiers held towards the Union ex-president Ulysses S. Grant as evidence of the fact that "the restored Union has deeper roots in the South than Northern people of strong sectional feeling imagine."[5] If the Union has roots in both South and North, then union is the nation's natural state, and any attempts to divide the national territory is equivalent to an uprooting and a disruption in natural cycles of growth.

Thus, the civil war: a disorientation and a disruption. And so, as the conflict between North and South had worsened, President James Buchanan had first discovered that the "firm rock on which he had always rested had crumbled beneath him," and he had then "found himself drifting helpless and alone on the seething waters of secession and civil war."[6] The collapse of the Union resulted, figuratively, in radical disorientation, a metaphorical construction that reflected the *Atlantic* view that the secession movement was the result of Southern leaders having simply lost their bearings – "nothing appeared to them as it really was, nothing had its true proportions; they lost in this way even the capacity to recognize existing facts."[7] Even as late as 1884, the pre-war years were "so murky with the tempests of passion and hate which raged through them," that it was "even now difficult to see them clearly."[8] This kind of confusion was both destructive of "place" and "out of place" itself, for "the normal condition of a healthy society is not the fierce conflict engendered by a great moral issue any more than a thunderstorm is the normal condition of any tolerable climate."[9]

The attempt at secession is presented most often in the *Atlantic* in the image of a gulf, or chasm, and is thus related first to the geographical separation or disruption of the organic whole of the union (nation) and second to a historical disjunction created in the continuity of national memory. Disturbingly, the chasm created by the national "earthquake" creates a metaphorical dislocation that is both spatial and temporal at the same time. The nation's geographical integrity is shattered as South divides from North, and its historical continuity is broken as past is cut off from present. The civil war was thus "a seething gulf" that separated North and South, then and now; it "had drawn [a] red line which had the effect of giving an air of obsoleteness to everything on the former side."[10] The antebellum period, only 20 or so years in the past, had become "history." "The intervening war," we are told, "has riven a chasm so deep and wide between that time and this that [previous] events . . . belong to a different era."[11] While *Atlantic* writers in the 1880s might sense "a postwar rush of optimism in New England about the national future," they were nonetheless aware "that the war marked a sharp break with the settled and traditional life of rural, small town America." (Sedgwick 1994: 95). At its most dramatic, the gulf/chasm image of the war is both a dividing line and a black hole. Within one article, for example, we find North and South "moving steadily in opposite directions, the gulf opening and widening between them," while the two opposing sides are at the same time *facing* a "chasm," "toward which they were blindly hastening."[12] To make sense of this odd image we have to take into account both spatial and temporal dimensions: the two halves of the nation are pulling apart, and so the united history of the American states is about to fall into the gap.

A short story that appeared in the February 1881 issue of the *Atlantic* provides an extended metaphorical version of the disruptive, disorienting effects the war had on the nation. But the narrative of the North–South love story finally carries the reader through the chaos of the war to a safe landing on solid ground. This kind of "only-just-historical" romance functioned as one of the ways in which the nation reassured itself that the war was over and that reconciliation and reunification had been achieved. These narratives move comfortingly from disruption and crisis through resolution to happy ending, often ending with the symbolic union of a marriage. "Is Anything Lost?" is one of the war's "stupid love stories" that were criticized for getting in the way of serious oral history, but after all they had their own serious effects; as one writer remarks, it is "an encouraging sign of progress" to have "reached (the) story-telling stage."[13]

"Is Anything Lost?" opens in Richmond, on the eve of its evacuation. Annie Somerville's Confederate fiancé is leaving, just as her future husband, a Union soldier, is arriving. Annie's fiancé has found himself in possession of a bag of army gold; the Confederate army is in total disarray – "no man is in his right place to attend to anything" – and so he has decided that his best course of action is to give it to his fiancée before he evacuates with the retreating army. She refuses it, but he buries it at their favorite meeting place, telling her that she can dig it up later. Annie never finds it. Their old meeting place is taken over by the Union forces; by the time she returns, it is a building site. Completely disoriented, like the Southern leaders, she fails to recognize the significant tree: she has "mistaken her moorings" because "it was impossible to find anything out there."

The Southern gold is lost forever, but Annie finds other hidden things. At first, as her Confederate fiancé leaves her, amid disorientation and disruption, she finds tragedy: "her trembling feet had slipped through this outer earth of bloom and verdure and had struck the heavy layer of tragedy which underlies it." But in the end, reoriented, she finds her lost happiness in marrying the Union officer who had secretly replaced her lost gold out of his own funds: " 'My happiness was, after all my grieving, only hidden like the gold,' she mused, 'and God brought it back to me.' " Annie learns to cross the "gulf" that Major Graham believes she has created in her mental map of North America. "In your geography," he tells her, "the large, though unimportant tract of land north of Mason & Dixon's line is labeled, 'The Yankee States – inhabited by a horrid race,' " separated from the South by "a great gulf fixed on your map between them and you." The story achieves its happy ending in 1876, a significant centenary year, with the newly-married couple arriving in Philadelphia at the end of a cross-country railroad journey. Major Graham's insistence that he "can't help feeling sorry about the 'gulf' " has finally persuaded Annie to cross it both literally and figuratively, and the narrator can tell us that in 1876 "one should not be surprised at any possible encounter or combination."

This story provides a large and reassuring answer to the post-war question "Is Anything Lost?" Some things have indeed been lost forever; but other things have been rediscovered. At the start of the story, Annie's "world was under-going...convulsive, transforming agonies," and the women left behind in Richmond were telling each other that "to-night the ground does seem to be shaking under our feet." But the disorientation that characterized the establish-ment of the narrative problem is replaced by its conclusion with images of reorientation and shared direction: with the opening of the Centennial

Exposition in Philadelphia, "extremes were meeting, en route to that city, every day now." The organic unity that should characterize the nation has been restored.

Virtue and intelligence

The *Atlantic*'s view of "nature" as a stable, harmonious system directed by some form of moral intelligence fits in well with national post-war emphases on patriotism, centralized government, and continental unity. It also, critically, makes the assumption of governmental responsibility by an educated, moral elite seem not only natural but imperative. The definition of nature that lies behind the *Atlantic*'s narrative of natural and unnatural wars thus not only created a particular version of US history but also endorsed a particular, elitist view of democracy and national leadership. The emphases on centralized government and national unity were responses not only to the past disruptions of the civil war period, but also to a new range of perceived threats to national coherence. This time the earthquakes included labor disputes and class conflict; the rising tidal waves were of immigrants speaking alien languages and the state-educated "masses" waving ballot papers. The three great Chicago crises that Carl Smith discusses in his *Urban Disorder and the Shape of Belief* – the fire of 1871, the Haymarket bombing of 1876, and the Pullman strike of 1894 – were all, in their turn, as Smith points out, linked with the threat of the "foreign" and the "unnatural" and perceived as new battles for the Union (Smith 1995). In the early 1880s the *Atlantic* could feel the ground shaking underfoot again, and again felt the need for the nation's intelligent leadership to take control.

At this moment of renewed threat, the theme of the nation's continuity with the history of its parent country was carefully rehearsed, and the identification of national leadership with the descendants of those who had fought in the war of American Development was reinforced. This reaction not surprisingly strengthened the *Atlantic*'s sense of organic identity at the expense of democratic inclusiveness. Common origins and continuing similarities with the modern British nation were insisted upon to the point that it could seem only natural that new immigrants and newly-emancipated slaves be marginalized if not regarded as overtly hazardous. Throughout the magazine, writers placed themselves within "the English race," made an inclusive "we" of the British and the Americans, implied (more or less directly) that (we) Americans still retain traces of "the colonial spirit in our modes of thought," and struggled to find an inclusive term for English speakers: "Anglo-Saxon," "English-speaking," "the Englishry."[14]

The conservative reading of the revolutionary war (as not really revolutionary at all) had played a critical role in the antebellum period in resolving the dilemma of a nation founded on the "consent of the governed" and created by rebellious resistance faced in turn by a non-consenting section that wanted to secede. Modern popular history may have made the two conflicts seem very different to us, but their similarities were deeply troubling at the time. Some people believed, for example, that the South had the same rights as the American colonies had had in 1776. As one northern abolitionist put it: "according to the fundamental principles of our government, the secessionists are right in their main principle." (Fredrickson 1993: 58) Pro-Union Northerners found their way out of this problem in various ways, one of which was simply to deny that it had really been a revolution at all. Charles J. Stillé insisted in 1863 that the founding fathers "had really been conservative Englishmen," and that "the glory of our system is, that there is nothing revolutionary about it." (Fredrickson 1993: 142)

And so, in the *Atlantic* version of American history, the revolution was not revolutionary and the civil war had not primarily been fought out of a moral aversion to slavery. Writing about the upcoming 1884 elections, a writer can dismiss the former slaves – "the public mind is no longer interested in the negro race" – and build his criticism of the Democratic Party on the point that it is the "opponent of an efficient centralized government."[15] Thus, while in retrospect the anti-slavery movement may have become regarded as "the cardinal fact in the spiritual life of this people" marking the founding of "the regenerated nation," it was recognized within the *Atlantic* that at the time of the Appomattox surrender the "common purpose of the most diverse minds" had simply been to "restore the States lately in rebellion to their normal relations to the Union."[16] In the 1880s, national unity was paramount: "local tradition impedes progress," as one writer put it.[17]

The *Atlantic* connected this post-war re-establishment of national unity with a natural tendency towards centralized intelligence and purpose, and connected *this* to assumptions about the political responsibilities of the educated elite. George Fredrickson shows by reference to the work of the historian John W. Draper how an emphasis on centralized intelligence developed very quickly after the war. The lesson of the conflict had been clear: it had been "a punishment for. . .fighting against the laws of nature," and it had "demonstrated the scientific law that 'centralization is an inevitable issue in the life of nations,' or, put another way, that 'all animated nature displays a progress to the domination of a central intelligence.'" (Fredrickson 1993: 201) The central intelligence, that was divine in nature, was to be provided in the nation by people like the readers of the *Atlantic*.

That the elite should accept its responsibilities was imperative – the next step in the natural progress of the nation that had just been preserved by the triumph of the Union forces in the civil war. The Declaration of Independence, the justification of the revolutionary stage of national development, had been "framed for a highly intelligent and virtuous society."[18] Social changes associated with emancipation and immigration meant, for the *Atlantic*, that society as a whole was rapidly becoming neither highly intelligent nor virtuous. As the nation grew and diversified, the need for a centralized intelligent governing elite seemed to become more and more apparent. Writers argued that "the successful shaping of our national life depend[ed]. . .upon the mutual understanding between different parts of the country," and that a "steady progression" towards the realization of "a broad ideal of national authority" meant that the theory of states' rights was now only "a sentiment associated with the beaten rebellion."[19] National authority was needed, and that authority had to belong to the elite, for while "a better understanding between the different stations and conditions of society" was as necessary as that between different parts of the country, it still had to be recognized "how low the capacity of the masses – the public-schooled [state-educated] masses – still [was] for right thinking."[20]

It is significant that when the "American Development" and the "Southern Rebellion" themselves came to be used in as images in the *Atlantic*, the goodness of the "good" war became associated with the politically responsible behavior of the elite classes, and the badness of the "bad" war became associated with the unthinking actions of mass human "ammunition." There was considerable slippage between the idea of the "natural" war and the idea that those who fought it were the literal progenitors of the nation's "natural" leaders. "We have become accustomed," R. R. Bowker remarks, in 1880, "to look upon a political 'campaign' as the grand battle of two opposing armies, with their officers and their generals disdainfully regarding their privates as ammunition to be fired against the enemy, and nothing more." Bowker argues that, on the contrary, "citizens go to the polls one by one; each casts his ballot by himself and for himself. . .and the very act of voting is the invitation to use his individual judgment as between the opposing forces." Out of context, this redefinition seems democratic, suggesting that in an election each citizen can make up his own mind, while in a battle "if he deserts he is rightly shot." But in the end, the emphasis on direct historical continuity which allows for the *Atlantic* version of the War of Independence as the "good" war means that while Bowker's title is "The Political Responsibility of the Individual," his "individual" is in fact a property-owning son of the founding fathers, the "voter of intelligence and education." "The independent voter must keep at his work," Bowker concludes, "as his grandsires fought the first battle of the Revolution, each from behind his own tree."[21] We are back in the orchard again, although this time, the Americans are not even throwing rocks, they are simply involved in an energetic form of democracy – voting, we might say, from behind trees.

The Revolution, far from being in any sense really revolutionary, was taken by the *Atlantic* to be a natural step in the historical development of an American nation characterized by steady progress and geographical unity. While it was the dropping of the ripe fruit from the tree, the fruit that dropped seeded a strain of true Americans who would form the moral intelligence of the nation. "The truly American spirit is only in the descendants of men who founded our institutions," M. H. Hardaker explained, in 1882, while it was "a matter of statistics that the primitive, self-reliant, and self-respectful revolutionary stock bears a steadily diminishing relation to that of more recent importation, and of inferior quality."[22]

The Southern attempt at secession was not revolutionary either: it was an unnatural rebellion with all the disorienting and destructive effects of a major earthquake. But the moral regeneration that followed the civil war buried the bad, unnatural desire for separation beneath a healthy, natural organic unity in the same way that the "immense and kindly recuperative force" of nature always in the end lets the "grass grows over her extinct volcano."[23] And this moral regeneration, this national expression of natural intelligence, depended entirely upon the descendants of those who had fought the earlier revolution from behind their own trees, for "even the foreigners who fought in our civil war," M. H. Hardaker insists, "were not lifted to a moral conception of its issues."

The War of Independence was thus, for the *Atlantic*, a "natural war" not only in the sense that it facilitated the birth of the organic nation, but also because it was a war fought by the nation's natural leaders, individual members of "a highly intelligent and virtuous society." The Civil War, on the other hand, was "unnatural" not only in the sense that it threatened the organic unity of the nation, but also because not all of the combative participants had been "lifted to a moral conception of its issues." It was initiated by men who had lost all sense of "true proportions" and fought by the masses: it was a war without intelligent direction, and thus unnatural.

Nature, nation and democracy

Two aspects of the *Atlantic*'s understanding of "nature" have a direct relationship to its understanding of the revolution and the civil war. First, it believes in American "nature," or a natural America: in other words, associating national self-fulfillment with a westward movement from coast to coast, it regards an east–west geographical section of the American continent as the national territory of a group of naturally united states. Second, it believes in a natural world – a "universal nature" – that is essentially stable, meaningful and directed by a moral (divine) intelligence. It believes that this natural world provides the model for national government and national progress; the nation is therefore, by analogy, understood to be at best (most naturally) a unified, purposeful state directed by a moral (elite) intelligence.

In the context of his discussion of just-war theory, Michael Walzer explains that it is probably impossible to tell the story of the American war in Vietnam "in a way that will command general agreement." This is partly because the official US version is "on its surface unbelievable." There is no faith in a controlling moral intelligence, and so Walzer is able to discount the official version immediately: "fortunately, it seems to be accepted by virtually no one and need not detain us here" (Walzer 1992: 97). But in the case of the War of Independence, which in its manifestation as the "natural war" is placed outside the bounds of just-war theory, there seems to be much greater popular consensus, founded, in part perhaps, on a lingering faith in the moral leadership of its individual heroes.

The problem with this consensus is that it can easily lead to the assumption that the particular national history resulting from that "natural war" and the particular version of national identity that seems to result from that history, are themselves in turn both natural and inevitable. This is surely problematic. At one time, the War of Independence was remembered as a "natural" war in part because of the way in which it was read as the expression of an intelligent, moral elite; the Civil War, in contrast, was remembered as "unnatural" first because of its association with the absence of intelligent, moral leadership and second because of its image as a war of massed armies rather than thoughtful individuals. The long-term effects of a nexus of meanings in national history that connects "natural" with "good" with "the educated individual," and "unnatural" with "immoral" with "unintelligent mass" are surely worth considering.

Notes

1 Quotations from the *Atlantic Monthly* 1880–1884 are cited in subsequent notes, which will list title, year, volume number, relevant page(s), and the author's name when it was included in the original Table of Contents. "Review" refers to an unsigned book review; "Contributors' Club" pieces come from a collection of unsigned comments and short essays printed at the end of each monthly issue.
2 "The Embryo of a Commonwealth" 1884, 54: 610–619 (Brooks Adams).
3 "American History on the Stage" 1882, 50: 313 (R. Fellow).
4 "Governor Thomas Hutchinson" 1884, 53: 662–676 (George E. Ellis).
5 "Republican Candidates for the Presidency" 1880, 45: 553 (unsigned).
6 "James Buchanan" 1883, 52: 710 (Review).
7 "The Rise and Fall of the Confederate Government" 1881, 48: 410 (Review).
8 "William H. Seward" 1884, 53: 694 (Henry Cabot Lodge).
9 "Julian's Political Recollections" 1884, 53: 562 (Review).
10 "Nathaniel Parker Willis" 1884, 54: 212 (Edward F. Hayward).
11 "Daniel Webster" 1882, 49: 228 (Henry Cabot Lodge).
12 "Webster's Speeches" 1880, 45: 98 (Review).
13 "Is Anything Lost?" 1881, 47: 262–277 (Fanny Albert Doughty); "Studies in the South: IX" 1882, 50: 633; "Studies in the South: V" 1882, 50: 101.
14 French and English" 1884, 53: 436, (Contributors' Club); "Foreign Lands" 1883, 52: 832 (Review); "Richard Grant White" 1882, 49: 220 (E. P. Whipple).
15 "The Political Field" 1884, 53: 129–30 (E. V. Smalley).
16 "Johnson's Garrison and other Biographies" 1881, 47: 560 (Review); "The End of the War" 1881, 47: 396 (Theodore Bacon).
17 "Local Patriotism" 1880, 46: 439 (Contributors' Club).
18 "Equality" 1880, 45: 24 (unsigned).
19 "The Progress of Nationalism" 1884, 53: 701 (Edward Stanwood); "Politics in Southern Life" 1882, 50: 426 (Contributors' Club); "The Strong Government Idea" 1880, 45: 273 (unsigned).
20 "A New Observer" 1880, 45: 849 (Review).
21 "The Political Responsibility of the Individual" 1880, 46: 320–328 (R. R. Bowker).
22 "A Study in Sociology," 1882, 50: 215 (M. H. Hardaker).
23 "Is God Good?" 1881, 48: 536 (Elizabeth Stuart Phelps).

References

Brosseau, Mark (1994) "Geography's Literature," *Progress in Human Geography*, 18: 333–353.
Fredrickson, George M. (1993) *The Inner Civil War: Northern Intellectuals and the Crisis of the Union*, second edn, Urbana: University of Illinois Press.
Graham, Gordon (1997) *Ethics and International Relations*, Oxford: Blackwell Publishers.
James, Louis (1982) "The Trouble With Betsy: Periodicals and the Common Reader in Mid-nineteenth-century England" in Joanne Shattock and Michael Wolff (eds)

The Victorian Periodical Press: Samplings and Soundings, Leicester: Leicester University Press.

Sedgwick, Ellery (1994) *The Atlantic Monthly 1857–1909: Yankee Humanism at High Tide and Ebb*, Amherst: University of Massachusetts Press.

Smith, Carl (1995) *Urban Disorder and the Shape of Belief: The Great Chicago Fire, the Haymarket Bomb, and the Model Town of Pullman*, Chicago: University of Chicago Press.

Smith, Henry Nash (1950) *Virgin Land: The American West as Symbol and Myth*, Cambridge, Mass.: Harvard University Press.

Smith, Neil (1996) "The production of nature," in George Robertson *et al.* (eds) *FutureNatural: Nature, Science, Culture*, London: Routledge.

Walzer, Michael (1992) *Just and Unjust Wars: A Moral Argument with Historical Illustrations*, second edn, New York: Basic Books.

13 Altered states
Nature and place in the Russian Far East

Jeremy Tasch

The geographical imagination is concerned with interpreting how places and the spaces in between are produced and reproduced, constructed and deconstructed, by whom and to what purposes. A place-influenced environmental ethics shares a common interest in how individuals and communities act upon and understand local to global environments. If we consider the Greek origins of the word "ethics," *ētrea*, i.e. "habitats," the historic disciplinary intersection of interest in place is evident. Whether places are known with affection, ambivalence, disdain, or complex combinations, they are distinguished in time and space in the context of their meanings to individuals, social groups, nations, and cultures. These meanings and social relations are undergoing rapid transformations in Primorie, a region within Russia's Far East. The geographical and ethical imaginations meet here, tracing the complex dynamic of how, and by whom, local places within Primorie are being "re-mapped."

Gorky's often repeated aphorism, "Man, in changing nature, changes himself," (1935) cogently expresses the former Soviet state's official view of nature–society relations. Further, the gender bias was deliberate: it was *homo sovieticus*, the new Soviet Man, who would tame taiga and tundra, dam lakes and rivers, and harvest forests to meet the demands of the Soviet economy (Josephson 1997). Both nature and society were considered engineerable, improvable, and both *should* be harnessed to serve the state.

Gorky's statement also underscores the intersection of natural, social, and cultural spheres, relationships which obviously shift and are expressed differently from place to place as well as within places. Different contexts, different places, constrain and enable our actions, and reciprocally, our actions construct and maintain places. The construction and maintenance of places involve decisions about how nature will be exploited, transformed, or preserved. Our actions stemming from these decisions have consequences; the entire process incorporates notions of motive, intent, desire, and one's own/others' interests. Traditionally ethics (a subfield of moral philosophy) has addressed these notions in terms of "rights" and "wrongs" in relations among humans. Geography, in its particular focus on nature–society relations, can redress the incompleteness of this conception of responsibilities. In its appreciation of context – place and space, cultures and multiculturalisms – geographical inquiry explores how physical and social landscapes emerge in a particular space through the operation of various processes functioning at different temporal and spatial scales. How we try to understand and participate in processes which change nature and create new worlds imbued with meanings produced through place-based social processes, involves ethical questions of what is, what could be, and what *ought* to be.

Whether overstating or not, Epstein (1995) convincingly argues that in the case of the former Soviet Union, Soviet-Marxist ideology became "the underlying force of all economic, political, and aesthetic movements in the USSR. . .ideas, not economics, determined material life and produced the 'real'" (1995: 157). Through its direct and indirect control of the principal organs of power, the Soviet state was engaged in dictating to the Soviet citizenry who they were, who they could be, and who they ought to be (Heller 1988). Similarly, in its treatment of nature, the state's goal was "the domination of nature by technology, and the domination of technology by planning, so that the raw materials of nature will yield up to mankind all that it needs and more besides" (Trotsky 1969: 171; see also DeBardeleben 1992). Both nature and citizen were instruments to be used by the state in the production of a new world. Through education, bureaucracy, coercion, terror, the state defined what was "natural," for people and nature, and worked to hold both to this norm. Alternative practices, difference, deviation, the local and the particular, were actively planned against.

Russia's glasnost years marked a denaturalization of Soviet cultural myths. A future-oriented ideology was replaced with a common apocalyptic vision of what was to come. Watching graphically televised images of contemporary crime and contending with first-time experiences of unemployment and currency devaluations, many Russians (from the old Stalinists to the young non-"bisnismenii") are reinventing a mythical, imaginary Soviet Union which corresponds little to less nostalgic reminiscences. From the vantage of a Soviet dormitory in 1990 Moscow, an apartment in Western Siberia in 1994, and life with a host family in Russia's Far East in 1998, I have glimpsed friends and acquaintances living through the disintegration of one-party domination and the loss of a master narrative. Phrases such as "perestroika destroyed our country," and "earlier was better," are popularly repeated cliché-like as summary analysis of the current state of affairs.

While the old system is dying, a new one remains unborn. Out of this "in between" arises a great variety of crises. The legacy of the Soviet past in confluence with the high expectations and disillusionment of the present creates tensions between the simultaneous pursuit of economic development, improved social welfare, and environmental protection. From this "in betweenness" a post-Soviet landscape is co-evolving with a post-Soviet *something*: an ethics in-process and affirmatively un-Soviet. Ethics here is neither a "thing" nor an order nor a regime. As the public detaches from a 70-year controlling ideology, there are examples – such as those discussed in this chapter – of locally-based, internationally-connected alliances attempting to reappropriate places (both physical and socio-political) in ways which express a reclamation of deteriorating ecological and socio-political lifeworlds. This reclamation, this affirmative *something*, may not be single, ordained, or agreed upon. But in the absence of institutional conduits between the state and the public, this *something* is a practice of ethics itself, a form of justice, created between the cracks of Soviet deconstruction and post-Soviet legitimization. As Caroline Nagel discusses (cf. Nagel, this volume, Chapter 10), justice here involves conceptions of what is fair and unfair, and the social arrangements necessary so that members of nature-society are treated accordingly or have recourse to work towards such treatment. The local expressions of re-mappings of local places in the wake of a disintegrated Soviet master narrative *are* a discourse of justice.

Like Russia's symbolic double-headed eagle facing in opposite directions, Pacific Russia is positioned at a cross-roads of political and economic allegiances and reconfigurations. This chapter contends that place, as a contextualization of socio-ecological relations, can serve as a site to locate self and community (e.g. geographical and ethical) within a moral space and through which to navigate post-Soviet anxiety. The first section briefly describes Primorie's geographical context within a wider Russian Far East. The following section then considers three sites within Primorie, where local mobilizations contend with the unequal interaction of global and local processes within a post-Soviet context. These examples show how citizens can usurp, redirect, redefine power relations to reclaim particular places as empowering sites for a historically oppressed locale to have a say in its own (re)production.

Geographies of Russia's Far East

There is a multiplicity of definitions and representations for what constitutes "Northeast Asia," "Russian Asia," "Siberia," and the "Russian Far East." The exact borders are drawn differently depending on the historical moment, the particular context, and what or whose purpose is being served. Cartographic-ally, the entire region of the Russian Republic east of the Ural Mountains can be drawn as part of Northeast Asia; and historically, this region can be considered in many ways an antipode to European Russia, that part of the country west of the Urals. Referred to collectively as "Siberia" by, for example, nineteenth-century Siberian regionalists, Eurasionists, and by many contemporary Westerners, the portion of Northeast Asia known as the Russian Far East (RFE) extends from the Arctic southward along the Pacific Ocean to the borders of China and North Korea.[1] Throughout its history of colonization by Russia (intensified during the twentieth century by Soviet industrial development schemes), this region and its larger geographic context has formed the periphery to Moscow's center in a traditional center–periphery resource colonial relationship.

Though slightly less than 1 percent of Russia's present territory, the Maritime Province of Russia's Far East, Primorie, nonetheless consists of a patchwork of maritime and continental climates, alpine and lowland regions, mixed evergreen and deciduous old-growth forests, villages, factory towns, and major cities. Primorie is also geographically closer to the American city of San Francisco than to Russia's capital, Moscow. Vladivostok, the region's capital and largest city, is situated along the "Golden Horn," an inlet of the Sea of Japan. With optimistic reference to both the horn of abundance (of Jason and the Argonauts fame) and Istanbul's Golden Horn Bay, this inlet today leads to a geographically diverse region attempting to navigate a tangled knot of historical, political, and economic relations. Temporally it straddles a moment in history between what was Soviet and what is becoming its successor. Geopolitically it shares borders with China and North Korea, and via the Sea of Japan it is connected to Japan and South Korea. Socio-economically it is situated between the industrializing Pacific Rim and the fault-strewn post-Soviet Russia, and is poised for fundamental reorientations of institutional arrangements, communications, and capital.

A specific Soviet-socialist approach to nature–society relations contributed to the valuation of this peripheral region (by government officials) insofar as it could serve both as waste repository (political and other) and bank of natural capital (Bassin 1991; Sinyavsky 1988). Other cultural understandings of nature were for the most part undermined or silenced.[2] In fact, the spatialities produced by dominant Soviet organizations of socio-ecological relations in this part of Russia maintain an enduring influence on local and extra-local interactions. And since the reopening of Vladivostok and the Golden Horn to foreigners in 1992, the region has found itself additionally and increasingly influenced by industrial capitalism and the unprecedented extensions of the world market. From Pacific Russia through the Russian Far East to Siberia in general, the entire region, combining a Soviet legacy with market influences, is largely perceived by regional government administrators, corporate foreign timber and oil conglomerates, and federal production ministries as an underutilized land of opportunity. The development of the region's natural wealth (including contestations over how and by whom) is the medium through which this place is being reshaped at both global and local levels. Yet, conflict exists between local residents' needs, an export-focused drive for hard currency, and an ill-formed political infrastructure unable to reconcile the conflicting demands of post-Soviet transition.

Within Primorie, local struggles, negotiations, and contests over control of resources intersect with globally-oriented strategies for resource access and development. The making of the discourses and practices through which nature is known and transformed is ecologically and socially connected in and to particular places. The Soviet discourses and practices through which nature was known and transformed produced asymmetrical and hierarchically organized power relations between monopolistic enterprises and government ministries on the one hand, and local communities on the other. The Russian state's present interest in Far Eastern resource development forms a strong historical and geographical connection with its past. And what puzzled previous regimes remains sought after today: how to finance and build the infrastructure necessary to move Siberian natural wealth out of Siberia and into the global market, presently via the Pacific Ocean. This continuation of an extensive development approach to resource access and control necessarily intersects with local places and their continuing navigation of post-Soviet transition. The significance of this confluence for the local requires taking stock of the socio-economic relations which come together in particular locations, linking with other places and multi-scale interactions.

Globalization of the Russian Far East

The Russian Far East is in general experiencing a burgeoning of economic ties to the Pacific Rim (Paisley and Lilley 1993; Shlapentokh 1995; Thornton 1995; Zochowska 1993), giving it potential entree into the center of the global political-economic system. This potential lies in its combination of tremendous mineral, fossil fuel, timber, hydro, and fisheries wealth, its geographic and increasingly economic distance from Moscow, and its proximity to industrializing Pacific Rim countries. Large-scale international banking, governmental, and corporate concerns are eager to gain access to this Siberian wealth while reciprocally Russia is eager to obtain hard currency. Within a 1,000-km radius of Vladivostok, Primorie, live 260 million people along the northeast and eastern coasts of China, 130 million in the Koreas, 120 million in Japan, and 8 million in the Russian Far East, a total of over 500 million. This greater area has a collective GNP of $3 trillion and accounts for nearly a third of the world's trade..

How nature is accessed, locally or globally, whether in the RFE or elsewhere, involves its discursive construction, incorporating power, capital, and social differences. Certainly the Tumen River Area Development Project (TRADP), the largest of the development schemes in Primorie, is representative. It is intended to create an international free trade zone on the borders of Russia, China, and North Korea. The project, as presented by the United Nations Development Programme (UNDP) and the United Nations Industrial Development Organization (UNIDO), is designed to transform the region into a low-cost processing and shipment zone, primarily for the export of Siberia's previously inaccessible timber, oil, gas, coal, and mineral resources. TRAPD, as envisioned by its powerful promoters in China, North Korea, South Korea, Mongolia, Japan, and Russia, the UNDP, and the UNIDO, will transform the Tumen River into the transportation and exporting hub for Northeast Asia.

The perceived inexhaustibility of the USSR's natural environment, the drive for economic growth, adherence to Soviet-socialist ideology, and technological Prometheanism fueled the drive for subduing and converting the environment (cf. Low and Gleeson, Chapter 3, this volume). But for the replacement of Soviet-socialist ideology with transnational capitalism, the situation with regard to the Tumen project shows little change. Despite the UNDP's vision of TRADP as "a model of people-centered, environmentally sustainable development" (1997a: 3), a variety of Russian Far Eastern scholars from various fields have expressed counter opinions.[3] Some objections to the Tumen Project are ecologically based: sewage, trash, and gas fumes from the proposed Tumen port potentially would be transported via warm water currents to Russia's only marine nature preserve, located 20 km to the north. Others object for logistical reasons: as the Tumen River is approximately 4 meters deep, the bottom would need to be dredged another 15 meters to handle ocean vessel traffic (*Vladivostok News*

1996). Consequently, local plants and animals would be harmed and regional ocean waters would experience additional pollution. Some in this southern part of Primorie suggest that the area's four currently operating Russian ports, the Trans-Siberian Railroad, and those depending on these facilities would suffer due to increased competition and subsequent redundancies. Deep-seated mistrust of China's intentions towards this part of Russia also contributes to local anti-project sentiments (see Hunter 1998).

While the UNDP acknowledges that the risk of environmental degradation will increase due to the effects of expanded trade and development brought on by the Tumen Project (Chinese Research Academy for Environmental Sciences, 1994), it concurrently hopes that the "welfare of people in TREDA (Tumen Regional Economic Development Area) and in its North East Asia hinterland will improve rapidly as a result of development of infrastructure, industry and trade in forms that are ecologically and economically sustainable and equitable" (RDS Sub-group, UNDP 1994; see also UNDP 1997a, 1997b). Environmental impact assessments and monitoring, and investment in infrastructure and community redevelopment, is each sovereign region's own responsibility (see also Rosencranz and Scott 1993). The intent is to encourage effective domestic environmental protection and economic development appropriate to each participating country. It assumes, however, both the willingness and ability to balance these concerns.

Russia has decentralized environmental monitoring and policy enforcement to regional administrations. Funding for resource and environmental management no longer dependably comes from Moscow. Regional administrations receive funding for resource management and monitoring from those firms which lease – from the state – the right to develop resources. Funding is inadequate and the power relations between the forces for protection and those for development thus become ambiguous (Bowles *et al.* 1996). The Primorskii regional government's underlying commitment to environmental protection is in any case questionable.[4] Regional control and protection of resources in Primorie, as in other regions of Russia, is often a function of personal interests and participants' personalities, not policy or free market activity (Kryukov and Moe 1993). Consequently, conflict and gridlock exist as new power relations evolve and are tested at institutional and personal levels. The result is a struggle over control, an attempt by governmental institutions and quasi-capitalist enterprises to consolidate and expand their positions within the evolving system with little regard for society's welfare in general (Mkrtchian 1994).

The general tendency of extractive export economies to experience initial gains in regional incomes followed by rapid collapse is well documented (Daly 1977; Furtado 1970; Georgescu-Roegen 1970, 1975; Levin 1960). When natural resources are extracted from one regional ecosystem to be transformed and consumed in another, the resource-exporting region can eventually experience a deceleration or impoverishment of its economy, in addition to demographic, social, and ecological disruptions (Amin 1976; Bunker 1988; Wallerstein 1974). Continued reliance on large-scale domestic and multination-ally sponsored resource development projects threatens to damage the complex and interdependent environmental and social systems of Pacific Russia (Altshuler et al. 1992; Bothe *et al.* 1993; Bradshaw 1991; Kuleshov *et al.* 1994; Peterson 1993). This is exacerbated by an absence of formally established procedures for public participation in natural resource allocation and land-use decisions. But these remain abstract challenges at this stage for the residents of southern Primorie, the majority of whom live in Vladivostok. Residents contend daily with water and electricity shortages, unemployment and unstable inflation rates, urban crime and pollution. The moment the downstream effects of Tumen Development negatively impact the lifeworlds of Primorie's residents is when the abstract will become materially transformed.

Nature, as it is discussed, exploited, and/or protected is an object of globalized treatise (e.g. Montreal Protocol, the Rio and Kyoto conferences, etc.). Commodities are the actual manifestations of local natures, signs pointing to the environmental alteration necessary for their manufacture. Ideologies, social practices, and the power relations which inhere in these create the conditions for resource development and commodity production, and together with the physical entity itself, are expressive of the globalization of the local. An effective place-based environmental ethics is one that serves as a necessary counterweight to the dominating narratives which too often marginalize local places and concerns for nature. The following examples suggest ways in which citizens are working to reappropriate physical and socio-political places in the wake of a disintegrated Soviet state in ways which express a reclamation of deteriorating ecological and socio-political lifeworlds.

Ethics in place

Dalenergo, Primorskugol, and popular protest

In May 1997, the Primorskugol mining company cut off coal supplies to Primorie's only energy producer, Dalenergo, to protest the non-payment of back wages. Primorskugol claimed that for several months Dalenergo had not made payment on its debt. On the other hand, Dalenergo maintained that it in turn was owed 209 billion rubles ($37 million) from the region's capital, Vladivostok, for past unpaid energy deliveries (Ogden 1997). The stand-off forced the closure of several of the city's factories, bakeries, and processing plants. City officials met at Vladivostok's city hall to discuss emergency measures. Mayor Cherepkov declared a state of emergency, and, unintentionally reflecting the unequal distribution of regional control over natural resources, the meeting was conducted without lights as power cuts of up to 12 hours a day were being experienced throughout the city. City workers predicted a litany of disasters, ranging from failed sewage systems to epidemics if the situation was not brought under control.

The mayor had attempted earlier to avoid such severe energy shortages by circumventing the authority of Dalenergo and buying coal directly from Primorskugol to deliver it to local power stations. Dalenergo, however, in a strategic move to retain control, prohibited generating stations from receiving coal from the city. The power dynamic which underlay this situation came about, in large measure, from the Soviet legacy of regional energy sector organization. Under the Soviet system, a fuel enterprise was more than a company which produced energy. Such an enterprise was an integrated production complex in control of most of the infrastructure found within its areas of operation, ancillary infrastructure which in the Western context would be outside the domain of the company (Kryukov and Moe 1993). These individual enterprises historically behaved as owner and monopolist; today these companies, such as Dalenergo, are reluctant to divest control within borders traditionally considered "theirs." Consequently, the energy enterprise which effectively controlled a region under the prior Soviet system may seek to maintain and consolidate its place within the currently evolving system.

The miners' protest shows, by example, how our actions can interrupt the daily flow of social relations and so potentially re-make our "places." To get at this, consider first that through a diverse and continuous flow of accidental, planned and potential social interactions, particular (for a given time and place) political, economic, and behavioral patterns (i.e. society) emerge. Metaphorically or cartographically, these patterns become mapped; sometimes by us, and sometimes – in a process manifest with power relations delineating territories, domains, and boundaries – for us. In this way we come to know our "places." The coal miners' strike can be viewed simultaneously as a social (i.e. ethical) and spatial (i.e. geographical) re-mapping of places.

Consider, for example, that energy production and distribution enterprises, and their employees, are part of a pattern of social relations. These relations, albeit exhibiting movement, are nonetheless relatively stable and through the interactions of authority, capital, habit, and inertia, form bounded, ordered, and semi-permanent (for a time and place) formations (material and figurative). There exists such an array of employees, departments, firms, production complexes, through to governing federal-level ministries such that people and the ordering institutions know their "places." The miners' strike was, in effect, an acknowledgement, or re-mapping, of new social relations. Geography and ethics are neither underlying context nor ground upon which we act or construct ourselves and our communities. Rather, they are *constitutive* of the process of making our social worlds. The miners, geographically and ethically, were making for themselves a new place and voice from which to be heard.[5] And it was in a similar vein, only four months earlier, that a thematically related protest took place.

Along Vladivostok's major downtown intersections and within the city's central square, a broad cross-section of residents, ranging from teenage girls to elderly veteran pensioners, gathered for protest. They joined together to express dissatisfaction with a decline in living conditions (including the on-going energy crisis) and to demand a no confidence vote on the region's governor, Yevgeny Nazdratenko.

In this collective expression of protest, economic, political, and cultural social-relations contemporaneously met in the central square in a moment of socio-spatial reordering. What actually does this mean? To begin with, the symbolic meanings with which particular places become imbued can incorporate, as Edward Said explores in *Orientalism* (1978), differentials in access to power, locations of specific social functions, and expressions of political authority. The making of certain places (police stations, city halls, and open-plan office suites are most obvious) is fundamental to socio-political control. Now consider the location for the May 1997 public protest. In several ways this has historically been a place of "real" and metaphoric political and social power. From the central square, which is located along Vladivostok's oldest avenue, many of the city's grandest examples of pre-revolutionary architecture can be seen. What also is observable is the degree to which they are in need of maintenance. But rising far higher (and whiter) than any of these pre-1917 buildings, and dominating the central square, is the Regional Government Administration Building, nick-named the "White House." It can be seen from various corners of the downtown; the reverse is also obvious.

Also within the central square is a major monument dedicated to "those heroes who fought for Soviet Power in the Far East." It is a large bronze revolutionary atop a pedestal. He is muscular, square jawed, gaze set determinedly on the horizon, foot placed firmly on a rock, a flag in one hand and a bugle in the other. In literature promoting the region, the statue is used (albeit with a twist of irony) as a reminder of the city's and nation's Leninist legacy. "Vladivostok is far away but it is (truly) ours" is an excerpt from a speech of Lenin's after the unification of the short-lived Far Eastern Republic with the rest of the country. The idea was that, despite the vast distances and cultures which constituted the newly formed Union of Soviet Socialist Republics, all cities were in fact united and belonged to the same "us."[6] In a material expression of a reordering of social and therefore power relations, the protesters in view of the "White House" were in fact eloquently transforming one revolutionary symbol and reclaiming it as "theirs."

Despite the important differences between the miners of Primorskugol and protesting citizens in downtown Vladivostok, both have sought to dismantle or bypass pre-existing and continuing regional power structures through decentralizing, community empowerment. Place-based movements and strikes can appeal to, reinforce, and build a locality's shared sense of community (see Herman and Mattingly, Chapter 15, this volume; see also Lipschutz and Mayer 1996). For protesters in Vladivostok, abstract notions of resource flows and control were put into very real and material terms relevant for a specific territory and group of people. Local groups are using the material changes in production, information flows and communication, in addition to the spaces of resistance opened by the disintegration of the Soviet state, to experience the viability of place as a realm for the exploration, contestation, and affirmation of citizen–state relations and of contextual ethical discourse.

Udegeitsy and the Nanaitsy

Consider a different case, one which began in the remote northern part of Primorie, inland from the Sea of Japan, along the eastern side of the Sikhote-Alin

Mountain Range. This struggle involved two local Siberian ethnic communities, an international joint venture, and eventually the Supreme Court of the Russian Federation. The focus of the dispute was on jurisdictional boundaries between national, regional, and local authority. Controversy resulted from the efforts of the Udegeitsy (and to a lesser extent the Nanaitsy) ethnic group to prevent the Hyundai-backed Russian–South Korean joint venture, Svetlaia, from expanding logging operations within their traditional lands in the Bikin River basin.

Svetlaia applied for and was granted a license by the then head of Primorie's regional government, Vladimir Kuznetsov, to increase its logging operations within the old-growth forest of the Bikin River valley. In response, regional ethnic groups and supporting scientists argued that forestry operations along the upper Bikin would reduce the areal extent of permafrost and reduce tributary flow into the river. Natural habitat for rare birds and sable would consequently be disrupted or destroyed, as would the habitat of the endangered Amur/Ussuri ("Siberian") tiger, a subspecies protected by international conventions.[7] Further, the resource base upon which the Udegeitsy and the Nanaitsy had traditionally relied would be destroyed.

Svetlaia's operations constituted violations of international environmental agreements and human and minority people's rights. With the help of the Socio-Ecological Union and the Far East Information Center,[8] the Siberian groups influenced the regional government to overturn the governor's previous decree which had granted Svetlaia the extra logging rights within the upper stretches of the old-growth forest. Svetlaia appealed the decision to Primorie's regional court, which favorably sided with the international joint venture. The court ruled that the regional governor's original decision on granting logging rights should stand. The case continued to the Russian Supreme Court where the lower court's decision was overturned in favor of the Udegeitsy and the Nanaitsy. The Supreme Court made a unilateral ruling based on a presidential decree, passed in 1992, that gave native Siberians the right of formal consent for any encroachment onto their traditional lands. As no consent had been granted by the Siberian groups, the governor had no legal standing for transferal of additional logging rights to Svetlaia. Despite the Supreme Court's decision, members of Primorie's regional government continue to look for a way to allow Hyundai access to the Upper Bikin River valley (Newell and Wilson 1996; Wishneck 1995; Chisholm 1998).

Legislation passed in Moscow or international conventions agreed upon in Geneva were not sufficient to prevent environmental violations. New and reconstructed legislative initiatives to protect the environment are often rendered ineffective in the chaotic political and economic turmoil of Russia. Federal enforcement of policy appears increasingly difficult as the political and economic distance between Moscow and the Russian Far East grows. Consequently, local authorities may pursue short-term economic and political gain in their navigation of post-Soviet resource access and control. As the Russian Far East increasingly faces away from Moscow and toward the Pacific Rim, the region's resources are becoming more accessible to exploitation, both international and regional.

Though physically still isolated from material centers of commercial activity, international capital's search for more profitable places of accumulation helped position this river valley and its residents more centrally within macro-scale political and communication networks. This underscores the notion that places (despite physical isolation) are often porous, open to the ebb and flow of international socio-economic relations. And in this case, it is the very porosity of place that aided local residents to build and strengthen supportive ethical networks with other locally-based environmental groups in order to defend their place. In contrast to the previous examples of protest, this is a case where participants used new networks of global communications combined with evolving post-Soviet socio-political relations in order to promote a *continuation* of their place in Primorie and the wider Far East.

The most powerful discourses in society are institutionally based (Weedon 1989), yet, as the Bikin logging controversy and the previous protest examples poignantly show, the institutional locations are themselves sites of contestation, and the dominant discourses structuring the functioning and relationships of these institutions are under continuous challenge and re-formation. This actually suggests that there exists the discursive space to challenge or resist environmental and social practices which have traditionally subordinated the health of the local and the environment.[9]

Places can be the sites of multinational corporations and their operations, but they are also sites of local mobilizations contending with the interaction of local and global processes. The plethora of locally-based, globally-connected environmental groups in the Russian Far East, forming coalitions with other environmental organizations situated in distant locations, against abusive environmental access by capitalized, multinational corporations or a hard-currency-seeking government sector, is representative. This suggests local adaptations and re-uses of the global spaces being created by transnational capital to reassert the vibrancy of their places. As de Certeau (1984) suggests, and these examples of protest hopefully demonstrate, local places in resistance to movements of global capital are developing strategies to re-use and recode capital's disciplinary structures. This stands in contrast to Dear's (1997) reading of Jameson (1991). Dear suggests that "*place* no longer exists," for in the "saturated space" of multinational capitalism, place has been "drowned by other more powerful abstract spaces such as communications networks" (1997: 58). In Russia's Far East, however, it is in large measure precisely the globalization of the local, and access to new global technologies of communication, which have enabled local environmentally concerned groups to mount resistance to international and transnational enterprises such as Exxon, Marathon, and Hyundai.

Prospects

A number of disparate themes have so far been raised: what remains to be done is to draw them together to uncover what may be a coalescing (if still disparate) contextual ethics of place. On the one hand, this discussion has emphasized the conflicting definitions by those involved of nature as site of extraction, preservation, culture, artifice, or morality. On the other hand, what is apparent is the lack of political infrastructure to reconcile these multiple constructions. As the Russian Federal Government's enforcement power in Primorie has weakened since 1991, struggle for access and control of regional resources has intensified, exacerbating a legacy of lop-sided power relations where the locality was typically on the disempowered side.[10] While today there may be a multivocality regarding Primorie and its nature, this does not mean that the voices have equal power. By recalling the agglomeration of transnational investment, influence, and infiltration represented by some of the institutions involved in accessing Primorie's resources – for example Exxon, Mitsubishi, Hyundai, UN, USAID – this point is made all the more obvious. Where, then, does this lead in identifying a nascent operant ethics of place?

Consider, first, that it is the confluence of social institutions organized at a variety of scales, symbolic and experiential associations, and material existence which interplay to express place (see Curry, Chapter 7, this volume). In other words, dynamic and shifting socio-economic relations come together in particular locations, contributing to a place's identity. Place, then, is not "bounded," somehow frozen in an absolute Cartesian grid, isolated and cut off from macro-scale socio-economic flows. Places are porous, in fact partially produced through inter-relations with other places, conditions, and multi-scale interactions (see Massey 1994). Places within Primorie, in a decisive break with a Soviet past, are not simply changing in relation to a one-way development process whereby communities and local cultures are overshadowed or absorbed by macro-scale forces (see previous section). Disregard for the vibrancy of place-based socio-ecological relations (e.g. by state and capitalist sponsored development) has provoked local protests which are extending outwards to a larger environmental politics.

Amid the encroachments of the global in the form of capitalized multinational interests and the dismantling of Soviet central authoritarian structures, communities are attempting to construct new local spaces from which to be heard within the evolving democratic political system. These communities take multiple spatial forms, depending on ethnicity, class, gender, age, continuity or break with tradition, identification with a particular place, or catalyzing social issue. Freed from the Soviet-era ban on open protest, confronted with the uncertainties of economic turmoil, and facing the accumulated environmental and cultural disruption brought about by Promethean Soviet modernization and continuing post-Soviet dubiety, Siberian communities are variously positioned to gain some measure of control over their land and resources (Dienes 1991; Mitchneck 1991; Wishnek 1995).

Pursuit of local interests, however, is sometimes nothing more than that. As laudable as local self-determination is (and most would agree that it is), it is not always appropriate to assume its moral glorification. But consider, for a moment, the context here. Grassroots public protests, regardless of their reasons (e.g. wage demands, quality of life concerns, regional autonomy from the center) for the most part simply were not allowed to happen under the Soviet regime. And take, say, the period from the first coal miners' strikes of 1989 until Boris Yeltsin's abolition of the Russian Parliament in 1993. It was still not uncommon for provincial leaders to condemn worker and citizen protests as negative displays of where lack of discipline and over-permissiveness on the part of government could lead (see Christensen 1995).

I do *not* wish to propose the rather simplistic polarity that all local resistance movements must be celebrated as positive reactions to the alienation and disempowerment that sometimes results from extra-local processes. I do want to recognize that the process of post-Soviet democratization (meaningfully used here in opposition to prior Soviet management of nature and society) largely requires that all major power-holders agree to equitable forms of negotiation and contestation. As long as there continues a Soviet managerial elite in control of formerly state owned enterprises, an indeterminacy of property rights, and *ad hoc* policy-making – regionally and nationally – perpetuating systemic instability, local expressions of self-interest (whether influenced by environmental, wage, or quality of life concerns) serve a positive moral purpose.

The future of this region, and of formerly socialist states in general, depends largely on what the local will tolerate and defend *vis-à-vis* relations to weakened central control and strengthened international capital flows (Best and Kellner 1991; Fiske 1989; Peet and Watts 1996). In this altered state of post-Soviet politics, newly liberated voices from the periphery strive for representation no longer structured according to a single official discourse. Under conditions of the disintegrated Soviet state and decentralization of power, there has been a multiplication of possibilities and necessities for political struggles, local politics, and regional interests.

Russia's Far Eastern landscape is currently being reshaped by a variety of economic, political, and social interrelations at all scales. The global influence of finance and international market competition, the spatial dissolution and transformation of national political power, and the changing social and economic relations between regional institutions and local communities reflect the social processes and circumstances which are shifting and reforming post-Soviet society. A new regional geography of the Far East is thus being created, where local spaces and identities are being reshaped by the confluence of macro-political and micro-political environmental access and control. As the Far East opens its borders and enters the export markets of the Pacific Rim, its identity in large part will be reformed by its specific interactions with the "outside" and the manner in which those interactions are manifested locally. This global/local interface is tied to the dynamics of how nature will be accessed and controlled, and by whom. When considering local resource development pressures and their underlying discursive (post-Soviet) re-formations, an effective place-focused environmental ethics must strive to subvert an imposed development agenda which ignores local nature–society relations and deep historical connections to place.

The practice of an environmental (or other) ethics, however, does not stop there. As Fraser and Nicholson (1990) and Bertens (1995) point out, power must not only be subverted, but exercised. How the places of the Russian Far East are reconstructed has materially much to do with how local desires, expectations,

188 Ethics and nature

longings, and investments (emotional, economic, discursive) are expressed in place. This has thus much to say in regard to undoing the traditional dominance of Moscow *vis-à-vis* the Far East, presenting an alternative to the contemporary encroachments of the transnational, and empowering a historically oppressed local to have a say in its own (re)production.

Local expressions of dissatisfaction or protest may not determine the content of power relations in a post-Soviet Russia. Nor will the struggle of local Siberian ethnic communities against transnational capital guarantee ecological sustain-ability. But, without such local places of expression, without a re-mapping of local places within post-Soviet transition, the likelihood of a democratic Russia ethically better in kind than a previous Soviet Union, is discouraging.

Acknowledgments

I would like to acknowledge the American Council of Teachers of Russian and the National Science Foundation for their generous support of my research in the Russian Far East.

Notes

1 In Russia, the Far East is generally considered to include the Amur, Kamchatka, Magadan, and Sakhalin Oblasts, the Primorie and Khabarovsk Krais, the Sakha (Yakutia) Republic, the Jewish Autonomous Oblast (Birobidzan), and the Chukotke and Koryak Autonomous Okrugs.
2 These other cultural understandings include early and mid-nineteenth-century Siberian regionalist views of nature as an entity abused by a far away and uncaring central Russian government (Ianovoskii 1983; Vilkov 1974; Kalashnikov 1905, in Diment and Slezkine 1993), Siberian nature as an embodiment of a "purer" Russia (Kropotkin 1988; Chekhov 1890), and alternative pre-Bolshevik visions of nature and society (see Timonov 1922 and Sementov-tian-shanskii, 1919, in Weiner 1988). This of course does not even begin to account for the multitude of ethnic Siberians' conceptions of nature.
3 Local researchers who have expressed opinions in opposition to the Tumen Project include, for example, the Director of the Pacific Geographical Institute (Russian Academy of Sciences), the Director of the Far Eastern Center for Economic Development (participating member of the Gore-Chernomyrdin Commission), the Deputy-Director of the Marine Biology Institute (Russian Academy of Sciences), the leading researcher of the Economic Research Institute (Russian Academy of Sciences), the Director of the Vladivostok Institute of Inter-national Studies, as well as local ecologists, students, and members of international NGOs located in Vladivostok.
4 One example of questionable environmental commitment involved a gift pre-sented by Primorie's governor to the president of Belarus. At a reception in Vladivostok in early 1998 for the Belarussian president A. Lukashenko, the skin of a rare Amur ("Siberian") tiger was presented by Primorie's governor, Y. Nazdratenko. The Governor hoped that the fangs of the tiger would give inspiration to Lukashenko in his fight against political opponents in Minsk and Moscow (Working and Chernyakova 1998). The presentation of the tiger skin was made in violation of both Russian environmental laws and international treaties.

5 In February 1998, striking miners took their protest inside the "walls of power" directly to the office of the mayor of Vladivostok. Simultaneously regional and municipal representatives were meeting in Moscow with Federal Ministry of Fuel and Energy officials. Shortly after, the Primorie regional government announced there would be a massive restructuring of the regional energy complex, including energy diversification and upgrading of the coal industry. What effect this will have practically is still an open question.

6 The strength of Lenin's statement may have been reinforced by the fact that he had never visited Vladivostok.

7 The larger geographical context of this region is sometimes referred to as Ussuriland. This region supports the last population of the Amur/Ussuri tigers, generally estimated at not more than 500; several thousand black and brown bears; and further down to the southwest, the last 30 (approximate) known surviving wild Amur (Far Eastern) leopards. The region is also habitat for 350 bird species, 50,000 insect species, 100 species of fish, and contains 25 percent of the Russian Federation's biodiversity.

8 The SEU, formed in 1988, is an international non-profit voluntary association of over 250 environmental groups active on local and regional levels in the former Soviet Union. The FEIC is a non-profit organization of journalists that provides independent information on Far Eastern politics, economics, environmental and resource-use problems.

9 "Discursive space" is used here to connote the site of expression where the processes of discourse production are known and made visible (see Dorst 1989: 125).

10 The year "1991" refers to the moment when the Soviet Union, as a recognized sovereign nation, no longer formally existed.

References

Altshuler, I., Y. Golubchikov and R. Mnatsakanyan (1992) "Glasnost, Perestroika and Eco-Sovietology," *The Soviet Environment: Problems, Policies, and Politics*, J. M. Stewart (ed.), Cambridge and New York: Cambridge University Press, 197–212.

Amin, S. (1976) *Unequal Development*, New York: Monthly Review Press.

Bassin, M. (1991) "Inventing Siberia: Visions of the Russian East in the Early 19th Century," *American Historical Review* 96: 763–794.

Bertens, H. (1995) *The Idea of the Postmodern: A History*, London and New York: Routledge.

Best, S. and D. Kellner (1991) *Postmodern Theory: Critical Interrogations*, New York: Guilford Press.

Bothe, M., Thomas Kurzidem and Christian Schmidt (eds) (1993) *Amazonia and Siberia: Legal Aspects of the Preservation of the Environment and Development in the Last Open Spaces*, London and Boston: Graham & Trotman/M. Nijhoff.

Bowles, F., J. Newell and E. Wilson (1996) *The Russian Far East: Forests, Biodiversity Hotspots and Industrial Developments*, Tokyo: Friends of the Earth-Japan.

Bradshaw, M. (ed.) (1991) *The Soviet Union: A New Regional Geography?*, London: Belhaven.

Bunker, S. (1988) *Underdeveloping the Amazon: Extraction, Unequal Exchange, and the Failure of the Modern State*, Chicago: University of Chicago Press.

Chinese Research Academy for Environmental Sciences (May 1994) "Tumen River Area Development Project. Preliminary Environmental Study: Draft" in G. B. Hayes, Team Leader, PDP Australia Pty Ltd. *A Regional Development Strategy for the Tumen River Economic Development Area*, Preliminary Draft Report.

Chisholm, B. J. Director of the Russian Far Eastern Office of ISAR, Vladivostok, personal communication. February 1998.

Christensen, P. (1995) "Property free-for-all: regionalism, 'democratization,' and the politics of economic control in the Kuzbass, 1989–1993," *Rediscovering Russia in Asia: Siberia and the Russian Far East*, S. Kotkin and D. Wolff (eds) Armonk, NY: M. E. Sharpe.

Daly, H. (1977, updated 1992) *Steady-state Economics: The Economics of Biophysical Equilibrium and Moral Growth*, San Francisco: W. H. Freeman and Co.

De Certeau, M. (1984) *The Practice of Everyday Life*, Berkeley, CA: University of California Press.

Dear, M. (1997) "Postmodern Bloodlines," *Space and Social Theory: Interpreting Modernity and Postmodernity*, Georges Benko and Ulf Strohmayer (eds), Oxford: Blackwell Publishers.

DeBardeleben, J. (1992) "Ecology and Technology in the USSR," *Technology, Culture, and Development: The Experience of the Soviet Model*, J. Scanlon (ed.), Armonk: NY: M. E. Sharpe, 149–169.

Dienes, L. (1991) "Siberia: Perestroika and Economic Development," *Soviet Geography* 32(7): 445–457.

Dorst, J. (1989) *The Written Suburb: An American Site, An Ethnographic Dilemma*, Philadelphia: The University of Pennsylvania Press.

Epstein, M. (1995) *After the Future: The Paradoxes of Postmodernism and Contemporary Russian Culture*, Amherst: The University of Massachusetts Press.

Feshbach, M. (1995) *Russia in Transition: Ecological Disaster, Cleaning Up the Hidden Legacy of the Soviet Regime*, New York: The Twentieth Century Fund Press.

Fiske, J. (1989) *Reading the Popular*, Boston: Unwin Hyman.

Fitzpatrick, S. (1994) *Stalin's Peasants: Resistance and Survival in the Russian Village after Collectivization*, Oxford: Oxford University Press.

Fraser, N. and L. Nicholson (1990) "Social criticism without philosophy: an encounter between feminism and postmodernism" in L. Nicholson (ed.), *Feminism/Postmodernism*, New York: Routledge, Chapman and Hall, 19–38.

Furtado, C. (1970) *Economic Development of Latin America: A Survey from Colonial Times to the Cuban Revolution*, London: Cambridge University Press.

Georgescu-Roegen, N. (1970) "The entropy problem and the economic problem," *The Ecologist*, July (2): 347–381.

—— (1975). "The entropy law and the economic process," *Southern Economic Journal* January (41): 347–381.

Gorky, M. *et al.* (1935) *Belomar: The Construction of the Great White Sea–Baltic Canal*, New York and London: Smith and Haas.

Heller, M. (1988) *Cogs in the Wheel*, New York: Alfred Knopf.

Hellman, L. (1955) *The Selected Letters of Anton Chekhov*, New York: Farrar, Straus and Company.

Hunter, J. (1998) "The Tumen River Area Development Program, Transboundary Water Pollution, and Environmental Security in Northeast Asia," text from talk given at The Woodrow Wilson Center, Environmental Change and Security Project, Working Group on Environment in US–China Relations.

Ianovskii, N. N. (ed.) (1983) *Literaturnoe Nasledstvo Sibiri*, Novosibirsk: Zapadno-Sibirskoe Knizhnoe Izdatel'stvo.

Jameson, F. (1991) *Postmodernism or the Logic of Late Capitalism*, Durham, NC: Duke University Press.

Josephson, J (1997) *New Atlantis Revisited: Akademgorodok, the Siberian City of Science*, Princeton: Princeton University Press.

Kalashnikov, I. T. (1905) "Zapiski Irkutskago Zhitelia," *Russkaia Starina* no. 123, July, in G. Diment, "Exiled from Siberia: The Construction of Siberian Experience by Early-Nineteeth-Century Irkutsk Writers," 47–68, in G. Diment and Y. Slezkine (eds) (1993) *Between Heaven and Hell: The Myth of Siberia in Russian Culture*, New York: St Martin's Press.

Korkunov, I. N. (1994) "O Proekte CEZ na Territorii Russii, Kitaya, i KNDR," *Problemii Dal'nevo Vostoka* 3: 13–17.

Kropotkin, P. A. (1988) *Memoirs of a Revolutionist; with a New Introduction and Notes by Nicolas Walter*, New York: Dover Publications.

Kryukov, V. and A. Moe (1993) "Controlling the Russian Oil and Gas Complex: The Regions vs the Centre," Nato Economics Directorate and Office of Information and Press.

Kuleshov, V., V. Seliverstov, V. Kharitoniva, V. Pushkarev, V. Shmat and V. Tchourachev (1994) "Final Report: Institutional Strategies of Sustainable Development: The Development of West Siberian Resources," Institute of Economics and Industrial Engineering, Siberian Branch of the Academy of Sciences of Russia, and the Siberian International Centre for Regional Studies, Novosibirsk.

Levin, J. (1960) *Export Economies*, Cambridge, Mass.: Harvard University Press.

Lipschutz, R. D. with J. Mayer (1996) *Global Civil Society and Global Environmental Governance: The Politics of Nature from Place to Planet*, Albany, NY: State University of New York Press.

Massey, D. (1994) *Space, Place, and Gender*, Minneapolis: University of Minnesota Press.

Mitchneck, B. (1991) "Territoriality and regional economic autonomy in the USSR," *Studies in Comparative Communism* 24(2): 218–224.

Mkrtchian, G. Dean and Chair, Department of Economics, Novosibirsk State University, Novosibirsk, personal communication, November 1994.

Newell J. and E. Wilson (1996) "The Russian Far East: foreign direct investment and environmental destruction," *The Ecologist* 26 (March): 68.

Ogden, K. (1997) "Blackouts anger mob." *Vladivostok News*, 9, 141 (May 15): 1.

Paisley, E. and J. Lilley (1993) "Economies: bear necessities." *Far Eastern Economic Review* 156(27): 40–43.

Peet, R. and M. Watts (eds) (1996) *Liberation Ecologies: Environment, Development, Social Movements*, London and New York: Routledge.

Peterson, D. J. (1993) *Troubled Lands: the Legacy of Soviet Environmental Destruction*, Boulder, CO: Westview Press.

Pryde, P. (1995). "Russia: An Overview of the Federation, *Environmental Resources and Constraints in the Former Soviet Republics* P. Pryde (ed.), Boulder, CO: Westview Press, 25–39.

RDS Subgroup, United Nations Development Program (1994) "A Regional Development Strategy for the Tumen River Economic Development Area," Preliminary Draft Report, April 22.

Rosencranz, A. and A. Scott (1993) "Siberia's threatened forests," *Nature* 355, January.

Said, E. (1978) *Orientalism*, New York: Pantheon Books.

Semenov-tian-shanskii, V.P. (1919/1988) "Svobodnaia Priroda, kak Velikii Zhivoi Muzei, Trebuet Neotlozhnykh Mer Ograzhdeniia," *Priroda* 199–216 in Douglas

Weiner, *Models of Nature: Ecology, Conservation, and Cultural Revolution in Soviet Russia*, Bloomington: Indiana University Press.

Shlapentokh, V. (1995) "Russia, China, and the Far-East – old politics or a new peaceful cooperation," *Communist and Post-Communist Studies* 28 (N3, September): 307–318.

Sinyavsky, A. (1988) *Soviet Civilization: A Cultural History*, New York: Little, Brown and Co.

Thornton, J. (1995) "Recent trends in Russia's Far East," *Comparative Economic Studies* 37 (Spring 1): 79–86.

Timonov, V. E. (1922/1988) "Okhrana Prirody pri Inzhenernykh Rabotakh." *Priroda* 72–86 in Douglas Weiner, *Models of Nature: Ecology, Conservation, and Cultural Revolution in Soviet Russia*, Bloomington: Indiana University Press.

Trotsky, L. (1969) *The Permanent Revolution, and Results and Prospects*, New York: Merit Publishers.

UNDP (1997a) "Preliminary Trans-Boundary Analysis of Environmental Key Issues in the Tumen River Area, its related Coastal Regions, and the North-East Asian Hinterlands," UNOPS/GEF-SAP Fact Finding Mission, Beijing (June).

UNDP (1997b) "Draft Project Brief for the Preparation of a Strategic Action Programme (SAP) for the Tumen River Area, its Related Coastal Regions, and its North-East Asian Environs," Global Environment Facility, Beijing (August).

Vilkov, O. N. (1974) *Goroda Sibiri: Ekonomika, Upravlenia i Kul'tura Gorodov Sibiri v Do Sovetskii Period*, Novosibirsk: Nauka.

Vladivostok News (1996) "Institute puts environment first in Tumangan Port Project," 1, 132 (January 10): 7.

Wallerstein, I. (1974) *The Modern World System: Capitalist Agriculture and the Origin of the European World Economy in the Sixteenth Century*, New York: Academic Press.

Weedon, C. (1989) *Feminist Practice and Poststructuralist Theory*, Oxford: Blackwell Publishers.

Weiner, D. (1988) *Models of Nature: Ecology, Conservation, and Cultural Revolution in Soviet Russia*, Bloomington: Indiana University Press.

Wishneck, E. (1995) "Whose environment? A case study of forestry policy in Russia's Maritime Province," in S. Kotkin and D. Wolff (eds) *Rediscovering Russia in Asia: Siberia and the Russian Far East* Armonk, NY: M. E. Sharpe.

Working, R. and N. Chernyakova (1998) "Reported tiger gift outrages ecologists," *Vladivostok News* 5, 162 (March 6): 1.

Zochowska, B. (1993). "The Far East in its time of transition," *Revue d'Études Comparatives Est–Ouest* 24 (1 March): 69–80.

14 The problem of pigs

Alice Dawson

Introduction: animals, ethics, and geography

> A sunny day in March. Within the birch tree's slender shadow on the crust of snow, the freezing stillness of the air is crystallized. Then – all of a sudden – the first blackbird's piercing note of call, a reality outside yourself, the real world. All of a sudden – the Earthly Paradise from which we have been excluded by our knowledge.
>
> (Hammarskjöld 1964: 71)

I begin with Hammarskjöld's poignant statement that simply being human sets us apart from the world around us. Yet humans long to be a part of this world. What are the meanings and consequences of this fact of human existence? In response to this self-awareness, we struggle with the meaning of our existence and world. This includes attempts to come to terms with who we are and how we act, and what we should be and how we should act. We ponder our relationship with our environment, and what this relationship should be. This reflection occurs within the human societal and cultural context, with collective acceptance of individual beliefs, thoughts, and ideas. Within the societal and cultural context, each individual develops knowledge of what is and what should be. We have come to expect and use an objective, empirical approach to the resolution of questions and problems. Yet ultimately the answers to many questions, particularly those of meaning and what should be, of values and beliefs, are in nature subjective and ultimately personal. Ironically, we must use that which sets us apart from our world – our knowledge, to use Hammarskjöld's word – to attempt reconnection with our environment and discover our place in the world.

Animals occupy a unique place in the human environment. They are at the intersection of the cultural and biophysical environments. Animals are living, sentient, biological components of the natural environment. Yet they are a part of human culture: food, clothing, cosmetic and pharmaceutical products, medical products, physical assistance to the disabled, vocabulary and symbolism, and friendship and companionship. Humans form deeply meaningful, personal relationships with animals that do not occur with any other component of our environment. Cultures and individuals hold strong beliefs about where different animal species should be placed in space and in relation to humans, and these reflect perceptions about the human and natural environment relationship. Humans recreate their environment, including other living beings, in the image that they see fit. To aid in our discovery of our place in the world, it seems appropriate to consider the consequences of these aspects of our relationship with this particular component of our environment – the animals – for the animals themselves. Our sense of separateness and difference from the other animals, of being outside of nature, is a critical factor in the ways in which we construct animals and our relationships with them. This plays out in our ethical consideration of animals.

While the formal study of ethics is primarily the realm of philosophers and theologians, life confronts humans with ethical choices daily. Birdsall (1996) speaks of the numerous choices that we each must make daily, and the cumulative consequences of these choices. Elsewhere in this volume (Chapter 18), Gormley and Bondi comment on the ethical dimensions and consequences of everyday activities. As a society, as individuals, and as geographers, we must determine those things that are important and reflect these in our everyday actions, including geographical inquiry. Unwin speaks elsewhere in this work of geography's "right and duty to be involved in social, economic, and political change" (p. 263), and much of this volume echoes the underlying conviction that geography does indeed have such a right and duty, "a moral responsibility to address issues of social suffering, injustice and oppression" (Kitchin, p. 233).

In Chapter 8 of this volume, Tuan considers evil, citing the traditional difficulty of addressing this issue in geography (and within the Western scientific framework more broadly). Particularly for physical geography, with its largely inanimate subject, issues of ethics have seemed far removed from physical earth processes and systems. Western moral philosophers struggle to determine if a rigorous philosophical argument can be constructed to allow the environment as a whole as well as the animals within the environment "rights." Yet humans have reflected, and will continue to reflect, on the morality of their interactions with their environment, including the animals.

Buttimer (1974) notes that values are an inherent component of an ethical approach, and that the practice of geography is based on personal and societal values. Values become individually and collectively what Proctor describes in his essay (Chapter 1 in this volume) as the basis of morality and systems of good and bad, of right and wrong. These values become the basis for an individual's ethical framework. Each of us brings something of ourselves – world view, values, and a consequent sense of "what should be" – to what we do as geographers. As we learn and increase knowledge and understanding of our world, re-examination and reflection on our values and our relationships with our world in light of our increased knowledge seems imperative. Where better than in geography can the ethical issues of the human/environment interaction be addressed? What is the human place within our environment, and what should it be? What issues of oppression and justice for animals, as sentient beings within our environment, arise within the context of interactions with our environment, and what should our response to these be?

My concerns

Relationships between human and animal continue as a matter of life-long reflection for me. Although as a child I could not kill insects to collect them "properly," I observed insects from the time I could walk. I read about animals and eagerly took advantage of any opportunity to see and interact with animals. As an undergraduate, I changed from my zoology major as the scientific study of animals had little to do with their interests. I worked for years in various capacities in animal welfare work, including animal and rabies control. One aspect of this work was the deliberate killing of society's surplus animals, an irony and source of considerable conflict for one who cannot kill insects to collect them. I continue my exploration of the relationships between humans and animals, and to share my life with many individual animals of assorted species.

Geography intrigued me with its concern with the human/environment interaction, and the human place within our environment. My reflections about the human–animal relationship as a geographer are within the realms of the intellect and academia. The basis for these thoughts comes from experiences with animals in my immediate and daily environment: the experiences previously mentioned, frustration with predations of a hungry raccoon on a small flock of chickens, and determining the responsibility of my actions in terms of the best interests of a handicapped, tamed wild opossum. These various dilemmas arise in allowing the different species of animals that are a part of my daily life to be as truly themselves as possible. What is and should be my response to these components of my environment? These dilemmas fuel my reflections as a geographer.

Birdsall (1996) urges a "moral geography of the everyday." "Day-to-day experiences are the phenomena of our existence, its raw material. They are what we 'know,' even without conscious awareness of that knowledge" (1996: 619–620). These experiences are all the more powerful if we attempt consciously to examine and reflect upon them, give them meaning, and integrate them into our ethical considerations. What should the human relationship with animals in the environment be?

The context

Humans have constructed within their intellectual traditions various understandings of nature and animals, and human relationships with animals. Exploration of these relationships continues to be very much a part of current intellectual and academic endeavors. Understanding of the human intellectual tradition concerning human–animal relationships is useful before examining the human–pig relationship.

Rationalism and a dualistic way of understanding the human place in the world are fundamental to the late twentieth-century American understanding of animals and our relationships with them. Tracing the development and role of these ideas in Western history is well beyond the scope of this essay, but as underpinnings of our understanding of the human relationship with animals, these ideas must be mentioned.

The divide between reason and emotion was delineated toward the end of the medieval period. Universality, generalization, and science gained primacy, and the importance of particularity and difference dwindled. During the Enlightenment, the Newtonian/Cartesian scientific approach emerged (Donovan 1993). The world was divided into the generalized and repeatable (that which can be expressed mathematically) and the particular, specific, and unpredictable (all that which is not man, i.e. nature, including the animals and women). Human males, with their exclusive rationality, determined themselves separate from and superior to the rest of their fellow creatures, including women and animals. Descartes carried this reasoning to what might be considered the logical conclusion. Because animals lack the supreme faculty of reason, they are unfeeling and mechanical; they are machines (in Regan and Singer 1976). Consequently humans need have no concerns about the nature of their relationships with animals. Birdsall (1996) describes the consequences of this understanding:

> Western culture is based on a conceptualization of humanity that places us in the world but not of the world. We are apart from the rest of the world, not a part of the world. At a conscious level, we are aware that our existence is one part of many in the knowable universe. But at the subconscious level, the assumptions that guide our every day behavior tells us that we are not merely one part among many, but the only part that really matters.
>
> (Birdsall 1996: 621)

Each species is indeed different from the others (cf. the definition of a species), and each has its own particular gifts. So too humans are different. Humans have a sense of self-awareness and a level of cognitive processes that set us apart. But we believe that because we are rational, and since it is our gift and therefore the ultimate gift, humans have rights and powers over everything else. (But as Simmons (1993) notes, we are also the only reference point we have.)

In *The Descent of Man*, Darwin acknowledged our connection to other animal species (in Regan and Singer 1976), including the commonality of "complex emotions." Darwin describes the difference in the human mind and that of the other animals as a difference of degree rather than kind, questioning human presumption of our exclusive possession of reason. Recall Hammarskjöld's words at the beginning of this chapter. The irony and imperative is that we must use this which makes us unique and separates us to recreate our connections with our world, including the animals.

Current academic research continues to focus on human/animal relationships. Issues of animal rights are a part of this intellectual context, and thoughtful debate on this topic has emerged within moral philosophy in particular. Even here, however, there are varying perspectives. For example, two of the seminal thinkers on animal rights take fundamentally different approaches. Tom Regan speaks of animals' rights based on animals' inherent value, while animal sentience and speciesism are the foundation of Peter Singer's concerns (Ryder 1989). In *The Case for Animal Rights* (1983), Regan carefully develops a rigorous philosophical argument for the rights of animals, drawing on philosophy's understanding of issues such as direct and indirect duties, consequentialist and nonconsequentialist ethical theories, inherent values, and concepts of moral and legal rights. He concludes that animals, like humans, have "certain basic moral rights, including in particular the fundamental right to be treated with the respect that, as possessors of inherent value, they are due as a matter of strict justice" (Regan 1983: 329). Conversely, Singer's approach within the philosophical methodology is based in utilitarianism. He maintains that human interests should not have greater weight over animals' interests and preferences. He notes Jeremy Bentham's oft quoted remark in response to Kant: "The question is not, Can they (animals) reason? nor Can they talk? but Can they suffer?" (Singer 1975: 211). Singer develops the concept of speciesism, and believes that it violates the principle of equality. According to Singer, it privileges the interests of one's own species (i.e. humans) over the greater interests of members of other species, even though one's own species' interests may be quite trivial in relation to the far greater interests of a member of the other species (Singer 1975).

Feminist thought, particularly ecofeminism, has also considered the human–animal relationship, and what this relationship should be. Rooted most firmly in radical feminism, ecofeminist thought concludes that the mechanisms that dominate and oppress women and nature, including animals, are the same. "Ecofeminism's basic premise is that the ideology which authorizes oppressions such as those based on race, class, gender, sexuality, physical abilities, and species is the same ideology which sanctions the oppression of nature" (Gaard 1993: 1).

Amongst late twentieth-century Americans, there is a spectrum of responses to animals and nature. Kellert (1993, 1996) developed a taxonomy of the primary views held by Americans about animals and nature. The categories are utilitarian (practical, material exploitation), naturalist (direct experience, exploration), ecologistic-scientific (systematic study), aesthetic (physical appeal, beauty), symbolic (nature in language and thought), humanistic (strong attachment, love), moralistic (spiritual reverence, ethical concern), doministic (domination, control), and negativistic (fear, alienation). These attitudes are expressed culturally in various ways, including those of human organizations concerned with animals. At one extreme are groups that want to ensure wild animals continue to exist so humans will be able to hunt them, and at the other, those who believe that any human interference in any way in an animal's life is oppression and exploitation.

A case in point: the pig

> Lord,
> their politeness makes me laugh!
> Yes, I grunt!
> Grunt and snuffle!
> I grunt because I grunt
> and snuffle
> because I cannot do anything else!
> All the same, I am not going to thank them
> for fattening me up to make bacon.
> Why did You make me so tender?
> What a fate!
> Lord,
> teach me how to say
> Amen
>
> Carmen Bernos de Gasztold,
> *Prayers from the Ark,* 1966

This prayer of a little pig summarizes the pig's situation in relation to humans. A pig does pig things because it is a pig, and finds its meaning in being what it is. But humans find it tasty (among other things), and so we attempt to modify this component of our environment to suit our various needs. It is implicit in this poem that humans have control over the pig, and human desire for use of the pig has priority over the pig being a pig. This is our response to the pig. Should our response be otherwise?

Human response to pigs is something of a "mixed bag," and because of this complexity the pig presents a useful case study of various aspects of the human/ animal relationship. Pigs are not perceived as embodying the spirit of wildness, nor virtues of selfless devotion to humans. Pigs are food, life's necessity. Because of the ways we perceive pigs, they are accorded treatment that would not be tolerated by American society towards another species. A wrecked truckload of injured and/or terrified pigs on a major highway provokes perhaps an attempted journalistic witticism about not bringing home the bacon (personal experience). A truck hauling dogs in a comparable fashion would create public outrage. In Kellert's taxonomy (1993, 1996), most of us approach the pig with a utilitarian and/or perhaps doministic viewpoint. This chapter offers no definitive answers to what our relationship with pigs and other animals in our environment should be, but rather calls attention to and reflects upon particular aspects of the pig–human relationship as an example of our relationship with one component of our environment. As we clarify our relationship with animals, including pigs, we can better contemplate what our relationship with animals, including pigs, should be, and envision a better response to this living being in our cultural and biophysical environments.

Pigs have been a part of human material culture for thousands of years and remain so: food most of all, an agricultural commodity, a working animal, some-times companion, and increasingly a medical resource. They are laden with symbolic meanings and are a part of our daily vocabulary, mythology, folklore, and religious traditions. Like other animals, they are the interface of the cultural and biophysical environments.

There are numerous varied perceptions of pigs through time and space. The pig's status as "unclean" in the Jewish tradition may have arisen because pigs were sacred to deities other than Yahweh; or because eating pigs was unhealthy, leading to trichinosis; or perhaps because pigs feed on garbage. Perhaps this anti-porcine sentiment may have been an ecologically adaptive behavior as pigs are unsuited to the nomadic lifestyles and hot arid environment of this region (Lawrence 1993). Lawrence also cites anthropologist Marvin Harris' view that disdain for pigs is ecologically adaptive because pigs are omnivorous and compete with humans for food. Prior to the Middle Ages, pigs were associated with divinity (Nissenson and Jonas 1992), yet pigs were burned at the stake during the European witch hunts. In addition, the wild boar was believed to exhibit the attributes of nobility: courage, power and cunning, while the domestic pig represented the sloth, gluttony, and lasciviousness of the common peasant (Kearney 1991).

Our use of pigs plays a critical role in our perception of pigs. Humans may feel a deep guilt for being so mercenary about pigs. "Hogs are generally only com-modities without respectability or identity, and they are harvested without a qualm. As repositories for our own fears of ourselves and the animal within us, pigs bear the brunt of our self-reproach" (Lawrence 1993: 324).

Humans feel guilty about pigs, and our relationship with them. Unlike other species such as sheep, who have wool, and cows that give milk, humans raise pigs simply to kill and eat them. The shame for this relationship becomes attached to the pig itself. Lawrence also notes anthropologist Edmund Leach's comments:

> Besides which, under English rural condition, the pig in his back-yard pigsty was, until very recently, much more nearly a member of the household than any of the other edible animals. Pigs, like dogs, were fed from the leftovers of their human masters' kitchens. To kill and eat such a commensal associate is sacrilege indeed!
>
> (Ibid.: 316)

Pigs were our "near neighbors" and backyard dwellers. They shared food and space with us, and are physiologically more similar to humans than perhaps any other species, save the primates. This familiar proximity fosters human identifica-tion with pigs and allows humans to make them "vehicles for human feelings" (Ibid.). Their nearness and similarities allow us to see too easily in them those qualities we would prefer not to recognize in ourselves. Kearney (1991) con-cludes that "the object of man's peculiar cultural disdain for the pig is less the beast itself than man's own speckled soul" (1991: 322).

If sharing space with pigs and our use of pigs results in "uncomfortableness" about pigs, how do these perceptions play out as we order our landscape? Kitchin's comments in Chapter 16 of this volume concerning the disabilist nature of society and its inherent spatialities and the consequences for disabled humans are applicable in this context concerning animals produced for food. "Forms of oppression are played out within, and given context by, spaces and places," (p. 223) and appear as messages on the landscape. We are uncomfortable with these places where pigs are rendered into neatly packaged pork products from live creatures. (Singer's "Down on the Factory Farm" in *Animal Liberation* (1975) or Serpell's "Of Pigs and Pets" in *In the Company of Animals* (1986), are difficult reading for most.) So pig farms and processing plants are located out of sight (and smell), actually hidden from the landscape. Adams (1991) speaks of the ways we remove ourselves from the discomfort of eating the flesh of other creatures by referring to meat as "pork" or "beef" or "veal." Paradoxically, pork products often are marketed with smiling, happy pigs, enticing us to eat this happy creature as pork. Shepard (1996) writes of the most essential relationships between early humans and animals: the animals ate us, and we ate them. A sense of respect for an animal because it would allow itself to be eaten was part of this relationship. Perhaps these happy pigs asking to be devoured are modern humans' rather feeble attempt to express respect for the pig for allowing us to live by eating it. It alleviates our feelings of guilt to know that by eating the pig we are giving it what it really desires, allowing it to be what it truly is, and therefore making it happy.

The concept of human power over nature, especially with our increasing technological abilities, is another aspect of the culture–environment interaction, and certainly pertains to the human–animal relationship. Tuan (1984) considers how humans change the face of the earth by looking at power and domination in relation to affection. Gardens are created by humans to shape their environment as humans believe nature should be. So too animals are modified. Whether innocent and benign or destructive and exploitative, the modifications resulting from human power are based on the needs and desires of humans. Genetic variations in dogs are institutionalized in human society (cf. the American Kennel Club), but these variations have no value to the dog and in many cases are detrimental to it. Tuan's comments in Chapter 8 of this volume and elsewhere concerning "telescope goldfish" are another example of this problem (Tuan 1984).

Domestication of animals is certainly one of the essential aspects of the human/animal relationship and is often viewed as a form of domination of nature. However, a broad spectrum of understanding about this relationship can be found. Domestication may be a terrible distortion of the original relationship between humans and the other animals, and not part of our evolution or "biological context" (Shepard 1996). It is yet another example of the imposition of human will on the world. In a completely different light, domestication may be a beautifully adaptive evolutionary strategy. Budiansky (1992) writes:

domesticated animals chose us as much as we chose them. And that leads to the broader view of nature that sees humans not as the arrogant despoilers and enslavers of the natural world, but as a part of that natural world, and the custodians of a remarkable evolutionary compact among the species.

(1992: 24)

Interestingly, several animal species, such as rats, raccoons, white-tailed deer, and coyotes, that can live in proximity to humans, are thriving. This lends validity to the idea that some animals might have chosen us as we chose them. Domestic animals too exist in great numbers, so great that we can kill vast numbers for food or simply because there are too many for our society to support.

Whether or not animals were willing participants in the process of domestication, or it was an intrusive intervention for our own ends, or the meshing of evolutionary development, the fact remains that in large degree we have created these animals. Domesticated animals depend upon human beings for their survival; we have created them to be dependent. Our evolutionary paths are now irrevocably linked. An important consequence of creating a dependent species is that our responsibilities and obligations to these creatures are far greater. As Shepard (1996) comments: "Experimental science, widespread pet keeping, the increasing application of technology to animal husbandry, and the marketing of livestock all produce their own forms of abuse and reaction against them" (1996: 305). All of these relationships between humans and the other animals present new ethical dimensions, with perhaps even greater claims on the humans involved as the situation is of our own making. The issue of domestication and its consequences presents another opportunity for reflection upon the nature of the human relationship with the other animals.

Porcine particulars

Three particular aspects of the pig–human relationship bear consideration and offer the opportunity for ethical reflection, particularly within the context of cultural and environmental interaction, including the impact of human activity on this particular component of the environment. These are the pig as agricultural commodity, the use of pigs as a medical resource, and the pig as pet.

In a sedentary agricultural setting, the pig may have many roles. In scavenging it finds its own food and cleans human living areas. Rooting clears the land and prepares the soil. Pig manure is used as fertilizer (Porter 1993). Pigs are no longer allowed this diverse lifestyle, nor are they "backyard" animals, common on our landscape. Agriculture has become agribusiness, and with this cultural change we modify our biophysical environment in numerous ways. Greater and greater numbers of pigs are kept in geographically more limited areas. In North Carolina, the number of hogs has grown from 2,000,000 in 1970 to almost 9,000,000 in 1996, yet these animals are concentrated in "smaller geographies," as the pig population "geographically implodes" (Furuseth 1997). During a seven-month investigation, the Raleigh (North Carolina) *News and Observer* monitored the development of this industry in the state. Environmental damage from hog waste lagoons and the extensive use of hog waste as fertilizer, legal restrictions to protect citizens' land values, issues of hegemony and equity for agribusiness corporations and small farmers, and legislative ethics are considered in the report of this comprehensive investigation. This is indeed a considerable array of concerns for ethical reflection within the context of the human–environment relationship and human modification of the environment. The well-being of the aspect of the environment most affected by all this system, the pig, is never mentioned.

Modern agricultural production of pigs creates an environment based on human convenience and economic profitability that is quite different from the image of Old MacDonald's farm. Part of the *News and Observer* report (Stith and Warrick 1995) is a full-page graphic illustrating "Manufacturing Hogs." The process is described as an "industrial assembly line" that produces "thousands of lean, carbon-copy hogs produced at the least possible cost" (1995: 8). Advertisements in pork producer publications refer to sows as "farrowing units" (*Pork* January 1996). If Descartes' description of animals as unfeeling automatons gives us pause, so too should these descriptions. Additionally, Curry (Chapter 7 of this volume) examines the making of places and human activities that define places. He concludes in part that in this process, "some people and actions and things belong in a place and some do not." Mason and Singer (1980) begin their examination of such industrial agricultural facilities with five lines from Milton's *Paradise Lost*:

> No light, but rather darkness visible
> Served only to discover sights of woe,
> Regions of sorrow, doleful shades, where peace
> And rest can never dwell, hope never comes
> That comes to all; but torture without end . . .

Animals perhaps do not share our sense of place, but it is appropriate to ask if the environment we have created for living beings is as it should be. Do pigs belong in the places we have made for them? What responsibility do we bear for the creation of these places? If humans can develop such differing senses of place about where animals are raised, it would seem a point of ethical reflection. We have changed the environment; what are the consequences? Such inquiry should not be limited to the realm of moral philosophers, but is open to anyone who sees human modifications in our world every day and reflects on what they may mean.

Not only the places but also the scale at which we raise pigs has changed. This contributes greatly to the environmental consequences mentioned previously, and the scale of the consequences to pigs increases with the scale of production. The individual relationship and actual "caring for" an animal is gone (Shepard 1996). Pigs are housed in industrial-type warehouses of hundreds or thousands of pigs. Humans breed pigs that are genetically predisposed to produce large litters (so large that sows cannot nurse all the piglets) and lean meat quickly (so quickly that bone and muscle development cannot keep up with the piglets' growth rate)

(Mason and Singer 1980). This environment that humans have created and deemed suitable for pigs does not permit normal social interactions and behaviors such as rooting and nest building. Instead, aberrant behaviors develop such as chewing on the tails and hindquarters of other pigs. Physical responses such as porcine stress syndrome and ulcers may occur.

Pigs are used increasingly for medical purposes. Pig heart valves repair human hearts, insulin for diabetics is developed from pigs, and pigs' skin may be used in burn treatment (Lawrence 1993). These products are primarily gleaned as by-products of the slaughter of pigs for food. With increased biomedical techno-logical sophistication, new ways to modify the pig emerge. Pigs are bred with retinitis pigmentosa (personal communication with Dr Charles Stanislaw 1996), a human eye disorder, in order to study this human disorder. Genetic engineering opens a world of possibilities. At least four biotechnology companies are working to develop genetically altered pigs as organ donors for humans (Fisher 1996). Xenotransplantation offers relief of the shortage of organ donors. The pig, the "foreigner," may become part of our physical body in new ways. It is not only as food that this part of our environment sustains us. With this, come a whole host of additional questions about our relationship with pigs and other animals.

In recent years, the pig has been advertised as a fashionable pet. The I Pig of North Vietnam, or Vietnamese Potbellied Pig, was introduced to the USA in the 1960s and now "suffers the indignity of becoming a city pet" (Porter 1993: 187). The advertised intent is a mature pig weighing 35–50 pounds but the average weight of these "hot new pets" is closer to 150 pounds. The pig's intelligence and natural behaviors require an outlet, and pet pigs will root in soil or vegetation, or carpet and linoleum. One owner said that her pig tore up the kitchen floor and ate it like fruit roll-ups (Jefferey 1995). Territorial behaviors are exhibited when the pig reaches puberty; many of these behaviors are deemed unacceptable by humans. These pigs are abandoned more and more at animals shelters because their owners simply are not prepared to deal with a real pig. The problem is not that a pig should not be a pet, but rather a pig is a pig first and then a pet. The perception of and resulting response to the pig does not jive with what the pig truly is. Our efforts to recreate pigs within our cultural context in the image we choose continue to conflict with what the pig in the biophysical environment is, with serious consequences for the pig.

Conclusions

It is difficult to delineate a "proper" or "right" relationship between humans and the animals that are a part of our environment. The issue of what this relationship is and what it should be has been a point of human reflection for much of our history, and will continue as a consequence of being human and our search for meaning and our place in the world. Pythagoras, Aristotle, Thomas Aquinas, Descartes, Voltaire, Darwin, Hume, Kant, Bentham, Schweitzer and Ghandi, as well as countless others before and after them, have struggled with the meaning of this aspect of human existence. It seems unlikely that the issue will be resolved in the immediate future. The critical point is that we continue to examine the nature of this relationship and continually ask if it is as it should be.

As humans, we accept our distinctiveness as a species; our cognitive processes and self-awareness set us apart. However, this becomes the foundation on which we construct a false dichotomy between ourselves and the other animals that strives to deny our commonalities. Within our intellectual context, humans have further constructed the animal–human relationship in such a way as to reify this separation from the other animals. The urgency for this separateness increases as we recognize characteristics that we have in common with animals but that we least want to accept in ourselves. Our commonality with the other animals threatens the very foundation of the false dichotomy, and drives us to insist upon our uniqueness and superiority. We cannot be Us without an Other. We cannot be superior without that which is inferior. With this false dichotomy as the premise for our relationships with animals, we may act with impunity towards animals. In fact, there is an urgency to act in such a way to confirm and validate this dichotomy and further establish our superiority and separateness.

Part of this process is to define individual animal species in such a way that furthers this separation. By creating an animal, such as the pig, as we want it to be, we can then use it as we chose. We define the pig's characteristics, its needs, and an appropriate environment. Therefore we can define the proper role and use of the pig in relationship with ourselves. As our needs for the pig change, we can change the pig. We can even say that we have considered the pig's interests, since we have determined those interests. It is far harder to see the pig as an individual, as it is, with much in common with ourselves. I suspect Pygmalion's relationship with Galatea became far more complex when she became who she was, not simply the form he created. If we see the pig as it is, and acknowledge that in it which is in us also, and how it is much like us, this calls into question all of our relationships with pigs. A restructuring of the previously mentioned relationships with pigs involves major restructuring of human society and activity. If perhaps this is a necessity in the relationship humans have with pigs, might it also not be required in our relationships with the other animals and the rest of nature? This becomes a daunting prospect, requiring a reordering of our individual and collective lives, as a consequence of seeing pigs, animals, and nature in a new light.

References

Adams, Carol J. (1991) "Ecofeminism and the eating of animals," *Hypatia* 6(1): 123–145.

Birdsall, Stephen (1996) "Regard, respect, and responsibility: sketches for a moral geography of the everyday," *Annals of the Association of American Geographers* 86(4): 619–629.

Budiansky, Stephen (1992) *The Covenant of the Wild*, Leesburg, VA: Terrapin Press.

Buttimer, Ann (1974) *Values in Geography*, Washington, DC: Association of American Geographers, Resource Paper No. 24.

de Gasztold, Carmen Bernos (1966) *Prayers from the Ark* trans. Rumer Godden, New York: Viking Press.

Donovan, Josephine (1993) "Animal rights and feminist theory" in *Ecofeminism: Women, Animals, Nature*, G. Gaard (ed.), Philadelphia: Temple University Press.

Fisher, Lawrence M. (1996) "Down on the farm, a donor: breeding pigs that can provide organs for humans," *New York Times*, 5 January (Business Section).

Furuseth, Owen (1997) "Restructuring of hog farming in North Carolina: explosion and implosion," *The Professional Geographer* 49(4): 391–403.

Gaard, Greta (ed.) (1993) *Ecofeminism: Women, Animals, Nature*, Philadelphia: Temple University Press.

Ginsberg, Susan (1994) "Plight of the pig," *Pet Product News* 48(7): 1 and 72–73 (July).

Hammarskjöld, Dag (1964) *Markings* trans. by Leif Sjöbert and W. H. Auden, New York: Alfred A. Knopf.

Jeffery, Clara (1995) "Pigstown," *Washington (DC) City Paper*, 21 July: 18–30.

Kearney, Milo (1991) *The Role of Swine Symbolism in Medieval Culture Blanc Sanglier*, Lewiston, NY: Edwin Mellen Press.

Kellert, Stephen R. (1996) *The Value of Life*. Washington, DC: Island Press.

Kellert, Stephen R. and Wilson, Edward O. (eds) (1993) *The Biophilia Hypothesis*, Washington, DC: Island Press.

Lawrence, Elizabeth Atwood (1993) "The sacred bee, the filthy pig, and the bat out of hell: animal symbolism as cognitive biophilia" in *The Biophilia Hypothesis*, S. R. Kellert and E. O. Wilson (eds), Washington, DC: Island Press.

Mason, Jim and Singer, Peter (1980) *Animal Factories*, New York: Crown Publishers.

Nissenson, Marilyn and Jonas, Susan (1992) *The Ubiquitous Pig*, New York: Abradale Press.

Pork Magazine (1996) *The Business Monthly for Pork Producers* 16(1): January.

Porter, Valerie (1993) *Pigs: A Handbook to the Breeds of the World*, Ithaca, NY: Comstock Publishing Associates (a division of Cornell University Press).

Regan, Tom (1983) *The Case for Animal Rights*, Berkeley and Los Angeles: University of California Press.

Regan, Tom and Singer, Peter (eds) (1976) *Animal Rights and Human Obligations*, Englewood Cliffs, NJ: Prentice-Hall, Inc.

Ryder, Richard D. (1989) *Animal Revolution*, Oxford, UK and Cambridge, Mass.: Blackwell Publishers.

Serpell, James (1986) *In the Company of Animals*, Oxford, UK: Blackwell Publishers.

Shepard, Paul (1996) *The Others*, Washington, DC: Island Press.

Simmons, I. G. (1993) *Interpreting Nature*, London and New York: Routledge.

Singer, Peter (1975) *Animal Liberation*, New York: Avon Books.

Stanislaw, Charles (1996) Personal communication, February 9, North Carolina State University, Raleigh, NC.

Stith, Pat and Warrick, Joby (1995) "Boss hog: North Carolina's pork revolution," Raleigh, NC *News and Observer*, February 19–26.

Tuan, Yi-Fu (1984) *Dominance and Affection: The Making of Pets*, New Haven: Yale University Press.

Part 4
Ethics and knowledge

How we come to know raises issues of personal professional practice with an ethical dimension. Indeed, professional ethics is a major growth area in contemporary geography. The chapters in this final part of the book exemplify from a variety of research contexts. Thomas Herman and Doreen Mattingly begin: they invoke a communicative ethics in engaging in participatory research practice. They describe the creation of a framework designed to empower people living in a particular neighborhood to communicate their own use of space and perspectives on their own community, by the use of art. They identify three insights from critical social theory which were found useful in guiding their research experiments: the emphasis on social relations as an object of study and an aspect of the research process, the importance of representation in defining and reinforcing social relations, and the analysis of the public sphere as a discursive process defining citizenship and democratic participation. They argue that art can play an important part in connecting critical thought with community action. Their form of applied critical geography expresses an ethical commitment to both the academy and society at large.

Rob Kitchin provides another example of participatory research. He examines two separate but related issues in geographical studies of disability. The first is the moral responsibility of (non-disabled) academics to undertake what he describes as critical emancipatory and empowering research on disability. The second concerns the epistemological and ethical basis of conducting such research, including the legitimacy of non-disabled persons writing about a group of which they are not members. One possible reaction, which the author himself employs, is to involve disabled research subjects in the role of co-researchers, a practice which need not prejudice what may be considered necessary scientific procedures. Such inclusive research might even claim scientific advantages over more conventional (exclusive) methods, by drawing directly on the experience of the disabled themselves, as well as being ethically superior in the sense of being emancipatory and empowering.

Robert Rundstrom and Douglas Deur raise other issues concerning relationships between researchers and research subjects. They address the articulation and reconciliation of what they refer to as the colonizing academic gaze and the institutional apparatus out of which it peers. Cross-cultural research in particular involves the geographer encountering alternative views of the world, including what constitutes ethical behavior. For these authors (like Gormley and Bondi, see the following paragraph), ethics is relational and contextual, requiring reciprocity between researcher and researched. Ethics and ethical behavior in research does not emerge from isolated reflection, but has to be negotiated as a form of what they term reciprocal appropriation, involving exchange of information as part of long-term relationships.

Nuala Gormley and Liz Bondi use a different kind of relationship, that between research student and supervisor, to explore some further ethical issues arising from everyday research practice. They recognize, with others, that research involves an attempt to steer a path between understanding and the exploitation of research subjects. And like others, their approach to ethics is situated and relational. Their focus is on issues arising in the transformation of "data" from one context to another. The student (Nuala) explores the problem associated with the balance to be struck between appropriating knowledge for her thesis from others (research subjects), and displacing their perspectives by removing data from its context. The supervisor (Liz) notes a similar problem, of displacing one agenda or objective (that of the student, to produce a thesis) with another (that of the supervisor, to publish a paper), in the joint authorship of their contribution to this book. Throughout, they reveal a critical self-conscious integration of theory and practice.

Finally, Tim Unwin addresses a fundamental ethical issue underlying much geographical research in recent years. He explores the proposition that we have a right and a duty to be involved in social, economic and political change. With the postmodern challenge to foundationalism in mind, he asks how it is possible to ground judgments concerning better or worse conditions and actions. He calls on the writings of John Locke, who played a crucial role in shaping the Enlightenment ideal that it is both possible and desirable to improve society, to reveal something of the complexity of the social activist stance. His personal conviction that the geographer should be engaged in changing the world raises the question of how we know what to do with our own praxis, which forces us to make and to justify moral choice. Unwin's chapter provides a fitting close to the volume by highlighting the tension between modernist and anti-/postmodernist conceptions of being, knowledge, and morality raised at the outset in Roebuck's chapter and alluded to in many other of the volume's chapters.

15 Community, justice, and the ethics of research

Negotiating reciprocal research relations

Thomas Herman and
Doreen J. Mattingly

Amid the diversity of interpretations of both geography and ethics, we concern ourselves in this chapter with the ethics of our professional practices of research and writing. Following Jane Flax (1993) and Iris Marion Young (1990), we view ethics as procedural, composed of sets of interrelated social practices. We evaluate "what is ethical" not in terms of absolute standards, but rather in terms of processes that bring about more just social relations (for similar theoretical perspectives, see chapters by Nick Low and Brendan Gleeson, Ron Johnston, Jeremy Tasch and Caroline Nagel in this volume). Accordingly, we evaluate the ethics of our own behavior in terms of our common sense responsibility for the influences of our actions on larger social processes. From this theoretical perspective ethics ceases to be something that can be objectively researched and becomes something that must be activated in social practices, including the practices of academic work. Thus, for us, ethics and research are both intensely personal and necessarily public. They are themselves composed of, and have an effect on, social relations.

In addressing issues of research and writing practice, our concerns parallel those of Nuala Gormley and Liz Bondi (Chapter 18 of this volume) and Rob Kitchin (Chapter 16 of this volume), who also seek to examine and make more just the relations of research and representation manifest in academic work. In particular, we are motivated by ethical concerns about the unequal exchange of knowledge and the power relations implicit in our relationships with the people we study. Feminist and postcolonial critical scholarship has developed critiques of traditional research practices that resonate with our own experiences of research (Clifford and Marcus 1986; Fonow and Cook 1991; Harding 1991). Like other social scientists we mine the lives of our research subjects for our own use and write stories that simplify, objectify, and at times misrepresent them. In return we offer them only token payment for their time (if they're lucky) and the vague promise that our work might some day change the academic discourse about their lives in a manner that might indirectly benefit them. In academic writing, the primary means of addressing these ethical concerns is reflexivity, which involves highlighting the social location of the scholar and scrutinizing the effects of that location on the research and analysis (England 1994). While reflexivity may satisfy our ethical responsibilities to our fellow scholars, it does not affect our everyday interactions with our research subjects. For us this ethical bind persists not only in theory but also in everyday life. Both of us are conducting research projects that require intimate involvement in a place-based community. We feel that the task of representing the community in our writing makes us a part of the community. Therefore, as we research and write, we must simultaneously negotiate two sets of social relations: those with other scholars and those with other community members (see also Katz 1994).

In this chapter, we describe our attempts to negotiate this ethical dilemma by pairing our research agendas with involvement in community arts projects. Our participation in community arts projects contributes to the community we study in two ways. The first contribution is straightforward: we help to provide and support interesting activities and opportunities for young people in a resource-poor neighborhood. The second is more abstract and concerns the content of the arts programs, which attempt to provide spaces of self-representation and community articulation. Our hope and hypothesis is that by encouraging marginalized people (specifically inner-city youths) to project their own voices and positions, we can increase their participation in the public sphere and thereby facilitate justice. Our participation in community arts projects also serves our own research interests by improving our access to our research subjects and hopefully contributing to more grounded and nuanced interpretations. Thus we view our community involvement not as a gift but as an exchange; part of a reciprocal – and therefore more egalitarian – relationship.

Our motivation for taking on the additional work of community projects is to establish relations of mutual support with the individuals and communities we are studying. We seek to ensure reciprocity by making a practical contribution to the life of the community we are studying "up front," rather than assuming that positive effects will trickle down in the long term. We suspect that such an approach is not uncommon, although largely invisible in academic discourse (for exceptions, see essays in Burawoy *et al.* 1991; Kobayashi 1994; Pulsipher 1997). We hope our efforts will lead to greater discussion about methods for navigating shared ethical dilemmas, and while we realize that these attempts may be flawed, they are products of our commitment to take seriously both our theoretical positions and our responsibilities to the people and places we write about.

Reciprocal research relations

Our reading of critical debates about the process of knowledge production makes it possible to identify three moments at which ethical issues pervade academic practice. We review them here in the reverse order that scholars encounter them. The third moment is when academic discourses are read and mobilized. In particular, critiques have drawn attention to the effects that the categories and language of academic scholarship have on the world we study. One example comes from development discourse, in which people with few material goods are depicted as victims of poverty. This representation has been crystallized in development practice in a manner that has frequently exacerbated and insti-tutionalized hunger in "underdeveloped" regions (Escobar 1995; Yapa 1996). The act of writing constitutes the second moment. Of particular concern is the way that we as scholars represent the people we study, and what we claim to know about them. The question of what shapes a researcher's ability to authoritatively represent the lives of others has been a persistent concern of feminist scholarship. For example, some feminists have argued that "objective" truth claims are in fact shaped by the social positions of scholars (Haraway 1991; Rose 1993). Others argue for "standpoint epistemology," emphasizing experience as the basis for authoritative knowledge (Harding 1993). The first moment is our interaction with the people that we study. Although less critical attention has been paid to this moment, we are interested in the ethics of our immediate relationships with research subjects.

The widespread institutionalization of university policies and committees for the protection of human subjects attests to the widely held belief that research projects should have no immediate negative effects on subjects. Our concern goes beyond mitigating or limiting negative effects, to establishing relations of reciprocity between ourselves and the individuals and communities we study. Whether through qualitative or quantitative techniques, we as researchers extract and use the details of other people's lives to construct our own stories, analyses, and ultimately, careers. It is this aspect of the research relationship that we seek to change. While we acknowledge that our research continues to depend on others as sources of "data," we believe that we have an ethical responsibility to offer something in exchange for what we receive. One way that scholars have attempted to establish more reciprocal research relations is through applied research (e.g. Kenzer 1989). Specifically, applied researchers focus academic analysis on practical questions generated outside of the academy. Traditionally, applied research has been an exchange of problem solving (provided by the researcher) for money or influence (received by the researcher). The most com-mon constituencies of applied work have been industry, government, and other institutions. These entities have been able to form productive partnerships with the academy because they present their needs in an organized fashion and financially support research conducted within universities.

Although traditional applied research addresses the problem of reciprocity, we find it a problematic model for three reasons. First, problem solving involves hierarchical relations of knowledge and power. Those who are defining the prob-lem and attempting to solve it are removed from and more powerful than those who are defined as part of the problem, although the ideal of "objectivity" disguises these power relations (Haraway 1991). Second, the deployment of categories in social practices can have a material effect on those phenomena, at times creating problems where none existed before. More often than not, the categories of social science reify the centrality of some positions (white, male, affluent) while marginalizing or "othering" positions categorized as different (Rose 1993). Third, the emphasis on institutions as recipients of applied research severely limits the potential for such work to do other than perpetuate hegemonic power relations. Beneficiaries of traditional applied research are still limited to those consciously and outwardly identifying themselves as a group with articulated goals and interests. Thus, serving institutions and other already existing constituencies means that we remain apart from practices through which identities and coalitions are formed, the very processes that many critical social theorists emphasize and investigate (see, for example, essays in Keith and Pile 1993; Pile and Keith 1997).

Ironically, despite the powerful commitment of poststructural social theories to engendering more just social relations, their deconstruction of social categories and relations often leads to paralysis rather than action. At times, well-intended efforts at reflexivity become endless chains of deconstruction, leading many to question not only the stability of social categories but also the rationale for researching and writing at all. Rather than a foundation for engagement with the social world, these understandings of social relations can create hesitancy and even withdrawal from direct forms of public participation. Richard Rorty (1989) seeks resolution of this postmodern dilemma by differentiating critical positions within private and public spheres. He insists that our private experiences of irony and doubt – "doubt about (one's own) sensitivity to the pain and humiliation of others, doubt that present institutional arrangements are adequate to deal with this pain and humiliation, curiosity about possible alternatives" – must not interfere with the liberal hope that allows us to make positive contributions to the communities with which we identify (Rorty 1989: 198). The spirit of Rorty's argument resonates with us. Further, we find that critical social theory, though itself composed of multiple and necessarily evolutionary intellectual projects, can provide more than deconstruction. All critiques, and especially those that endeavor to make possible more just social relations, suggest practical paths for engaging and changing the world.

We find ourselves balancing private irony and liberal hope by coupling our own critically informed research projects with participation in community arts projects. We suspect, although we have no more than anecdotal evidence, that many researchers also negotiate their relations with research subjects by offering some form of service or community participation. That this practice takes place despite the stubborn fact that most scholars barely have enough hours in their day to do what absolutely must be done, much less take on additional and institutionally unnecessary service projects, speaks to the ethical dissatisfaction many researchers feel with allowing their analytic roles to stand as their only form of public participation. In the next section, we describe how integrating community participation, in this case with arts projects, into our activity schedules, advances our objectives of negotiating reciprocal research relations. To begin, though, we offer a cursory sketch of the place in which we are involved.

Our projects

The context in which we conduct research, participate in community arts projects, and think about ethics, justice, and research practice is City Heights, a neighborhood often described as the "Ellis Island" of San Diego. Lying directly to the southwest of our campus neighborhood, City Heights is home to a population of 70,000. It is the most diverse neighborhood in San Diego. The 1990 census reported that people of color comprised 60 percent of the population, and 1996 school enrollment data revealed that 26 languages and close to 100 dialects were spoken there, and over four-fifths of public high school students were nonwhite (38.5 percent Hispanic, 26 percent Indochinese, 21.8 percent African American and African refugees).[1] It is also the most densely populated and one of the poorest neighborhoods in the city. The vast majority of residents are non-property owners; in 1990, 78 percent of the dwellings were renter-occupied and 28 percent of family households lived below the poverty level. Business activity in the area is limited, as are employment opportunities. City Heights suffers from street crime and gang violence, and the influence of that activity is magnified by media representations in which the neighborhood is depicted as a hotbed of criminal activity. Nonetheless, recent redevelopment projects initiated and funded by charitable, commercial, and civic organizations have begun invigorating the neighborhood and its image. The programs we have worked with are both examples of this recent wave of interest and investment in the neighborhood and its institutions.

Doreen has been involved with "Around the World in a Single Day" (hereafter referred to as AWISD), a community theater project by and about City Heights' residents. Initiated by the San Diego Repertory Theater (The Rep) in 1996, the project is now a partnership between The Rep, San Diego State University (SDSU), Crawford High School, and the City Heights Community Development Corporation (CHCDC). It is funded through grants from corporate foundations and local initiatives for after-school arts programs. The end result is a multimedia theater piece performed by local high school students, adult community members, and professional artists. The script for the play is based on research about the community, interviews with community members, and community writing workshops. The play is performed at several venues in the San Diego/Tijuana area, including City Heights' annual street fair. Along the way, students and adults from the neighborhood receive training from artists, activists, and scholars in theater, writing, and community development. Doreen's role has been directing the research process, which involves coordinating interviews with community members by high school and college students, collecting basic data about the neighborhood (such as crime statistics and local history), sharing that information with students, artists, and community members, and helping them incorporate the data and interviews into the script.

Doreen's motivations for working on "AWISD" were twofold. First, she has a long-term interest in researching the relationship between public schools and their surrounding communities. Before designing a formal research agenda, she wanted to experience issues "on the ground" through community involvement that would help her to focus her research. Second, she was committed to ways of establishing more reciprocal research relations. Playing a leadership role in a community arts project gave her an opportunity to make a contribution to the community and form relationships with neighborhood institutions and gate-keepers. Her hope is that "paying her dues up front" will set the stage for research relations that are more reciprocal and more enriching both for her work and the community. Thus, the reciprocal relations embedded in Doreen's involve-ment with "AWISD" enframe her ultimate research goals. Her engagements are consecutive in time, with community involvement preceding research.

For Tom, conducting research and participating in a community arts project were simultaneous. Tom began with a research proposal in which he planned to work with 10- to 12-year-olds to investigate the lives of children in urban neigh-borhoods and the status of children's roles as innovators and improvisers within processes of social and cultural reproduction. His research objectives required him to negotiate access to his subjects through the protective shield of children's institutions. The interpretive nature of the investigation also required him to overcome barriers to communication and understanding that separate adults from children (Graue and Walsh 1998). The institution through which access was achieved was an after-school arts and literacy program that was being offered to fourth, fifth and sixth grade students in City Heights. "City Moves" brings kids together with visual and performing artists to develop individual skills and create original art that reflects children's perspectives on life in the city. The 14-week program culminates in public performances of the music, dance, and art that the children have created.

Tom faced two challenges in achieving his research goals. He needed an approach that would provide him with a comfortable role within the community institution and he needed the kids to be invested in the project. Meeting these challenges required negotiating reciprocal relations, not only with institutions and the people who administer and staff them, but also with the kids themselves. Tom committed to being a long-term volunteer in the program and to expanding his involvement beyond the pursuit of specific research objectives. He designed part of his research methodology around an enriching arts project, giving the "City Moves" kids cameras to photograph their local environments and providing materials to make scrapbooks featuring their photographs. His participation helped the program staff and the kids to meet their own objectives, as he aided in supervising the kids, producing the culminating show, and coping with a wide variety of issues that affect those participating in the program. At an inter-personal level, Tom also worked to transcend the authority relations that normally accompany adult–child relations in order to forge friendships with the kids.

We will return to these projects, but in the following paragraphs we first discuss the connections between community, public discourse, and justice that we hope to activate both in our research and through our community involvement.

Affecting the social relations of community

At their most basic level, our research relations are reciprocal in that we invest time and energy in "being there" with the kids and contributing to projects in which they choose to be invested. But we also have tried to make our contributions to the community in ways that make sense to us theoretically and ethically. In this section, we discuss the potential contribution of our work, which we think is not necessarily community building, but rather the promotion of an inclusive public discourse about community. We want to contribute to widening the local discourse about community to include kids, cross language boundaries, recognize and respect cultural diversity, and generally take account of the fact that community and social relations are generationally reproduced. We understand this as an ethical project in that opening the sphere of discourse promotes justice. Art is useful to this endeavor because it is a particular way of entering into discourse that has a lot of potential for diverse communities. Art is not age or language dependent. Even more importantly, art encourages and enables people to contribute to public discourse by speaking from within their own cultural spaces and social locations (Májozo 1995). Therefore, art is able to valorize difference while seeking mutual understanding and compassion.

Our projects make use of the communicative power of art both to promote inclusive public discourse (i.e. community) and to produce and ground our own stories about the neighborhood. Through his work with "City Moves", Tom hopes to encourage children to produce self-narratives and represent themselves in discourse by allowing them to play with modes of self-representation. With "AWISD," Doreen seeks to build a space of encounter for the community by providing an initial context for communication and expression and by assembling an audience. In each case, reciprocal research relations are constructed at two levels. First, the community arts projects provide interesting and engaging activities for those who participate in them. In that sense they couple our immediate needs with those of our research subjects in a straightforward exchange that occurs at the point of contact. Second, the range of interactions that we have with our research subjects facilitates the authoring of academic knowledge that is informed by and sensitive to our analyses of social relations and politics of representation.

Our approaches to promoting a discourse about community are shaped by the specific contours of social relations in City Heights. Community exists there in many forms that support people in their everyday lives and make the place functional. Ethnic and religious groups, extended families, neighborhood associations, and informal support networks provide important contexts for identification and social participation. Despite this myriad of connections, communication among the various groups within the space of the neighborhood is truncated by many immediate issues faced by residents: negotiating ethnic identity, economic survival, fear and concerns for personal safety, language and cultural differences, and high residential mobility.

We have observed that children's communication and interactions transgress social boundaries more fluidly than those of more completely socialized adults. Freedom to invest their time in play allows kids to experiment with multiple identities and roles. The assimilating environments of schools and other institutions routinely manage cultural difference within collective pursuits. Experiences are commodified and standardized by entertainment media and merchandizing campaigns that recognize and fuel a universalizing "kid culture" (Davis 1997). While these dynamics bring children together in inclusive communities, they do so in a way that minimizes the importance of place and specific cultural identities (Massey 1998). Among kids, therefore, there is a disjuncture between their cross-cultural interactions and awareness of the forms of community that already hold the social environment of City Heights together. There is a need for discourse about community among kids that can help to bridge that gap.

At a more abstract level, the issue of pan-ethnic, pan-generational discussion about community can be seen as an issue of the inclusiveness of the public sphere. The public sphere is an "institutionalized arena of discursive interaction" (Fraser 1992: 110), where the norms and ideals of citizenship and participation are constantly being contested, debated, and transformed, ever in the process of changing as positions are communicated and instituted. Theoretical analyses of community as process are in many ways predicated by Jurgen Habermas' (1979; 1987) emphasis on communicative ethics in the public sphere.[2] Iris Marion Young explains the ideal of communicative ethics in the following terms:

> For a norm to be just, everyone who follows it must in principle have an effective voice in its consideration and must be able to agree to it without coercion. For a social condition to be just, it must enable all to meet their needs and exercise their freedom; thus justice requires that all be able to express their needs.
>
> (Young 1990: 34)

Building a more inclusive community, then, requires an inclusive public discourse, where all have voice and the authority to represent themselves (Young 1990). Young also argues that conditions of participation in the public sphere cannot assume transparency; all participants cannot assume that they fully understand or share experiences with other participants. Thus, justice requires an inclusive public sphere that assumes difference but provides a shared arena for contention and consensus.

The dimension of community that is lacking in City Heights is the place-based experience of the public sphere. Our hope is that our contributions to community arts projects will help to foster that experience. Art is a form of self-representation and an access point to public discourse open to all, that allows for communication across generations and ethnic and language differences. The association of art with community activism can be seen in both a long history of community arts and the articulation of motivations behind "new genre public art" (Lacy 1995). Suzanne Lacy (1995) identifies this new mode of public art by its engagement, social intervention, and sensibility about audience, social strategy, and effectiveness. In her writing on public art, Patricia Phillips (1995: 65) refers to the "evacuation of the public domain" within rational society, and argues that art can and must work toward animating the idea of the public. Within the public sphere, we believe that art occupies a privileged position that makes it a particularly potent means of connecting critical thought with community action. In the next section we further discuss how the work that we have done in community arts projects is informed by theoretical analyses of representation, community, and the public sphere.

Cultivating the authority of self-representation

The desire to create contexts and structures for self-representation forms the crucial link between our academic interests and the ways in which we have become involved in the neighborhood of City Heights. For the purposes of producing insightful and useful accounts of the communities that we study, we perceive the promotion and facilitation of self-representation to be a strategic means of addressing the issue of objectification. By encouraging our research subjects to value their own stories and embrace a narrative space, we invest in a valuable resource for our investigations. At the same time, we understand the development of skills and self-awareness to offer our subjects expanded possibilities for self-expression, engagement with the discourse of community, and direct participation in the public sphere. We are therefore hopeful that the academic capital that we produce/extract from the act of research is reciprocated by enriching the personal capital of our research subjects.

"AWISD" uses research to collect stories and background information about the neighborhood with the ultimate aim of representing the community to itself. One of the goals of the project is bringing the local community together as an audience for the piece. The project includes several workshops where community members can participate in writing about their community and creating artwork about the community to be exhibited with the play. The research for the show also creates community and audience. By beginning from factual information about the neighborhood, and interviews with residents, and then engaging other residents in the interpretation of those interviews and data, "AWISD" is a vehicle by which residents of City Heights can take part in learning about the spaces they share and in claiming responsibility for representing themselves within the community. One way this happens is by our encouraging people who are interviewed or contribute to the project in any way to attend the performances, but we also form an audience by performing the show at the annual neighborhood street fair. Ideally, we create an audience that is also a community by virtue of the fact that they are all reflected in and engaged by the show.

Audience plays a second role when the show is performed to San Diego residents outside City Heights. These audiences often approach the show knowing City Heights only as a ghetto, and hopefully leave with some human connection to the place and its residents. Thus they become an audience for the students and adults who are on stage. The project forms an empowering social relation that gives the perspectives and voices of the community a new place within a public that extends beyond the local context.

A team of high school students plays a central role in the artistic process. They help determine how to (re)present the information that has been collected in their neighborhood. They become the characters in the show that communicate what the neighborhood is and what its residents need, want, hope, and fear. One example of this process revolves around a student fight that occurred at Crawford High School. The fight started just as school ended at the only exit gate on campus, and dozens of students were soon involved in what one participant termed a "rumble." Although no weapons were used and no one was seriously hurt, the local news media descended upon the fight in droves and reported the event as a "race riot." We decided to use the "riot" as a topic for the show. The students wrote their own experiences of the fight and interviewed other students about their experiences. In the performance the juxtaposition of allegedly "object-ive" and institutionally authoritative narratives of the event with the experience-based student narratives not only gives an audience to the voices of students, it also provides an implicit critique of the ongoing silencing of those voices in the public sphere. Even in less controversial scenes, the interweaving of student voices with "factual" material about the neighborhood allows for self-representation for young people in the context of a larger discussion about issues of concern to all residents. In this manner, the show contributes to a critical and inclusive public discourse about community.

Tom's principal concern with the kids in "City Moves," who were younger than the participants in "AWISD," was to develop voice. The structure of the after-school program guaranteed an audience for scheduled performances, and he was interested in empowering the children to tell their stories of life in the neigh-borhood. The extent to which Tom was successful would affect not only the quality of his research, but also the degree to which participation in "City Moves" classes and performances offered kids opportunities to invest something of them-selves in a form of public discourse. The major obstacle to achieving his goal was the social boundary by which children are separated from adults and constructed as irrational and generally lacking the competencies required to participate con-sequentially in public life (Bardy 1994). That differentiation suspends children and their principal activity of play within a holding pattern from which influence and authority are withheld pending further socialization. To encourage the "City Moves" kids to consider their voices valid and important, Tom needed to actively counter the assumption that rationality, an adult vocabulary, and conformity are the bases for claiming any authority in the public sphere. He encouraged the children to use play and individual expression as the starting points to self-representation. He also asked the children to consider the uniqueness of their own perspectives and experiences and how they might interact with other perspectives present in the diverse community.

The activities that Tom provided for the children emphasized individualized narration. The experience of taking photographs was new to most of the kids and the necessary commitment of resources communicated the value and legitimacy that Tom attached to each individual's potential contributions. He formally identified the kids as collaborators on his own project, which was explained as authoring a book about "what it's really like to be a kid in this neighborhood." Their role was that of photojournalists, insiders who could provide the stories he was after. In addition to handing over their accounts to Tom, they also created scrapbooks in which they independently authored some representation of their perspective on the neighborhood. While the legitimacy of the children's efforts at self-representation needed to be consistently communicated and performed within the context of "City Moves" in order to provide encouragement, the importance of their stories was firmly established within Tom's methodological strategy as well as his notions of what constitutes justice in the public sphere. The fluidity with which kids relate to, take up, and abandon positions within their social and physical environments, through play, is central to their subjectivities (Aitken and Herman 1997). That reliance on play, rather than rationality, to produce local, situated, and embodied knowledge, also demonstrates a potentially powerful means of negotiating diversity and multiple identities in the way that is necessary to generate more inclusive forms of community and ultimately a more just public sphere.

The pictures the children took and stories they told emphasized two aspects of their lives. First, they reflected the kids' social and geographic networks of accessibility. They recorded the occupation of both designated and appropriated spaces, including schoolyards, gated courtyards of apartment complexes, the area's steep and undeveloped canyons, and other in between spaces of the neighborhood. They also depicted the communities in which children located themselves, groups of classmates, extended family, and friends. Second, the photographs and narratives of the children identified potential roles within the larger community. Children recorded their explorations of adult contexts and the relationships they were forging with non-familial adults. A picture of a young Vietnamese girl posing with a local store owner, or an African-American boy sitting on a neighbor's motorcycle, or a Puerto Rican girl exhibiting her friendship with local police officers are all visions of the places the children might occupy as adults. They also speak volumes about the ways in which they relate to others as a means of defining themselves.

We have found community arts projects to have multiple benefits as spaces of encounter between ourselves and the communities we study. Art is one of the few existing socially acceptable means of mutual communication open to people separated by class, age, race, and authority. In particular, within a community where language and cultural differences may impede communication, art provides a common ground. Art is one of the few ways that we as white academics can interact with inner-city youth and have the kids, their families, and the supporting institutions all agree that it is a positive interaction. One of the reasons that we have been successful in engaging kids in a discourse about their community is because we have used the medium of art as a vehicle. Both photography and theater offer the possibility of transcending language and allowing children to play with communicating their perspectives. Performance, photography, illustration, interviewing, and writing have been deployed as mechanisms that reduce barriers to the expression and communication of knowledge and identity and give each voice a little amplification. These various media provide us with inside information about how these particular young people perceive their environments, and at the same time serve as more fluid bases for the types of

communication and encounters that generate community. When we are successful, our research becomes applied as it is immediately taken up and used by young
people in the neighborhood.

Conclusions: implications for academic research and discourse

We have argued in this chapter that ethical conduct within the academy requires
confronting and dealing with the contradiction between academic forms of participation and discourse, on one hand, and our responsibilities within the communities that we study, on the other. As we are both acutely aware, negotiating
reciprocal research relations in the manner we describe here is time-intensive
and infrequently rewarded within the academy; it is our own sense of ethical
responsibility that motivates us to take on the added work. We struggle to achieve
recognition and valorization from academic institutions at the same time that we
position ourselves as members of a broader society and as participants in local
communities. There are, however, many ways in which a researcher can reciprocate with the community upon which s/he relies for access to data. We are certain
that many researchers already contribute to the communities they study in supportive ways, but see their participation as "behind-the-scenes" work separate
from the sphere of academic knowledge and discourse. While we commend our
colleagues who are reflexive about the effect of their social position on their
research findings, we are even more interested in those who develop reciprocal
relations with their research subjects. We hope this chapter will contribute to a
critical discussion of reciprocal research relations not only as an ethical practice
but also as a means of incorporating a valuable reflexive feedback loop within our
intellectual endeavors.

For us, community participation significantly contributes to our theorizing and
the way we represent people. The repopulation of the public sphere is an objective
that enriches community and facilitates justice, but it is also an objective that
involves moving academic knowledge production into a dynamic interchange
with the social relations of community and the public sphere. We believe such a
move to be crucial to the contribution of the academy to the pursuit of more just
social relations.

Notes

1   Data for fall, 1996, from San Diego Unified School District.
2   We nevertheless share important critiques of Habermas' theory, agreeing that his
    understanding of the public sphere is both historically specific and excludes many
    discussions of inequality by bracketing off the "private" sphere (see Calhoun
    1992).

References

Aitken, S. and T. Herman (1997) "Gender, power and crib geography: transitional spaces and potential places," *Gender, Place and Culture* 4,1: 63–88.

Bardy, M. (1994) "The 100-years project: towards a revision," in J. Qvortrup, M. Bardy, G. Sgritta and H. Wintersberger (eds) *Childhood Matters: Social Theory, Practice and Politics*, Aldershot, UK: Avebury Press.

Burawoy, M., A. Burton, A. A. Ferguson, K. J. Fox, J. Gamson, N. Gartrell, L. Hurst, C. Kurzman, L. Salinger, J. Schiffman and S. Ui (1991) *Ethnography Unbound: Power and Resistance in the Modern Metropolis*, Berkeley: University of California Press.

Calhoun, C. (ed.) 1992 *Habermas and the Public Sphere*, Cambridge, Mass.: MIT Press.

Clifford, J. and G. Marcus (1986) *Writing Culture: The Politics and Poetics of Ethnography*, Berkeley: University of California Press.

Davis, S. G. (1997) "'Space Jam': family values in the entertainment city," paper presented at the American Studies Association Annual Meeting, Washington, DC.

England, K. V. E. (1994) "Getting personal: reflexivity, positionality, and feminist research," *The Professional Geographer* 46: 80–89.

Escobar, A. (1995) *Encountering Development: The Making and Unmaking of the Third World*, Princeton, NJ: Princeton University Press.

Flax, J. (1993) "The play of justice: justice as a transitional space," *Political Psychology* 14,2: 331–346.

Fonow, M. M. and J. A. Cook (eds) (1991) *Beyond Methodology: Feminist Scholarship as Lived Research*, Bloomington: Indiana University Press.

Fraser, N. (1992) "Rethinking the public sphere: a contribution to the critique of actually existing democracy," in C. Calhoun (ed.) *Habermas and the Public Sphere*, Cambridge, Mass.: MIT Press, 109–142.

Graue, M. E. and D. J. Walsh (1998) *Studying Children in Context: Theories, Methods and Ethics*, Thousand Oaks, Cal.: Sage Publications.

Habermas, J. (1979) *Communication and the Evolution of Society*, Boston: Beacon Press.

—— (1987) *The Structural Transformation of the Public Sphere: An Inquiry into a Category of Bourgeois Society*, Cambridge, Mass.: MIT Press.

Haraway, D. J. (1991) "Situated knowledges: the science question in feminism and the privilege of the partial perspective" in *Simians, Cyborgs, and Women: The Reinvention of Nature*, New York: Routledge, Chapman and Hall.

Harding, S. (1991) *Whose Science? Whose Knowledge? Thinking from Women's Lives*, New York: Cornell University Press.

—— (1993) "Rethinking standpoint epistemology: what is strong objectivity?" in L. Alcoff and E. Potter (eds) *Feminist Epistemologies*, London: Routledge.

Katz, C. (1994) "Playing the field: questions of fieldwork in geography," *The Professional Geographer* 46: 67–72.

Keith, M. and S. Pile (eds) (1993) *Place and the Politics of Identity*, London: Routledge.

Kenzer, M. S. (ed.) (1989) *Applied Geography: Issues, Questions, and Concerns*, Dordrecht, The Netherlands: Kluwer Press.

Kobayashi, A. (1994) "Coloring the field: gender, 'race,' and the politics of fieldwork," *The Professional Geographer* 46: 73–79.

222 *Ethics and knowledge*

Lacy, S. (1995) "Cultural pilgrimages and metaphoric journeys" in S. Lacy (ed.) *Mapping the Terrain: New Genre Public Art*, Seattle: Bay Press.

Májozo, E. C. (1995) "To search for the good and make it matter" in S. Lacy (ed.) *Mapping the Terrain: New Genre Public Art*, Seattle: Bay Press.

Massey, D. (1998) "The spatial construction of youth cultures" in T. Skelton and G. Valentine (eds) *Cool Places: Geographies of Youth Cultures*, London: Routledge.

Phillips, P. (1995) "Public constructions" in S. Lacy (ed.) *Mapping the Terrain: New Genre Public Art*, Seattle: Bay Press.

Pile, S. and M. Keith (1997) *Geographies of Resistance*, London: Routledge.

Pulsipher, L. M. (1997) "For whom shall we write? What voice shall we use? Which story shall we tell?" in J. P. Jones III, H. Nast and S. M. Roberts, (eds) *Thresholds in Feminist Geography: Difference, Methodology, Representation*, Lanham, MD: Rowman and Littlefield.

Rorty, R. (1989) *Contingency, Irony, and Solidarity*, Cambridge: Cambridge University Press.

Rose, G. (1993) *Feminism and Geography: The Limits of Geographical Knowledge*, Minneapolis: University of Minnesota Press.

Yapa, L. (1996) "What causes poverty?: a postmodern view," *Annals of the Association of American Geographers* 86,4: 707–728.

Young, I. M. (1990) *Justice and the Politics of Difference*, Princeton, NJ: Princeton University Press.

16 Morals and ethics in geographical studies of disability

Rob Kitchin

We ... are not interested in descriptions of how awful it is to be disabled. What we are interested in is the ways of changing our conditions of life, and thus overcoming the disabilities which are imposed on top of our ... impairments by the way this society is organized to exclude us.

(UPIAS 1976: 4–5)

Introduction

Disabled people have long been labelled as Other. Across the globe, ableist prejudice, ignorance and institutional discrimination is rife (Barnes and Mercer 1995). As a consequence, disabled people generally occupy inferior positions within society, marginalized to the peripheries. Disabled people are more likely to be unemployed, occupy poorer housing, and have restricted access to education and transport than their non-disabled counterparts. As noted by Gleeson (1996), Imrie (1996) and Kitchin (1998) the oppressive experiences of disability are rooted in specific socio-spatial and temporal structures. Forms of oppression are played out within, and given context by, spaces and places. Spaces are currently organized to keep disabled people 'in their place' and places written to convey to disabled people that they are 'out of place'. For example, urban space is implicitly and explicitly designed in such a way as to render certain spaces 'no go' areas. Implicit or thoughtless designs include the use of steps with no ramp, cash machines being placed too high, and places linked by inaccessible public transport. Explicit designs include the segregationalist planning including separate schools, training centres and asylums. Even within public spaces disabled people are separated and marginalized to the peripheries with separate and often shared-sex toilets and restricted access to theatres and other entertainment establishments. The messages written within the landscape by such designs are clear – disabled people are not as valued as non-disabled people. Finkelstein (1993) thus contended that disabled people occupy a 'negative reality'.

This 'negative reality' has to a large extent been ignored by academia and other institutions. Moreover, as with poor people in relation to poverty discourse (see Beresford and Croft 1995), disabled people have largely been excluded from disability discourse; marginalized from the political process and the media structures which influence public and policy discussion; and excluded from academic and institutional research, think tanks, charity and pressure groups. Instead, disability discourse has been, and to a large degree still is, overwhelmingly dominated by people who are not disabled.

In this chapter, two separate but related sets of moral and ethical issues are examined in relation to geographical studies of disability. In the first half of the chapter the moral responsibility of (non-disabled) academics to undertake critical emancipatory and empowering research concerning disability issues is examined. In the second half of the chapter the epistemological and ethical bases of conducting such research are explored. Central to, and linking, the two halves of the discussion are the questions: 'Can academics representatively address the exclusion and marginalization of disabled people within society?'; and 'Can an academic adopt and enact an emancipatory and empowering position in relation to both societal oppression and the research process?' These questions have come to the fore in my own research as I have increasingly questioned my positionality and motivation as a non-disabled researcher studying issues of disability. In particular, I have two main concerns: first, to find an approach that is emancipatory and empowering, and which is representative of the disabled people taking part in my research; and second, the legitimacy of acting and writing on behalf of a group of which I am not a member. These reservations clearly have currency beyond geographical studies of disability to include other excluded and oppressed groups within Western society and research on other cultures (see chapters by Deur and Rundstrom, and Gormley and Bondi, Herman and Mattingly, in this book).

Moral responsibility

> Justice in modern industrial societies requires a societal commitment to meeting the basic needs of *all* persons.
>
> (Young 1990: 91, my emphasis)

> Academics must . . . be prepared to answer what they believe the role of the academy should be in promoting social change, and what they envision – in real, substantive terms – as the means to achieve a more just society.
>
> (Nagel, Chapter 10, this book)

Smith (1994) and Sayer and Storper (1997) recently argued that geography tends to be positive in nature, avoiding questions about whether something is good or bad, right or wrong. As such, geography often focuses upon what actually exists and avoids normative ethics: the attempt to discover some acceptable and rational views concerning what is good and what is right. In relation to disability, normative ethics concerns social justice, the fair and equitable distribution of things that people care about such as work, wealth, food and housing, plus less tangible phenomena such as systems of power and pathways of opportunity, and specific moral issues such as how people should be treated (Smith 1994). Social justice, in essence, concerns human rights. A right is an 'obligation embedded in some social or institutional context where expectation has a *moral* force' (Smith 1994: 36). In other words, moral rights are those things that we as members of a society expect as members. In our society they include things such as freedom of expression, choice, access to accommodation, to vote in elections, full recourse to the law, and access to education and medical treatment.

Social commentators universally agree that disabled people suffer social injustice. They are systematically denied their moral rights to social relations and interactions that 'able-bodied', 'mentally competent' people take for granted. Given that scholars of disability studies recognize the social injustice that disabled people face, the mechanisms by which it is perpetuated, and appreciate that disabled people are largely marginalized and excluded from positions of power and influence to change their own conditions, two questions arise. First, do academics have a moral responsibility – an obligation – to disabled people to expose ableist practices and seek social change? Second, should academics become politically involved in disability issues (or other aspects of societal oppression relating to gender, race, sexuality, etc.) and engage with direct action? The answers to these questions are contested and five basic positions adopted.

In the first position are academics who view their role as voyeurs, objectively and neutrally studying society. They argue that it is not the role of academics to try and influence decision- and policy-making. Instead it is for others, in democratically elected or institutional positions, to interpret research findings and influence future policies. Academics occupying the second position recognize their own subjectivity and positionality in relation to a researched group, but likewise feel it is not their place to be suggesting and seeking societal change. A third group recognizes the need for change but seeks alternative futures through implicit means such as raising consciousness. Here, discourse is itself seen as an action, and writing and lecturing as mediums in which to engage fellow members of society and alter world views. As such, traditional research methods are still adopted and no explicit action is taken. Academics occupying a fourth position recognize the power imbalances in their own research and seek research strategies that will empower their research subjects either to be able to seek justice themselves or to seek justice through the research (see Herman and Mattingly, this book). The fifth group comprises academics who recognize the need for change and who explicitly seek change through their own political and social actions.

Those occupying the fourth and fifth positions argue that by not *actively* seeking change through empowerment or emancipation that will improve the human condition, academics are guilty of averting their gaze from oppression and human suffering. Not actively engaging with the group which is oppressed or their respective politics would be the academic equivalent of what Dickson (1982, cited in Mohan 1995), in relation to student education, termed 'systematized selfishness' – the study of a subject without giving anything in return. He suggested that unapplied knowledge is knowledge shorn of its meaning. Oliver (1992) contended that this has been the common model of disability research.

He described this model as 'the rape model of research' because whilst the researchers benefit from respondents' knowledges or experiences, the research subjects remain in exactly the same social situation. As a consequence, Routledge (1996) has questioned the current marginal, social responsibility of academics, given their training, access to information, and freedom of expression. He suggested that by not joining their work with political practice or imparting their knowledge to empower oppressed people academics are complicit in oppressive practices. Mohan (1996) has similarly lamented that the current focus upon identity, culture and difference is failing the research subjects and there is a need for critical geography to become more critical.

In other words, geographers should be engaged in an emancipatory project aimed at improving the lives of disabled people (and other groups) in both practical and political ways. This involves bridging the chasm that still exists between radical, academic theorists and 'on-the-ground' activists (Pfeil 1994) and engaging with what Touraine (1981) termed 'committed research', Katz (1992) a 'politics of engagement', and hooks (1994) described as an 'ethics of struggle' both within the academy and beyond. Here, there is a recognition that people are not merely subjects to research 'but lives to be understood in the interests of working for a redistribution of wealth and justice' (Deur and Rundstrom, this volume).

Routledge (1996) has demonstrated that there is a 'third space' between academic and activist that researchers can occupy. An uneasy space where respective roles have to be balanced and negotiated through a dialectical relationship, but nonetheless a space from where committed research can be practised. He does not, however, envisage that this space should necessarily be occupied by all academics. Indeed, it can be argued that the occupancy of this 'third space' does not necessarily make a project any more emancipatory although it may provide more insights through social interaction and personal experiences than might be gained from formal research strategies.

Even if the academy is willing to accept that it has moral responsibilities (whatever they might be) to engage in social and political action on behalf of, or with, oppressed groups, new questions concerning the positioning of the academic and the situatedness of knowledge are encountered. Here, two forms of ethics identified by Proctor in the Introduction to this volume, become entwined:

> In science, ethics typically involves reflection upon moral questions that arise in research, publication, and other professional activities . . . yet philosophical usage is broader than this prevailing scientific interpretation. Ethics, also known as moral philosophy, is commonly understood as systematic intellectual reflection on morality in general, or specific moral concerns in particular.
>
> (this volume, page 3)

Questions concerning the ethical nature of research practice become enmeshed in questions concerning whether the researcher should be trying to change societal relations. By trying to bridge the gap between academic and activist a tightrope is walked in relation to whether an oppressed group is gaining the representation it seeks (or in some cases does not seek). Indeed, as recent debates in the disability literature have illustrated, some critics would be dubious about non-disabled academics forwarding visions for disabled people, questioning both the motivation and positionality of researchers. Given that academic research has perpetuated, reproduced and legitimated the marginalization of disabled people, justifying segregation, eugenics, and the denial of civil rights, it is little wonder that disabled people are suspicious of research by non-disabled researchers including those who claim to be allies (Rioux and Bach 1994). As such, there is a need to seek paths that allow societal oppression to be tackled, but are also representative of those people potentially being liberated. It is to finding such a path that the discussion now turns.

Emancipatory and empowering research strategies

Central to finding a path that is emancipatory, empowering and representative are epistemological debates concerning how knowledge is derived or arrived at; and the assumptions about how we can know the world (what can we know? how can we know it?). Such debates are currently taking place in the disability literature, particularly in respect to how we gain knowledge. As noted, debates within the disability literature have increasingly questioned the relationship between (non-disabled) researcher and (disabled) researched. Protagonists on one side of the debate (predominantly academics who are disabled) have argued that it is only disabled people who can know what it is like to be disabled. They question the legitimacy of (non-disabled) experts to draw conclusions about disabled people's lives and experiences. They argue that research concerning disability is invariably researcher-oriented, based around the desires and agendas of the (non-disabled) researcher and able-bodied funding agencies rather than subject(s) of the research (disabled people). Indeed, Oliver (1992) argued that current expert models of research are alienating, and disempower and disenfranchise research participants by placing their knowledge into the hands of the researcher to interpret and make recommendations on their behalf; that researchers are compounding the oppression of disabled respondents through exploitation for academic gain.

Hunt (1981) illustrated, in a much cited critique, the experiences of being a 'victim of research'. He described how as a resident of Le Court Cheshire Home he and other residents became disillusioned with 'unbiased social scientists' who followed their own agenda and ignored the views of the people they consulted. Oliver (1992) suggested that continued academic 'abuse' is leading to a growing dissatisfaction amongst disabled research subjects who view academic research as unrepresentative. Indeed, disabled activists and organizations have declared that existing research has largely been a source of exploitation rather than liberation (Barnes and Mercer 1997); that current expert models of research, where disabled people are the subjects and academics the experts, controlling all aspects of the process from initial ideas to the contents of the final reports, reproduce current social relations. As such, critical research adopting an expert model is paradoxically seeking change at one level, while at the same time reproducing exploitation at another.

Drawing on work within feminism in particular, these disabled academics argue that power relations within the research process need to be destabilized and the research agenda wrestled free from academic researchers still using traditional research methodologies. Indeed, Finkelstein (1985 – cited in Barners and Mercer 1997) has called for 'no participation without representation'. Such a reformulation, they argue, will close the emerging credibility gap between researchers and researched, provide a 'truer' picture of the experiences of disability and strengthen policy-making formulation whilst moving away from a social engineering model (Oliver 1992; Sample 1996). Stone and Priestley (1996: 706) suggested that the core principles of a new research agenda should be:

- the adoption of a social model of disablement as the epistemological basis for research production;
- the surrender of claims to objectivity through overt political commitment to the struggles of disabled people for self-emancipation;
- the willingness to undertake research only where it will be of practical benefit to the self-empowerment of disabled people and/or the removal of disabling barriers;
- the evolution of control over research production to ensure full accountability to disabled people and their organizations;
- giving voice to the personal as political whilst endeavouring to collectivize the political commonality of individual experiences;
- the willingness to adopt a plurality of methods for data collection and analysis in response to the changing needs of disabled people.

French and Swain (1997: 31) suggested that one way to approach these issues is for researchers to ask themselves three principal questions before undertaking work on disability:

1   Does the research promote disabled people's control over the decision-making processes which shape their lives?
2   Does the research address concerns of disabled people themselves?
3   Does the research support disabled people in their struggle against oppression and the removal of barriers to equal opportunities and a full participatory democracy for all?

Consequently, disabled academics argue that there needs to be a change in the way that we research and come to understand the world; a shift to emancipatory and empowering approaches.

Not surprisingly, not all researchers agree on the path to emancipatory and empowering studies and three alternative approaches have been forwarded. The first seeks to retain the expert model of research but to enforce a strict code of ethical practices that are designed to try and make the research process fair and non-exploitative. These are usually designed and enforced by professional bodies whose members are meant to comply with the ethical codes adopted. For example, The Council for Exceptional Children (CEC)[1] has developed the CEC Code of Ethics for Educators of Persons with Exceptionalities, which states that special educators should:

- adopt procedures that protect the rights and welfare of subjects participating in the research;
- interpret and publish research results with accuracy and a high quality of scholarship;
- support a cessation of the use of any research procedure which may result in undesirable consequences for the participant;
- exercise all possible precautions to prevent misapplication or misutilization of a research effort, by self or others.

Guidelines within research manuals, and those issued by representative bodies, generally advocate a professional approach to research and focus upon issues such as privacy, confidentiality, and anonymity. They suggest that the researcher should carefully weigh the potential benefits of a project against the negative costs to individual participants. Such individual costs might include affronts to dignity, anxiety, embarrassment, loss of trust, loss of autonomy and self-determination, and lowered self-esteem (Kidder 1986). This is clearly a subjective exercise, but one that can be approached in an informed manner. As a general rule a deonto-logical approach is advocated which judges actions according to whether the researcher would wish them upon herself/himself, whether the participants are treated with the respect due to them, and seeks to adopt a non-parasitic position (see Stone and Priestley 1996). The basis of such an approach is the development of genuine trusting relationships, where researchers respect the situated nature of their interpretation and their authorial power (see Crang 1992). Here, a system of relational ethics is adopted:

> where (the researcher) is/are committed to working with the differences between (them)selves and those with whom (they) interact, without seeking either to erase difference – that is to presume similarities or identifications that do not exist – or to treat difference as representing something essential and exotic.
>
> (Gormley and Bondi, this book, p. 252)

Feminist analyses in particular have highlighted the situated and produced nature of research accounts, the asymmetrical power relations at play between researcher and researched, and the authority expressed in research accounts (see WGSG 1997). Feminist reassessment of conducting research has led to the formulation of a feminist methodology which is characterized by a search for a mutual under-standing between researcher and researched (Katz 1994). This methodology focuses thought upon four issues: ways of knowing; ways of asking; ways of interpreting; and ways of writing. Within each of these issues researchers are encouraged to reflect upon their own position, as well as that of the researched, and to acknowledge and use these reflections to guide the various aspects of the research process (Dyck 1993). For example, Robinson (1994) in discussing white women representing 'others' in post-apartheid and postcolonial times, contends that there is a need for researchers to continually question their social location (gender, class, ethnicity), their political position, their disciplinary stance, and the physical location of the research. Each is key in shaping the research and the relationship between the researcher and researched. The same is true for those conducting research on disability. This includes non-disabled and disabled researchers alike. Academics who are themselves disabled do not occupy privileged positions where they can speak on behalf of their fellow disabled people. Admittedly, the disabled academic has the benefit of personal experience but this does not provide him or her with the platform to speak for all disabled people – their knowledge is also situated and they should pay similar respect to their research subjects. Reflexivity is particularly important when researching and writing upon a group that is unable to represent itself adequately (such as severely mentally impaired people).

The second approach seeks to alter the expert model of research so that it becomes more representative. Here, feedback (empathetic) loops are inserted into the research process so that the whole process is monitored by the research subjects who provide constructive criticism at all stages (see Barnes 1992; Oliver 1992; Sample 1996). As such, the academic retains control of the research process and the questions being asked but the participants get the opportunity to correct misinterpretations and influence the direction of the research. By using such feedback loops the researcher aims to make her/his research more representative of the subjects' views and experiences. This is the approach advocated by Deur and Rundstrom in Chapter 17 of this book, in relation to cross-cultural studies.

The third approach, and the one I am currently using in a study of the measurement of disabling environments, seeks a radical departure from the expert model of research, forwarding a partnership approach. This approach seeks to integrate research subjects more fully into the research process so that they take on the role of co-researchers (see Lloyd *et al.* 1996; Kitchin 1997). Here, the research process is 'collectivized amongst its participants' (Priestley 1997: 89) with disabled people taking an active role in the whole research process from ideas to hypotheses to data generation to analysis and interpretation to writing the final report. In this approach, the role of the academic is not as expert but as enabler or facilitator. As such, the academic takes an emancipatory position which seeks to inform and impart her/his knowledge and skills to the disabled people who are co-researchers in the project, and provide an outlet to inform the policy-makers. The academic's role is primarily to provide specific technical advice to co-researchers to help them make informed choices. Second, it is to provide a relatively privileged position through which the co-researchers can speak. Cocks and

Cockram (1995: 31) detail that emancipatory and empowering (participatory) research is premised upon five factors:

1 An acknowledgement that oppression within society creates oppressed groups and this leads to a need to engage in some transformation of the larger society to counter it.
2 Knowledge generation, control and application is central to the effort to emancipate and liberate people who are oppressed.
3 People have the capacity to work towards solutions to their own problems.
4 There is a vital link between knowledge generation, education, collective action and the empowerment of oppressed people.
5 Researchers should act in accordance with an explicit values position and should become actively involved in the process of liberation.

Many researchers would reject inclusive, partnership-based research because scientific principles (e.g. separation of researcher/researched) are clearly being compromised. However, collaboration does not mean a radical departure from the procedures of conventional positivistic or interpretative science, just that such science is carried out with and by the participants. In other words, there is a renegotiation of the relationship between the researcher and researched rather than a radical overhaul of the scientific procedures underlying the research: the study still aims to be professionally administered. However, in contrast to the standard expert model of research where research subjects have little opportunity to check facts, offer alternative explanations or verify researcher interpretations, inclusive approaches facilitate such interaction. As a result, inclusive approaches, far from diminishing the academic rigour of research, enforce a rigorous approach that is cross-checked at all stages of the research process through the participant co-researchers. Consequently, Elden and Chisholm (1993) argued that inclusive approaches provide more valid data and useful interpretations and Greenwood *et al.* (1993) contended that this increase in validity is due to a democratization of knowledge production giving the participants a stake in the quality of the results.

Current indications highlight that disabled people do want to be involved in disability discourse. The growth, politicization and radicalization of disabled people's movements over the past two decades demonstrates a desire by disabled people to take charge of their own lives; to wrestle free of control by professional services and charity organizations. Disabled people and organizations run by disabled people have been commissioning their own research and actively undertaking their own projects (see Ward 1997). Disabled people are becoming more involved in academic research as valued consultants, research students and research assistants (see Vernon 1997; Zarb 1997). Emancipatory and empowering research is another step towards independence, self-advocacy and self-determination. Involvement also provides a rational and democratic basis for disability discourse, shifting discussions and policy from tolerance, charity and common humanity to diversity, difference and rights (Beresford and Croft 1995). This provides a more effective basis for the campaign for civil rights and the fight for self-organization, independent living and anti-discrimination legislation (Beresford and Wallcraft 1997).

As I have argued elsewhere (Kitchin 1997), involving disabled people in the research process is important academically for two principal reasons. In the first instance, participation by disabled people is the only mechanism by which disability research can truly become emancipatory and empowering. Zarb (1992) described emancipatory research as being defined by two principles: reciprocity and empowerment. Whilst many methodologies might claim to fulfil these two principles, in general, reciprocity is a by-product of research aimed at increasing knowledge rather than directly addressing a real-world problem and empowerment is largely illusionary as the researcher ultimately designs and controls the study (Sample 1996). Empowerment is not something that can just be bestowed by those in power (researcher) to those who are disenfranchised (subject) (Lloyd *et al.* 1996). Empowerment is a process of gradual changes which, although they might be instigated by the researcher, must be accepted and built upon by the subject. To be fully empowering, the study needs not only to be designed in conjunction with the research subjects but to be conducted with them in such a fashion that they learn from the process and gain some semblance of power, either politically through the research results or through the learning of research skills.

In the second instance, an inclusive research approach allows the research to become more representative and reflexive by addressing the issue of unequal power arrangements within the research process and recognizing the 'expertise' of disabled people in their own circumstances. Inclusion acknowledges and signifies a respect that the contributions of disabled co-researchers are valuable and worthwhile. Here the co-researchers' expertise is acknowledged as equal but from a different frame of reference than the academics' (Elden and Levin 1991):

- Disabled people occupy insider positions. Their knowledge on a particular subject is often individual, tacit, practical led, from first-hand experience;
- Academics occupy outsider positions. They have specialized skill, systematic knowledge, are theory led, and based upon second-hand experience.

Here there is the development of a mutual sharing of knowledge and skills (Lloyd *et al.* 1996). This is not to say that an expert/lay-person relationship between researchers and co-researchers does not exist but rather that such a position can be re-worked into a more favourable, emancipatory position. Emancipatory studies thus address some of the problems of representativeness, reciprocity and reflexivity that plague both interpretative and positivistic studies. As Routledge (1996) suggested, it is all too easy for academics to claim solidarity with the oppressed and claim to act as relays for their voices. Inclusive studies are designed to negate such criticism and allow disabled people to speak through the research rather than have voices in it.

Whilst emancipatory studies are demanding, it is suggested that the shared benefits to researchers, policy-makers and disabled co-researchers outweigh costs in terms of time and organization. Involving disabled people in disability discourse as controllers or partners then offers practical and social gains for disabled people. It is only with their active involvement that disability discussions will reflect their needs, concerns and interests. Through participation and partnership, research will become more reflexive, reciprocal and representative. It will provide a platform from where disabled people can speak for themselves, to seek the services and support they want, explicitly to influence social policy, and fight for disabled rights. In short, research will become enabling and empowering.

Conclusion

Whether an academic feels (s)he has a moral responsibility to address issues of social suffering, injustice and oppression is clearly a personal issue. When researchers do, however, make the decision to fight for civil and material rights through their research and writings, a new set of problems are encountered concerning research ethics, positionality and representativeness. In this chapter I have discussed these new problems in relation to geographical studies of disability issues. Recent debates within the disability literature have led some researchers to question the ethical basis and validity of traditional expert models of research. Instead, they suggest that research should become more reflexive and, where possible, inclusive in design. Such a reformulation of research design, they contend, will lead to empowering and emancipatory research that will improve the social position of disabled people both within academic studies and society. I am currently trying to use one particular reformed approach, namely participatory action research, to address some of the concerns raised. In this study disabled people from Belfast and Dublin are designing and undertaking their own research into measuring disabling environments. They have complete autonomy and control over the process, deciding on the topic to be investigated, the methods of data collection and analysis, and writing the final report. My role is one of advisor or facilitator. The projects are action-led, aimed at confronting ableist practices (e.g. inaccessible public transport) and seeking change. Although the study is in its preliminary stage, early indications suggest that the projects will be a success and vilify the arguments of many disabled academics calling for a change in the social relations of research.

Note

1 http://www.cec.sped.org/home.htm

References

Barnes, C. (1992) 'Qualitative research: Valuable or irrelevant?', *Disability, Handicap and Society* 7, 2: 139–155.

Barnes, C. and Mercer, G. (1995) 'Disability: Emancipation, Community Participation and Disabled People' in M. Mayo and G. Craig (eds) *Community Empowerment: A Reader in Participation and Development*, London: Zed Books, pp. 46–59.

—— (1997) 'Breaking the mould? An introduction to doing disability research' in C. Barnes and G. Mercer (eds) *Doing Disability Research*, University of Leeds, Leeds: Disability Press, pp. 1–14.

Beresford, P. and Croft, S. (1995) 'It's our problem too! Challenging the exclusion of poor people from poverty discourse', *Critical Social Policy* 44/45: 75–95.

Beresford, P. and Wallcraft, J. (1997) 'Psychiatric system survivors and emancipatory research: issues, overlaps and differences' in C. Barnes and G. Mercer (eds) *Doing Disability Research*, University of Leeds, Leeds: Disability Press, pp. 67–87.

Cocks, E. and Cockram, J. (1995) 'The Participatory Research Paradigm and Intellectual Disability', *Mental Handicap Research*, 8, 1: 25–37.

Crang, P. (1992) 'The politics of polyphony: reconfigurations and geographical authority', *Environment and Planning D: Society and Space* 10: 527–549.

Dickson, A. (1982) 'Study service: for other countries, other institutions, other people's children – or for ourselves?' in S. Godland (ed.) *Study Service: An Examination of Community Service as a Method of Study in Higher Education*, Windsor: NFER-Nelson, pp. 7–33.

Dyck, I. (1993) 'Ethnography: a feminist method?', *The Canadian Geographer* 37, 1: 52–57.

Elden, M. and Chisholm, R. F. (1993) 'Emerging varieties of action research: introduction to the special issue', *Human Relations* 46, 2: 121–142.

Elden, M. and Levin, M. (1991) 'Cogenerative learning: bringing participation into action research' in G. F. Whyte (ed.) *Participatory Action Research*, London: Sage, pp. 127–142.

Finkelstein, V. (1985) Unpublished paper presented at World Health Organization Meeting, 24–28 June, the Netherlands.

—— (1993) 'The commonality of disability' in J. Swain, V. Finkelstein, S. French and M. Oliver (eds) *Disabling Barriers – Enabling Environments*, London: Sage, pp. 9–16.

French, S. and Swain, J. (1997) 'Changing disability research: participating and emancipatory research with disabled people', *Physiotherapy* 83, 1: 26–32.

Gleeson, B. J. (1996) 'A geography for disabled people?', *Transactions of the Institute of British Geographers* 21: 387–396.

Greenwood, D. J., Whyte, W. F. and Harkavy, I. (1993) 'Participatory action research as a process and as a goal', *Human Relations* 46, 2: 175–192.

hooks, b. (1994) *Teaching to Transgress*, New York: Routledge.

Hunt, P. (1981) 'Settling accounts with parasite people', *Disability Challenge* 2: 37–50.

Imrie, R. F. (1996) 'Ableist geographies, disablist spaces: towards a reconstruction of Golledge's geography and the disabled', *Transactions of the Institute of British Geographers* 21: 397–403.

Katz, C. (1992) 'All the world is staged: intellectuals and the projects of ethnography', *Environment and Planning D: Society and Space* 10: 495–510.

—— (1994) 'Playing the field: questions of fieldwork in geography', *The Professional Geographer* 46: 67–72.

Kidder, L. (1986) *Research Methods in Social Relations*, London: Harcourt Brace.

Kitchin, R. M. (1997) 'A geography of, for, with or by disabled people: reconceptualising the position of geographer as expert', SARU Working Paper 1, School of Geosciences, Queen's University of Belfast.

—— (1998) '"Out of place", "knowing one's place": towards a spatialised theory of disability and social exclusion', *Disability and Society* 13, 3: 343–356.

Lloyd, M., Preston-Shoot, M., Temple, B. and Wuu, R. (1996) 'Whose project is it anyway? Sharing and shaping the research and development agenda', *Disability and Society* 11, 3: 301–315.

Mohan, J. (1995) 'Geographies of welfare and the welfare of (British) geography' in R. Gachechiladze and D. Smith (eds) *Proceedings of the Second British–Georgian Geographical Seminar.* 28 June–5 July, Birmingham.

Oliver, M. (1992) 'Changing the social relations of research production', *Disability, Handicap and Society* 7: 101–114.

Pfeil, F. (1994) 'No basta teorizar: in-difference to solidarity in contemporary fiction, theory and practice' in I. Grewel and C. Kaplan (eds) *Scattered Hegemonies*, Minneapolis: University of Minnesota Press.

Priestley, M. (1997) 'Who's research?: A personal audit' in C. Barnes and G. Mercer (eds) *Doing Disability Research*, University of Leeds, Leeds: Disability Press, pp. 88–107.

Proctor, J. (1998) 'Ethics in geography: giving moral form to the geographical imagination', *Area* 30: 8–18.

Rioux, M. and Bach, M. (1994) *Disability is not Measles: New Paradigms in Disability*, North York, Ontario: L'Institut Roeher.

Robinson, J. (1994) 'White women researching/representing "others": from anti-apartheid to postcolonialism' in A. Blunt and G. Rose (eds) *Writing, Women and Space*, New York: Guilford Press, pp. 197–226.

Routledge, P. (1996) 'The third space as critical engagement', *Antipode* 28, 4: 399–419.

Sample, P. L. (1996) 'Beginnings: participatory action research and adults with developmental disabilities', *Disability and Society* 11, 3: 317–332.

Sayer, A. and Storper, M. (1997) 'Ethics unbound: for a normative turn in social theory', *Environment and Planning D: Society and Space* 15: 1–17.

Smith, D. M. (1994) *Geography and Social Justice*, Oxford: Blackwell Publishers.

Stone, E. and Priestley, M. (1996) 'Parasites, pawns and partners: disability research and the role of non-disabled researchers', *British Journal of Sociology* 47, 4: 696–716.

Touraine, A. (1981) 'An introduction to the study of social movements', *Social Research* 52: 749–787.

UPIAS (1976) *Fundamental Principles of Disability*, London: Union of Physically Impaired People Against Segregation.

Vernon, A. (1997) 'Reflexivity: the dilemmas of researching from the inside' in C. Barnes and G. Mercer (eds) *Doing Disability Research*, University of Leeds, Leeds: Disability Press, pp. 158–176.

Ward, L. (1997) 'Funding for change: translating emancipatory disability research from theory to practice' in C. Barnes and G. Mercer (eds) *Doing Disability Research*, University of Leeds, Leeds: Disability Press, pp. 32–48.

Women and Geography Study Group (1997) *Feminist Geographies: Explorations in Diversity and Difference*, Harlow: Longman.

Young, I. M. (1990) *Justice and the Politics of Difference*, Princeton: Princeton University Press.

Zarb, G. (1992) 'On the road to Damascus: first steps towards changing the relations of disability research production', *Disability, Handicap and Society* 7: 125–138.

—— (1997) 'Researching disabling barriers' in C. Barnes and G. Mercer (eds) *Doing Disability Research*, University of Leeds, Leeds: Disability Press, pp. 49–66.

17 Reciprocal appropriation

Toward an ethics of cross-cultural research

Robert Rundstrom and Douglas Deur

For most people, serious learning about Native American culture and history is different from acquiring knowledge in other fields. One does not start from point zero, but from minus ten.

Michael Dorris (1987: 103)

"Enjoy yourself, and never, never be an embarrassment to the administration." Anonymous faculty advisor.

(Clinton 1975: 199)

Peoples of the Northwest Coast of North America speak of sisiutl, the two-headed sea serpent, guardian of supernatural beings – one head masculine, the other feminine; one head hot, the other cold; one head good, the other evil. If you flee the sisiutl charging from the fjord you will be devoured, but if you stand firm before it, some say, its two heads will see one another at the very last moment as it lunges at you. Opposing forces collide: good will meet evil. The sisiutl will achieve a form of enlightenment and back off into the water. You will not be eaten. And you will find a form of enlightenment yourself, a spirit power of great magnitude.

The sisiutl story serves as a metaphor for the task before us in this chapter, the articulation and reconciliation of what often appears as two opposing forces: the abusive, colonizing academic gaze and the institutional apparatus out of which it peers; and the world of colonized peoples on which that gaze is frequently trained. Cross-cultural geographic researchers have long served as "cultural brokers," translating across cultural divides, representing – intentionally or otherwise – each group to the other (Szasz 1995). Particularly during the late twentieth century, these cultural borders have been subject to perpetual re-negotiation, as non-Western peoples challenge the authority of European institutions and question the veracity of past scholarly depictions of themselves (Deloria 1995). Today, geographers must confront the colonial legacy directly, interacting with people who often define their identities in opposition to the colonial world (a world of which, more often than not, the researcher is a part). In the process, geographers encounter alternative views of the world which must be recognized and engaged if their research is to continue, including alternative views of what constitutes "ethical behavior."

Perhaps, as in the sisiutl tale, steadfastness does not always produce a positive outcome, but – as cross-cultural researchers – we find truth in the story's recognition of the power wielded by contending forces outside ourselves, and in how that power compromises individual free will. Here, we emphasize the ethical issues surrounding cross-cultural research in "Indian Country" in North America because it is the place with which we are most familiar. And it is out of that experience, and our understanding of the experience of others, that we have developed a sense of the reciprocities involved when deliberating matters of ethics.[1]

Who decides ethical behavior?

For us, ethics is a relational and contextual matter, a topic requiring recognition of the social relationship shaping interaction between researcher and those who are researched. Ethical research does not simply involve an isolated individual seeking to "do good" or to "do the right thing." Such personal reflection is necessary but not sufficient to facilitate cross-cultural understanding and insightful research. In cross-cultural contexts, mere reflexivity is an even less credible foundation for ethical research behavior. Anyone who has worked extensively in Indian Country will recognize that reliance upon isolated reflection in developing ethical relations is a peculiarly non-Indian concept. In short, it is not enough in cross-cultural situations.[2]

Although we emphasize contextuality, we are not willing to argue against the existence of ethical universals. All people deserve respect, privacy, equitable treatment, and freedom from intrusion and oppression. But we know too that such rights are negotiated differently among different peoples. And the groundwork for an ethics of cross-cultural research rests upon a recognition of how these rights are considered and given meaning in the particular cultural contexts of both the observer and the observed.

We have found another valuable metaphor in the Kiowa writer N. Scott Momaday's concept of "reciprocal appropriation." Reciprocal appropriation provides a means of understanding the relationship between humans and the environments in which they dwell. In Momaday's terms, humans make investments in landscapes while simultaneously incorporating landscapes into their personal fundamental experience. These actions are "moral acts of the imagination" in which places and people use and motivate each other through innumerable acts. The relationship is not always a kind or honorable one; places and people make demands on each other. But if there is a balance struck that is deemed equitable on all sides, the synthesis or interpenetration produces a result that is pleasing to all involved (Momaday 1976: 80; Rundstrom 1993: 25).

Like an encounter with the sisiutl, it may be best to see the research relationship as a form of reciprocal appropriation. Ethical cross-cultural research involves: relinquishing control; placing both the observer and observed on the same plane of risk; recognizing that demands are made in both directions; understanding that demands made are not always kind or honorable; and, underlying all of it, realizing that all research is appropriation. Fundamentally, it means knowing that others will play a substantial role in one's research and one's life for the duration of a cross-cultural research program.

More specifically, the practice of cross-cultural research is complicated because evaluations of what constitutes ethical behavior may differ among the observer and the observed. Concepts such as 'confidentiality' and 'benefit' are defined differently, and possess different prescriptive implications, both between cultures and between individuals (Norton and Manson 1996). A gestalt shift is required of the researcher who hopes to engage and understand others intelligently and empathetically. In evaluating behavior, researchers may prioritize academic freedom or 'detached rationality' while Indian communities, for example, may value long-term personal involvement and the practical uses arising from research results. In the interest of reciprocity, researchers embarking on cross-cultural research projects can seek help among members of the group under study in accommodating these seemingly opposed needs.

Personal and ethical adaptation is therefore key to successful cross-cultural research. The extent of this adaptation will vary within different research contexts. But we share Byron's (1993) experience of contextual ethics, where – even after years of experience – field research situations arise that simply cannot be anticipated, and which require evaluation and accommodation.

Access and sovereignty

Cross-cultural researchers often are motivated by very real pressures to climb the tenure-and-promotion ladder of the academy, a not-so-idle curiosity, perhaps a vague sense of an ill-defined responsibility to "society," and to be fair, a sense of commitment to the people under study. Within the context of the Western scientific tradition, such motives spawn the presumption that one bears certain a priori rights to information regarding non-Western societies. The circle is completed in the view widely held by empiricist field geographers that the world is a laboratory inviting or even demanding investigation (Curry 1996). But as Sitter-Liver (1995) demonstrates, the right of powerful peoples to lay claim to the world for the seemingly benign purposes of examination and inventory has no defensible intellectual basis, and thus no foundation in the academy. We conclude on intellectual grounds that academic researchers possess no a priori right to conduct research in Indian Country.

But this issue of access is also a practical political matter. Academic interests are superseded legally by the rights of individuals and communities to privacy and self-determination. This is underscored emphatically in Indian Country by claims to legal and territorial sovereignty, claims derived from longstanding political traditions, the US Constitution, numerous international treaties, and nearly 200 years of US Supreme Court decisions. Thus, perhaps nowhere is the concept of reciprocal appropriation more revealing than when one engages this matter of native sovereignty.

Winchell (1996a) has claimed that both inherent and constitutional sovereignty are so fundamental to Indian experience that they form the foundations of individual and collective identities. But where does sovereignty reside? And how can researchers position themselves accordingly? Winchell advises that research should be done for and with tribal governments because they are the bearers of political sovereignty. Further, in the USA, the federal government recognizes the right of tribal governments to exclude researchers and other outsiders (American Indian Law Center 1994). For reasons that are neither surprising nor wholly unwarranted, these sovereign tribal governments increasingly seek something in return for tribal members' assistance, for their loss of privacy and proprietary knowledge, and for educating academic neophytes attempting to navigate their world. Researchers often must reciprocate for access privileges by investigating issues of local value (papers in Paine 1985). Thus, geographers increasingly have worked on applied topics of value to standing tribal governments, such as land claims, community planning, tourism, economic development, and resource inventories.[3]

But in truth, tribal governments are a form of institutionalized duplicity; they are comprador governments. In 1934, these governments were implanted within the USA by Act of Congress to allocate political authority along the lines of the majority culture. They usurped indigenous tribal authority structures in favor of a single, typically male "chief," a legislative council, and an "independent" judiciary branch. Consequently, members of tribal government often represent political interests residing in the US Congress and its subordinate, the Bureau of Indian Affairs (BIA), more than the interests of tribal members, who frequently adhere to an indigenous system where political authority is allocated quite differently (Rundstrom 1995). It is not surprising that the federal government considers sovereignty to reside in tribal governments of seriously compromised independence. The resulting political fragmentation leads to contests among different factions over which collective body oversees tribal sovereignty (Norton and Manson 1996). Often, researchers rely on information from a single individual or group with a particular set of agendas within this factionalized context, resulting in a host of conflicts and biases.

Researchers must choose their contacts and allegiances accordingly, and carefully. Rundstrom's work with the Mvskoke (Creek) of Oklahoma is a case in point. The government of the Creek Nation has been ruled by individuals termed "mixed-bloods,", who have severely compromised allegiances as described above. The rights of "full-bloods" or "traditionals" – designations more political than biological – have been uncoupled from and ignored by tribal government, resulting in full-blood opposition to that government. Rundstrom developed a medical-geography project of particular interest to full-bloods alongside individual full-blood decision-makers, people who had authority within their respective social units (e.g. ceremonial grounds, churches, speech communities) but who had very little within tribal government. Indeed, the Creek Nation has no interest in representing full-bloods, and has sometimes denied medical assistance to these members of the tribe. Had initial contact been made with the federally sanctioned tribal government, in recognition of national sovereignty, the concerns of full-bloods would never have been recognized in this project. Researchers, of course, risk aggravating internal divisions. In this case however, internal fragmentation could hardly be worse; it is a condition of everyday life and the two factions function more or less independently.

Ethics of representation

Functionary, activist, or . . . ?

The role of cross-cultural research has changed dramatically in recent years. Social scientists increasingly have become minor functionaries in a research bureaucracy, through which research progresses on a contractual basis. As Clinton (1975) suggested, contract researchers are increasingly accountable to a complex array of actors, each with different goals: funding agencies; consulting firms; academic tenure committees or departments; a professional discipline; and research populations. All are capable of manipulating researcher performance, which, cumulatively, can put the researcher in an ethical and political vice. Contract research is beset by additional issues: ultimate purposes of the work; relations between researcher and funding agency or contractor; relations between the researcher's title or official capacity and what the researcher is actually doing among the people being studied (e.g. is the researcher hiding behind an assumed neutrality?); how information will flow into and from the project; and how seriously the researcher regards members of the study population.

To combat the effects of an increasingly bureaucratized academic world, a growing number of geographers have adopted the role of political activist. Some "stand in the fire" to support the rights of the oppressed in response to competing demands on the researcher's loyalties. For example, Katz (1992) seeks to build a postcolonial geography through a "politics of engagement." She argues that places and people are not merely subjects to write about but lives to be understood in the interests of promoting equitable redistributions of wealth and justice. And she recognizes that a researcher's "neutrality" is a delusional construction. We agree, and endorse such an approach. However, the Eurocentric social and political theory that informs her work, and that of many cultural geographers, emanates from a global perspective on culture that presumably differs from that shared by the rural Sudanese women and children Katz studies. Certainly it is impossible to fully shed one's world view, and we are alert to the analytical strength of cross-cultural structures and patterns. But is it preferable to serve the ideological needs of particular theoretical camps within academe, rather than those of academic bureaucracies? Perhaps so, but in our experience in Indian Country, the activist is often greeted with suspicions and obstructions similar to those encountered by bureaucratic functionaries. We contend that neither role serves the interests of the people studied very well, people whose own voices usually are better suited to assessing their own world. Unfortunately, few in geography have sought suitable alternative research methods.

Intellectual property

In many respects, research remains fundamentally dependent on intellectual appropriation. Researchers have recorded the words and deeds of people and published them under their own names for their own professional advancement. Ethnographic facts are commonly decontextualized, and depicted as representative of a culture's collective worldview rather than being the words of a person or the intellectual property of an individual, a family, or a village. Many indigenous peoples view these depictions not only as invasive and ethnocentric, but as overtly plagiaristic. Still, few participants have been aware of these transgressions, fewer have sought a means of recourse until quite recently, and fewer still have successfully used what legal, economic, or political clout they may have. Two examples will underscore the reciprocity of intellectual appropriation rather emphatically. Both involve regrettable events, but in so doing may be instructive for those who still imagine that the isolated researcher defines ethical behavior.

In a native community on the Northwest Coast visited by Deur, most residents recall with vivid clarity the names of researchers who had passed through the village. Several were despised for taking proprietary knowledge (including dances and tales owned by families) and artifacts (including masks and other totemic art), and distributing them in publications and museum displays without permission. Although willingly shared in good faith with befriended researchers, losing such property to a mass audience was an unanticipated outcome of cordial exchanges in remote tribal towns. Within this village, as within most Northwest Coast societies, such losses may be considered debt demanding payment through reciprocal exchanges of equal value. Yet, not prioritizing their personal relationships over academic prerogatives, many researchers had not reciprocated. In residents' views, this added grave insult to past injuries, and the researchers were derided as vile profiteers. Lacking formal avenues for recourse, residents schemed to lure offending researchers back to the community to pay for their transgressions. Some spoke of hiring Vancouver lawyers; others advocated acts of violence.

Because of such concerns, many groups have developed mechanisms for controlling dissemination. For example, when both native and non-native executives working for the Inuit Cultural Institute (ICI) in northern Canada were accused of embezzlement by the federal government in 1988, elders simply and effectively locked up information associated with all ICI projects. This was possible because ICI cultural research is conducted in the interests and with sole permission of a council of local elders. Rundstrom's ICI-funded participatory place name mapping project ceased immediately, and all data were legally declared elders' intellectual property. Use and dissemination was cancelled until further notice. All projects were dismissed, and researchers were sent home with little to show for their effort.

Dissemination of information

Researchers' desires for publicity, material rewards, and prestige from publication conflict with a view of written documents and other material goods often held in Indian Country: these are containers of human dignity, empathy, and community strength. Their representation can have unexpected and adverse consequences. For example, among Northwest Coast communities, researchers invariably are struck by the visual aspects of totemism and ceremonialism. Increasingly, formally educated members of these communities encounter the printed results of past ethnographic studies. They recoil as they encounter photographs of ceremonies labeled "savage customs," or see images of totems – familial burial markers – that were taken by researchers and reprinted in books, museum catalogs, or even tourist postcards. Over time such representations have diminished or otherwise altered the cultural significance of these artifacts and traditions.

As a prescriptive note, the practice of cross-cultural research can be redeemed somewhat if researchers seek permission to publish potentially proprietary forms of information, and simply ask those providing information whether they wish to remain anonymous. Norton and Manson (1996), working among Alaskan Eskimos, recommended that individual identities must be held confidential. But in our own experience among Inuit and Northwest Coast peoples, participants often expect recognition and may actually delight in it.[4] Moreover, one subject area may invite recognition (e.g. environmental knowledge) for people who want anonymity on another (e.g. effects of alcoholism). In sum, the decision is highly contextual and should be made primarily by those who provide the information.

Cross-cultural research also can have adverse impacts through secondary representations. Writing on American Indians is characterized continuously by extractive representations buttressing non-Indian social, cultural, economic, and military agendas. Thus, Indians could be dangerous and brutal savages during the era of military conquest, yet seen as environmentalists and philosophers, the consummate noble savages, as non-Indians confront late twentieth-century cultural malaise (Churchill 1992; see also papers in Doty 1996; Moore 1993). Even among critical self-reflexive researchers alert to unbalanced power relations, representation remains a potentially extractive and destructive affair.

Researchers must attempt to understand the contents of their own cultural baggage and avoid assessing indigenous practices negatively, basing their assessments solely on Eurocentric conceptions of what constitutes ethical behavior. For example, in 1979, investigators released information to the *New York Times* regarding a study of alcoholism among the Inupiat in Barrow, Alaska. The resulting article was entitled "Alcohol Plagues Eskimos," and United Press International issued a story linking the "epidemic" of alcoholism with funds coming to the village from the Alaska Native Claims Settlement Act of 1971. These articles created great conflict between and within Inupiat communities, between Inupiat and many researchers, and between the Inupiat, researchers and local agencies who commissioned the study. The Standard & Poor's bond rating for incorporated Inupiat villages dropped immediately after publication, villages could no longer fund important municipal projects, and the overall quality of Inupiat life diminished. As of 1995, Alaskan native communities prohibited research on this topic, an unfortunate result because the study of alcoholism need not have fallen prey to issues of misrepresentation (Norton and Manson 1996).

As a result of such infringements, many tribes and native communities now have review boards that regulate both research and publication on health, customs, religion, and other topics. Many Institutional Review Boards within funding agencies and universities now require both individual and collective consent from participants and leading decision-makers. Photographs or quotations are sometimes screened, and a growing number of tribes now demand that they be allowed to review manuscripts before they are released. This requires completion of a manuscript long before final editing or publication, and a willingness to subject work to additional levels of scrutiny. This may sound chilling to researchers who prize academic freedom and fidget under tight deadlines. Increasingly, however, there is little choice in the matter.

Concerns about publication should also include consideration of audience responses. Among the most important objectives of academic representation is the construction of empathy, the ability to project one's personality into the personality of another to break down barriers to understanding. Cross-cultural researchers seek (or should seek) to achieve this in writings and other products viewed by people removed from the research process. However, audience reaction is often difficult to predict. Viewers or readers will often register those aspects of people and places they have been conditioned to expect from news and other media, other scholarship, college coursework, and social experience. For example, when viewing a film of her work on Nevis migrant communities, Byron (1993) found that students registered a priori images emanating from exactly such sources, while unanticipated facets of the film failed to generate much interest or discussion. Byron feared she actually might have reinforced negative stereotypes. Her suggestion – one we endorse – is to engage audiences (students particularly) both before and after the presentation of research, in order to replicate in these audiences the empathy and contextual knowledge realized by researchers during the process of field research. An empathic "bracket" may accompany books or articles as an integral component of the introduction and conclusion, thereby leading readers through an empathic "loop."

Further, through both research and teaching, researchers can engage the "crisis of representation" by aiding non-Western participants in their attempts at self-representation. Many obstacles face indigenous peoples who seek their own voices in postcolonial cultural and economic discourses (Clifford 1988; Clifford and Marcus 1986; Spivak 1985). Working alongside non-Western peers (both possessing or lacking formal university training) cross-cultural researchers are in a unique position to bring the tools and resources of the university to assist these peoples in the study of topics shaped by indigenous concerns, views, and agendas.

Codes of ethics

Geographers embarking on cross-cultural research might learn from the experiences of anthropologists. Acting as cultural brokers, they have had to adapt to changing circumstances within the larger context of relations between colonial and indigenous peoples. Typically, geographers have operated at different levels of resolution, assessing landscapes and geographical patterns rather than the intricacies of mental and social life. Thus, geographers have not been compelled to engage indigenous peoples' concerns (and objections) so directly. Geographers also are often an unknown quantity because they have been less visible than anthropologists in native communities throughout the former colonial world (save possibly in Latin America). At present, geography's relative anonymity is beneficial, providing an opportunity to go about our work differently.

Anthropological codes of professional ethics reflect the diverse and increasingly contentious issues discussed in this chapter. Anthropological codes of ethics from the early twentieth century tended to equate analytical and intellectual honesty with ethical behavior. By the 1960s, however, these codes began to manifest a search for a balance, addressing participants' rights and interests as an equally significant concern. This trend likely reflected anthropologists' growing moral concerns with minorities' rights, their practical concerns about growth in indigenous peoples' sovereignty, and the increasing ability and willingness of indigenous peoples to put that sovereignty into action, i.e. to exclude unwanted researchers from their communities (Fluehr-Lobban 1991). The solution stated in the current American Anthropological Association (AAA) Statement of Ethics is that, when the interests of researchers, participants, and other institutions collide, the "paramount responsibility," of anthropologists should be "to those they study."[5]

Whereas codes of ethics may be written at the level of national professional organizations, some universities and academic institutes have also established codes of ethics that shape the research programs of affiliated geographers. For example, Devon Mihesuah (1993), a Choctaw on the faculty at Northern Arizona University (NAU), served on a university committee formed by the president to establish university-wide ethical guidelines for cross-cultural research. The NAU committee recommended augmenting external strictures emanating from tribal governments and funding agencies like National Institutes of Health and National Science Foundation with the following formal objectives: researchers should safeguard the trust and rights of others; researchers should describe the project's purposes early and clearly to everyone involved; individual participants must be given the right to decide whether to remain anonymous or to be named and acknowledged in any concluding documents (also see Clinton 1975); a "fair and appropriate return" should be given to individuals; all anticipated consequences of the research should be described to individuals; researchers should "cooperate" with members of the host society in research design and implementation; "representative bodies" should be able to review materials prior to release by researchers; results should be given to elected or traditional leaders; and the research program, cumulatively, should embrace cultural pluralism and diversity.

Articulated codes of ethics for cross-cultural research of the sort produced by the AAA or NAU represent two possible options. Yet Curry (1996) has expressed doubt as to whether written codes are successful, and whether anthropologists, geographers, or anyone else have any reasons to adhere to them. And Winchell (1997) questioned whether codes too often ease individuals' concerns with ethical issues while demanding little individual responsibility. Certainly, the existence of so few mechanisms within academia for monitoring and disciplining violations supports both these claims. It is too easy to imagine that without reference to the specific individuals and arenas in which research is conducted, written codes may drift from the moorings of actual academic practice into treacherously ambiguous or self-celebratory waters. Accordingly, they can become ritualized means of assuaging academic guilt, rather than devices for improving research practices. Finally, pre-written statements may hinder research in Indian Country because they often ignore local concerns and raise the suspicions of potential participants, local governments, and others. Indeed, Daniel Wildcat (1996), Dean of Social Sciences at Haskell Indian Nations University, has reminded academic geographers that outsiders have presented many Indian communities with written statements of intent – from treaties to "binding" resolutions to codes of ethics – for over three centuries, and their frequent violation has left many indigenous peoples today with considerable warranted suspicions about the efficacy of most exogenous written declarations. To be sure, AAA, NAU, and other large academic organizations may be applauded for their sensitivity and good intentions in trying to induce greater reflexivity among cross-cultural researchers in their jurisdiction. After all, the number of people who would argue outright with the nine NAU guidelines is surely small. And researchers will continue to affirm their trust in such written codes, just as Byron (1993) does.

Yet ethical issues require attention at levels of resolution beyond faceless and politically inconsistent professional organizations and universities. Winchell (1996b) suggested that individuals who arrive bearing the code of ethics of a larger bureaucracy are less effective than those researchers who have sought to establish long-term relations between American Indian communities and smaller academic units like departments, centers, or institutes. Such a unit – seldom so large that its constituent individuals can remain faceless or wholly detached from group agendas – can become an active stakeholder in responsible research. The institutional permanency of such a relationship can help stabilize and smooth the eccentricities associated with individual research programs.

Conclusion

We hope to have raised key issues for consideration when geographers contemplate cross-cultural research, especially for those working in Indian Country. In our view, ethics and ethical behavior do not emerge from isolated reflection in a social vacuum, by adopting either bureaucratic or purely activist roles, or through pre-written codes. The issues surrounding us, including sovereignty, intellectual property, and the effects of dissemination of information in writing and teaching, cannot be addressed so easily. Indeed, they require the constant involvement of others. Thus, ethical behavior may best be negotiated as a form of reciprocal appropriation, wherein valued exchanges of information are made in more than one direction and as part of long-term relationships. And these relationships are best supported by mutual personal understanding between specific researchers (whether acting as individuals or as members of smaller academic organizations) and the people and places under study. Given the emphasis we place on social reciprocity, the task of producing an ethical research process is clearly relational and contextual.

Yet, if ethical behavior, broadly defined, is essential to the research process, but no single code of ethics is appropriate for the diverse range of contexts in which geographers find themselves, what option might be pursued productively by a committee of the AAG or other disciplinary body with an interest in "doing the right thing?" Instead of an articulated code of ethics, geographers might be aided by the creation of a set of guidelines for negotiating ethical issues within research-based relationships, that is, by a list of topics and ideas for consideration during specific cross-cultural negotiating processes. It seems self-evident that the members of a committee formed to develop such guidelines should manifest the spirit of reciprocity espoused in the document they would produce. Although we remain skeptical of written texts altogether, it may be that such guidelines could serve as an important reference tool or instructional device for geographers seeking cross-cultural connections.

Attempts at self-regulation may only reduce the speed with which the sisiutl approaches each of us. A researcher might stand a better chance simply by giving up pretenses to sole control of a project. For too long, within American Indian communities and elsewhere, academic research has meant a one-way extractive exchange in which European and Euro-North American social and academic agendas have been advanced at the expense of the dignity and continuity of non-Western peoples and places. Not only is this practice ethically indefensible, but as political and economic sovereignty are asserted in a postcolonial world, such one-sided control is simply impossible.

Notes

1   Both a social and legal expression today, "Indian Country" first appeared in print in the British Proclamation of 1763, had its legal basis affirmed in the Articles of Confederation, and has been defined and elaborated further in congressional legislation and decisions of the US Supreme Court ever since. It refers to the aggregate of numerous sovereign enclaves where American Indian land title has not been extinguished (Deloria and Lytle 1983: 58–65). This chapter represents, in part, one outcome of a larger project organized by the authors on the ethics of cross-cultural research. This project included two panels focused on ethical concerns surrounding field research in American Indian communities. These panels were held at the 1996 and 1997 meetings of the Association of American Geographers (AAG); both panels were organized and chaired by the authors. This chapter incorporates some ideas from all panel participants, not just those of the authors. The 1996 panel was entitled "Reciprocal Surveillance/Reciprocal Obligations: The Ethics of Research in American Indian Communities." Besides the authors, panelists included Dick Winchell (Eastern Washington University), Michael Curry (UCLA), Hedy Levine (San Diego State University), and Steve

Schnell (University of Kansas). Though not a panelist, Daniel Wildcat (Haskell Indian Nations University) contributed substantially to the 1996 panel. Participants in the 1997 panel, "Reciprocal Surveillance/Reciprocal Obligations: Continued Dialogue on the Ethics of Geographical Research Among American Indians" included Beth Ritter (University of Nebraska-Lincoln), Deur, and Winchell (who spoke briefly on behalf of invited panelist, Cecil Jose, Director of Native American Studies at Eastern Washington University and a member of the Nez Perce Tribe). Deur has conducted both ethnogeographic and archeological research among "Kwakwaka'wakw" (or Kwakiutl) and "Nuu-Chah-Nulth" (or Nootka) First Nations of British Columbia, as well as the Makah, Klamath/Madoc and Tillamook of the American Pacific Northwest. Rundstrom has worked among the Canadian Inuit, the Mvskoke (Creek) of Oklahoma, and the Tewa people of San Ildefonso Pueblo in New Mexico.

2   Anthropological treatments of ethics have centered on the "crisis of representation" and ethical conundrums that emerge when researchers serve as advocates of their host communities. See Doty 1996; Deloria 1995; Szasz 1995; Moore 1993; Fluehr-Lobban 1991; Clifford 1988; Clifford and Marcus 1986; Paine 1985.

3   One tribe may plead for research assistance, while another may vehemently prohibit almost identical research proposals; these differences seem to be a function of past histories with researchers, minor customary differences, varying levels of economic independence, and the need for information for cultural revitalization, land claims, or tribal petitions for federal recognition or assistance. In our experience, dire social, political and economic conditions produce greater interest in academic research and the possibility of a long-term symbiotic relationship it affords.

4   Indeed, those very same people on the Northwest Coast who objected strongly to depictions of graveside totemic art sometimes seemed overjoyed to show their own carvings to researchers. Some not only allow, but encourage photography of their more mundane or secular works. To distinguish the acceptable from the unacceptable, researchers need more than passing familiarity with the central assumptions of the society under study. Indeed, casual research efforts by non-specialists appear to be at fault for many tensions found in this region.

5   When the researcher is viewed as an "insider," a member or partial member of the community under study, the link between access and responsibility changes accordingly. Recording stories of London emigrants from Nevis, an eastern Caribbean island where she was born, Byron (1993), a geographer, remained part of her own subject. Though socially removed from her Nevis informants, she was granted access to information that might otherwise have been denied, only because island residents recognized that she might sympathize with their concerns and agendas, and might understand better than a complete "outsider" the specific issues and responsibilities implied by their disclosure of information. But this intimacy also meant she possessed a power to affect their lives that was greater than what an outsider could wield. The line Byron had to walk in her quest for information must have been thin indeed.

References

American Indian Law Center (1994) *The Role of Tribal Government in Regulating Research*, Albuquerque: American Indian Law Center.

Byron, Margaret (1993) "Using audio-visual aids in geography research: questions of access and responsibility," *Area* 25: 379–385.

Churchill, Ward (1992) *Fantasies of the Master Race: Literature, Cinema, and the Colonization of American Indians*, Monroe, ME: Common Courage Press.

Clifford, James (1988) *The Predicament of Culture: Twentieth Century Ethnography, Literature, and Art*, Cambridge, Mass.: Harvard University Press.

Clifford, James and George E. Marcus (1986) *Writing Culture: The Poetics and Politics of Ethnography*, (School of American Research Advanced Seminar), Berkeley: University of California Press.

Clinton, Charles (1975) "The anthropologist as hired hand," *Human Organization* 34: 197–204.

Curry, Michael (1996) "Panel presentation. Reciprocal surveillance/reciprocal obligations: the ethics of research in American Indian communities," annual meeting of the Association of American Geographers, Charlotte, North Carolina.

Deloria, Vine Jr (1995) *Red Earth, White Lies: Native Americans and the Myth of Scientific Fact*, New York: Scribners.

—— and Clifford Lytle (1983) *American Indians, American Justice*, Austin: University of Texas Press.

Dorris, Michael (1987) "Indians on the shelf" in C. Martin (ed.) *The American Indian and the Problem of History*, New York: Oxford University Press, pp. 98–105.

Doty, Roxanne (ed.) (1996) *Imperial Encounters: The Politics of Representation in North–South Relations*, Minneapolis: University of Minnesota Press.

Fluehr-Lobban, Carolyn (ed.) (1991) *Ethics and the Profession of Anthropology: Dialogue for a New Era*, Philadelphia: University of Pennsylvania Press.

Katz, Cindi (1992) "All the world is staged: intellectuals and the projects of ethnography," *Environment and Planning D* 10: 495–510.

Mihesuah, Devon (1993) "Suggested guidelines for institutions with scholars who conduct research on American Indians," *American Indian Culture and Research Journal* 17: 131–139.

Momaday, N. Scott (1976) "Native American attitudes to the environment" in W. H. Capps (ed.) *Seeing With A Native Eye: Essays on Native American Religion*, New York: Harper and Row, pp. 79–85.

Moore, Sally Falk (ed.) (1993) *Moralizing States and the Ethnography of the Present*, Arlington, VA: American Anthropological Association.

Norton, Ilena M. and Spero M. Manson (1996) "Research in American Indian and Alaska Native communities: navigating the cultural universe of values and process," *Journal of Consulting and Clinical Psychology* 64: 856–860.

Paine, Robert (ed.) (1985) *Advocacy and Anthropology: First Encounters*, St Johns, Newfoundland: Institute of Social and Economic Research, Memorial University of Newfoundland.

Rundstrom, Robert (1993) "The role of ethics, mapping, and the meaning of place in relations between Indians and whites in the United States," *Cartographica* 30: 21–28.

—— (1995) "GIS, indigenous peoples, and epistemological diversity," *Cartography and Geographic Information Systems* 22: 45–57.

Sitter-Liver, B. (1995) "Against the right of the stronger: ethical considerations concerning cultural property," *European Review* 3: 221–231.

Spivak, Gayatri C. (1985) "Can the subaltern speak?" in C. Nelson and L. Grossberg (eds) *Marxism and the Interpretation of Culture*, Urbana: University of Illinois Press, pp. 157–175.

Szasz, Margaret Connell (ed.) (1995) *Between Indian and White Worlds: The Cultural Broker*, Norman: University of Oklahoma Press.

Wildcat, Daniel (1996) "Panel presentation. Reciprocal surveillance/reciprocal obligations: the ethics of research in American Indian Communities," annual meeting of the Association of American Geographers, Charlotte, North Carolina.

Winchell, Dick (1996a) "The consolidation of tribal planning in American Indian tribal government and culture" in K. Frantz (ed.) *Human Geography in North America*. Innsbruck: Department of Geography, University of Innsbruck, pp. 209–224.

—— (1996b) "Panel presentation. Reciprocal surveillance/reciprocal obligations: the ethics of research in American Indian Communities," annual meeting of the Association of American Geographers, Charlotte, North Carolina.

—— (1997) "Panel presentation. Reciprocal surveillance/reciprocal obligations: continued dialogue on the ethics of geographical research among American Indians," annual meeting of the Association of American Geographers, Fort Worth, Texas.

18 Ethical issues in practical contexts

Nuala Gormley and Liz Bondi

Introduction

This essay arose in the context of a student–supervisor relationship. One of us (Nuala Gormley) brought to this relationship concerns about the presentation in her doctoral thesis of ethnographic evidence gathered during the course of a 21-month period during which she lived and worked in a mission place in sub-Saharan Africa.[1] The other one of us (Liz Bondi) found that these concerns resonated closely with some of those arising for her in various aspects of academic life including doctoral supervision. In very different contexts, both of us have deliberated over decisions we have had to make in the course of our practice as academic researchers. Moreover, for both of us it has not been a case of facing isolated moments of crisis; rather we have been aware that our everyday practices entail commitments that have ethical dimensions.

This emphasis on everyday practices highlights close connections between questions of ethics and research methods. It follows from this that what we have to say owes much to writings on methods, most especially to discussions among feminist geographers about fieldwork and about qualitative methods (for reviews and discussion see for example Nast 1994; Rose 1997). Our approach shares a good deal with those who argue against an objectifying separation between the research and the field or the researched (Sparke 1996), and for an understanding of our positions as always somehow in the field (Katz 1994). In this context, we share an interest in destabilizing distinctions between positions inside and outside the field (Kobayashi 1994; Moss 1995; Nagar 1997). While some of these issues have been explored in existing work on qualitative methods in and beyond human geography (see for example Eyles and Smith 1988; Jarvie 1982) much of this earlier work was more concerned with epistemological than ethical issues (but see Fabian 1983; Seiber 1992). But, like several other feminist researchers, we have become acutely aware of the risk that our efforts to listen to and understand those we interact with in the conduct of our research may turn into, or coexist with, an exploitative appropriation of other people's experiences (England 1994; Farrow 1995; Finch 1984; Gilbert 1994; Patai 1991; Shaw 1995). We hope that the reflections we offer on our attempts to steer a path between understanding and exploitation will open up this particular issue for further discussion.

Our practical orientation, with its sensitivity to context, means that we tend to find ourselves in sympathy with those who argue that ethics must necessarily be situated (Slater 1997). But, while our primary concern lies with practical issues rather than normative principles, we are aware of appealing to notions that may be of universal scope. Consequently, we agree with Sayer and Storper's (1997) argument that universal and situated perspectives on ethics may not be so far apart, and we hope that this discussion illustrates something of their interweaving in research and related academic practices.

A key aspect of situatedness concerns our relationships with others, and we find the notion of a relational ethics particularly useful. What this means is that we experience and engage with ethical issues in terms of our relationships with others, whether we are consciously aware of the connections or not (Whatmore 1997; also see Gilligan 1982). In this context we are committed to working with the differences between ourselves and those with whom we interact, without seeking either to erase difference – that is to presume similarities or identifications that do not exist – or to treat difference as representing something essential and exotic (Gilbert 1994; Katz 1994; Kobayashi 1994; Nast 1994; Slater 1997). This is no simple task; indeed the pitfalls are numerous (see for example Dyck 1997; Gibson-Graham 1994; Rose 1997). We do not claim to have avoided all such pitfalls, but our discussion can be understood as a working through, using ideas that we have found useful and which may also turn out to be of use to others.

In what follows we focus on issues arising in the transfer of 'data' from one context to another. Research of all kinds involves multiple contexts, and the movement of information, insights and so on from one context to another inevitably raises ethical concerns. We provide illustrations from Nuala's field research in rural Africa, and then point to connections with academic practices 'at home', which we illustrate through reflections on our experience of co-authorship. We draw attention to questions of integrity in relationships, arguing that this is central to the ways in which both of us address ethical issues. Before going further we relinquish the first-person plural in order to acknowledge and write about our different experiences and perspectives.[2]

Nuala's story

My (Nuala's) ethnographic material was gathered in a milieu initially unfamiliar to me, in which my presence was always assumed to be temporary and to which I may never return. Since I was not studying academia, my research engaged with a world and with ways of understanding life outside the institutional framework in which it would be processed and presented as a doctoral thesis, as conference papers, or as academic publications. For many personal and professional reasons, my interest lay with people at the receiving end of missionary endeavour in the Third World. The context was 'there', so with the intention of gathering data, there I went. Once there I began to engage with people experiencing mission in different ways. At least on the surface, they, together with the place itself, became the researched, while I was the researcher. Twenty-one months later, I returned home and gradually worked with the material I had collected in order to produce a doctoral thesis.

Together with my husband Michael, I went to Africa as a volunteer with a Catholic lay mission organization. We both worked full-time at a mission place in a rural area. I taught at the mission school, while Michael worked as an engineer at the mission hospital. I had permission to conduct research during my period as a volunteer from both the sending agency and the diocesan bishop by whom I was employed.[3] I endeavoured to be open about my research to those with whom I came into contact, but in practice much of the material I gathered arose when other roles were to the fore (compare Scheper-Hughes 1992).

Interconnections between researcher, researched, audience and text have been discussed by several commentators (see for example Keith 1992; Nast 1994). I draw upon one particular day towards the end of my period in the field, where my troubled relationship with a woman who was later to figure significantly in my analysis, provides a point from which to consider some ethical issues arising in this web of relations. My presentation of this moment is excerpted from field notes recorded in July 1994.

> After weeks of stories about the relationship between Sister Matron[4] and Mr Kaba,[5] a trader in the town, Michael[6] was driving her and several hospital employees into town (I was also there). Sister Matron was bringing the hospital safe to the diocesan workshop to be opened because the keys had been lost. The hospital accountant instructed Michael to remain with Sister Matron to witness the amount of money removed from the safe, since she suspected that Sister Matron would give it directly to Mr Kaba. Sister Matron was very agitated as Michael lingered at the workshop, and after they counted the cash together, she asked to be left in the town centre. Normally the hospital vehicle would wait outside the shop of Mr Kaba, who supposedly kept an eye on it, but since the stories had broken, the hospital driver had avoided that area of town. Sister Matron told Michael to drive to Mr Kaba's shop, but he refused, saying 'There? No way!'. At this point she realized that Michael (and I) had heard the rumours about her and Mr Kaba, and if *we* knew so did the entire county. She became very angry. She demanded to know why Michael would not park at Mr Kaba's shop, and Michael refused to be drawn. She stormed from the vehicle.
>
> I was relieved to get back to the mission station that evening, after a long, hot, frustrating day in town. Then Sister Matron arrived on the verandah, and she and Michael started a very heated discussion about the day's events. It was becoming increasingly loud and quarrelsome and I worried that he would say something he would later regret. So I intervened and told them that I was too tired and too pregnant to put up with such stubborn childishness, and told them both to leave me in peace. They were both stunned at my outburst and their immediate argument dissipated.

In my thesis I am interested in portraying, for a particular Western audience, something of the way in which the undercurrents and complexities of mission life, especially in its development projects,[7] contradicted the veneer that was presented to visiting field officers from NGOs and charities. I also want to explore some of the pressures and conflicts experienced by mission women. These were graphically illustrated by Sister Matron's relationship with Mr Kaba, which contravened her religious vows and therefore attracted enormous interest and comment when it became widely known to local people. Consequently, Sister Matron, who occupied a powerful administrative role in the mission hospital emerged as a very significant figure in my analysis of the contradictions underlying development in a mission place, especially in relation to sexuality. This chapter was prompted in part by the difficulties I encountered in writing about her in my thesis. The observations that follow are the product of my own reflections and those of my supervisor and co-author, generated individually, in discussion, and through the processes of writing.

While in Africa, I witnessed and absorbed the minutiae of everyday mission habits and events. Incidents which initially seemed bewildering to my European sensibilities usually came to make sense to me as I became more familiar with and integrated into the particular place of my research. In this process the boundary between myself as researcher and those I positioned as the researched became less clear (compare Katz 1994). The excerpt from my field notes makes it abundantly clear that I was deeply involved, and positioned in the field in complex ways. As in all ethnographic research, much of the material I gathered consisted to some degree of my experiences (compare Evans 1988). In what follows I comment on some of the issues that arise from this, concerned with appropriations and displacements from the field.

Appropriations and displacements

With time I grew to anticipate local understandings of much that I observed, but certain elements of local life remained alien and unresolvable to me. On my return to the Western academy, the place of my field research is represented through my memories, together with the tapes, field notes and artifacts I brought home. I continue to try to make sense of incidents and habits I observed and experienced. I find that it is issues that remain unresolved for me that gain prominence in my analysis and with which I engage relentlessly. These issues may be of little significance in the mission lives of others, illustrating forcefully how my research inevitably entails both an appropriation of what I find interesting in my relations with others and a displacement from their concerns to mine.

I have returned repeatedly to the events described in the excerpt from my field notes presented above, in part because they seemed to encapsulate for me something of the unresolvably complex and contradictory relationships between Western development agencies and local people. While Sister Matron was the main character in the story presented above, it was not really hers; rather it was a story about her impact upon others and also about relationships and communications within a rural community that included outsiders. Moreover, what may have meant something altogether different to Sister Matron came to signify for me some of the pressures experienced by mission women and some of the dilemmas they face. In this way, my appropriations from the field are informed by their displacement as I move between several contexts – an African mission, the Western academy, and a Western sending agency to name but three.[8]

The people who provided me with all my data have had no direct participation in what I choose to say or write about them. Although most were aware that I was engaged in research, they almost certainly did not anticipate that they would feature (however anonymously) in the way that they now do. Indeed, during my period in the field, I had very little idea of which incidents and which people would become prominent in the representations I later crafted. While in the field I did not ask Sister Matron for permission before I recorded in my field journal the stories I had heard about her, or incidents involving her, such as the one I have just described. I never explicitly sought her informed consent (compare Jarvie 1982). Neither did I inform her of what I wrote. Although she knew that I was conducting research, it seems very likely that on the day in question (and on many other days) she viewed me primarily as an irritable teacher, married to an infuriating engineer (and of course both of us as white Western foreigners in the mission place temporarily). In this context I have worried a good deal about using Sister Matron's story (and others), but I have in the end decided to write about her both in my thesis and in this chapter. Other material I gathered I have decided should not be used in my thesis or in papers, whether oral or written. Whether I set the right limits between what I have decided is available for use and what is not is, of course, open to question; here I explore the issues guiding my decisions.

Researchers sensitive to the relations of power between themselves and those they research have attempted to reduce power inequalities in a number of ways.[9] For example, some researchers have sent interview transcripts and/or drafts for their respondents to check, amend or comment upon (see for example Painter 1979). Some have discussed their interpretations with those they have researched (see for example Skeggs 1997). Some have sent copies of theses, research reports or publications to research participants (see for example Madge 1997). And some have decided against writing up aspects of, or even all of, their research (see Madge 1997; also see England 1994). These practices acknowledge that research inevitably entails appropriations: researchers acquire knowledge from others and take it elsewhere.

Increasing the involvement of the researched and emphasizing the importance of dialogue in the production of knowledge in the academic domain may reduce some of the inequalities integral to ethnographic research, but only to a limited extent. As Skeggs (1997) argues, the researcher is, in the end, the one whose interpretation is at issue and who makes the final decisions about oral or written (academic) products. Moreover, as Madge (1997) observes:

> Academic geographers are part of a system of knowledge production that has systematically undermined and dislodged Third World thought systems . . .

and created powerful institutions of knowledge which are located in the First World, and often (but not always) viewed from the perspective of the First World.

(Madge 1997: 116; also see Ake 1979; Ngugi Wa Thiong'o 1992)

Thus, any attempts I might make to convey Third World perspectives in a First World context will always be contradictory given my positioning as a white Western woman. Whatever I do I cannot avoid being part of an exploitative system of relationships between the First World and the Third World. In undertaking my research, I view my challenge to be about appropriating knowledges from the field in ways that make connections across difference that have the potential to unsettle aspects of this system, albeit in modest ways (compare Dyck 1997; Nagar 1997; Patai 1991).

In practice, my decisions about what to include and what to exclude in my thesis, and about how to involve the people with whom I interacted in Africa in its production, have been guided principally by two interrelated issues. The first concerns the qualities of relationships I developed with particular individuals (compare Gilligan 1982; also see Facio 1993; Jarvie 1982; Nagar 1997). In instances where friendships have been sustained beyond my time in the field, I have corresponded and in some of my letters I have sought to check or clarify particular interpretations. But in other relationships, which were more fraught – as in the case of Sister Matron – or which I felt could not be sustained over distance, such actions would have felt intrusive and demanding, as if I would have been attempting to 'take' even more, and so intensifying the exploitative quality of my appropriations. My anxiety about transferring what I learnt about Sister Matron in the context of our overlapping lives into the text of my doctoral thesis is not something I can easily resolve, but understanding the issues relationally clarifies the tensions I seek to sustain.

The limits I have placed on involving others in the production of ethnography highlight the extent to which I have been thrown back on my experiences in the field and my subsequent processing of those experiences as the basis for making decisions about what constitutes ethical practice. This raises issues about who my research is about. In writing my thesis I am faced with a problem both perennial and ubiquitous, namely, how to acknowledge the enormous significance of my personal experiences in this research, without displacing those who made these experiences possible. The story I have presented, in which Sister Matron figures prominently, takes the form of an account of my experience of interactions and interrelationships on a particular day. It also suggests something of the impact of Sister Matron on myself and others. I cannot tell her story – her version of the events I have described would no doubt be very different – but I want to keep her at the centre of my account, while representing my experiences of her and of stories circulating about her.

This suggests that there is a balance to be struck between appropriating knowledge from others and displacing their perspectives. In order to find this balance I think it is useful to consider the processes of ethnography in terms of a complex web of relationships. While I apply my own values in thinking through what it means to treat others with respect and integrity, I also open myself to the ideas and concerns of others both as a volunteer/ethnographer in the field, and as a doctoral student at home. These contexts frame my relationships with others, and it is by reflecting on these contextualized relationships that I make decisions about what constitutes ethical practice.

This brings me to the second main issue guiding my decisions about what to include and what to exclude. I have come to understand the 'bottom line' I have drawn in relation to such narratives, and in the broader context of my analysis, in terms of its relevance to the development projects taking place in the mission. Only when the personal lives of influential people (whether local or foreign) had a direct and detrimental effect on the efforts being made to improve the health and well-being of local people did I include stories of this kind. In such cases, I have endeavoured to present the story in its fullest context and to allow events and their consequences to illustrate any implicit judgments made by myself or by others. In the case of Sister Matron, her relationship with Mr Kaba eventually and detrimentally affected the moneys available to the hospital for medication. It was this connection that I considered to be important and only for this reason does the story of her relationship, considered 'illicit' by the institutions framing her life, feature in my thesis. It features in this chapter partly because it also serves to illustrate my sense of unease in making such decisions. But it also demonstrates how my research has been framed by many personal relationships, and how one small and apparently inconsequential encounter between myself and Sister Matron can encapsulate the methodological and ethical 'open wounds' of my ethnography.

Liz's story

My (Liz's) research experience has been based exclusively in Western, urban contexts. As Nuala knew from the outset of our supervisory relationship, I could bring no specialist expertise to my role.[10] At first, it seemed that there was very little overlap between our research projects beyond a shared interest in experiential knowledges that brought issues of gender to the fore. This situation has been unique in my experience of graduate supervision. While it has disadvantages it also has certain advantages. In particular, because I have brought few preconceptions to my task, I have, perhaps, anticipated less, and listened with particular care. But also, I have looked for connections between her research interests and my own.

The issues about ethical practice Nuala brought to our supervisory meetings fascinated me and prompted me to reflect further on aspects of my experience as a researcher. In due course, I recognized resonances with my experience as a supervisor, and now it seems that our sensibilities in relation to ethical issues overlap considerably although our research contexts differ. This was crystallized powerfully in the co-authoring of this chapter.

Processes of appropriation and displacement, which Nuala has discussed in relation to her ethnographic research, have echoes within the production of this chapter. To elaborate, the inspiration for this chapter came principally from Nuala's research, but the act of imagining how it might be presented as a chapter in a book on geography and ethics, was at least as much mine. As in many co-authoring partnerships between a student and a supervisor, I, the supervisor, have taken a lead role in turning text from, and discussion about, a doctoral thesis into text for a book chapter. This entails processes of appropriation: I have taken Nuala's writings, evidence and insights, incorporated them into my own ways of thinking, and created something to which I lay (partial) claim. Regardless of who is listed as an author, and in what order, such acts may be experienced and under-stood as exploitative. Likewise, processes of displacing one research agenda (that of the doctoral student) with another (that of the supervisor) are at work. Indeed, in the preparation of this chapter I spent a good deal of time impersonating Nuala: that is, I edited and reorganized much of her text, contributing to the production of the preceding section by pretending, as it were, that I could write from her position, and so displacing her in the act of writing. In so doing I selected from the text Nuala prepared for this chapter and amended it in ways that I judged to constitute a more focused and effective manner of addressing the audience of this volume.

It seems to me important to acknowledge these processes.[11] Further, while the power relations (and therefore the risks) in play are particularly clear when a supervisor and student co-author, the dynamics are similar in any co-authorship project, and also, arguably, in the academic review process (compare Berg 1997; Curry 1991). Again, it seems to me useful to consider the ethical issues at stake in terms of relationships. Within the supervisory relationship I can check out whether my drafts are acceptable to my co-author, and I can invite her to amend and edit the text. I can be alert to our different interests in seeking publication.

However, it is at least as important to acknowledge the enormous differences between Nuala's relationships with members of the community in which she lived, and mine with her. In the writing of this chapter there was a good deal of exchange between the two of us, and in our shared setting of the Western academy there is scope for mutual agreement on the text.[12] We are, in a sense, players in the same field, albeit differently positioned. In sharp contrast to this, Nuala must move between very different fields. Where I rely upon mutual agree-ment she cannot, because those who inform her ethnography share only one of the contexts of its production.

Clearly, therefore, the appropriation and displacement of knowledges rooted in relationships with people outside the Western academy raise far more problematic issues of ethical practice than occur within it. But drawing this parallel directs attention to the inter-subjective production of all knowledges. In these terms ethical practice is therefore necessarily about relationships.

Concluding remarks

In telling our stories, we have endeavoured to illustrate something of what it means to reflect on the ethics of academic practices in particular contexts. In particular we have attempted to foreground some of the dilemmas that layer the pursuit of academic knowledge across widely varying forms of interaction, and operating at widely varying intensities, within which questions of appropriation and displacement in relationships recur. With respect to ethnographic research in a cross-cultural context, we have drawn attention to the importance of reflecting upon relational dimensions of the processes involved, including both the qualities of particular relationships and the complex web of relationships binding the researcher to those who become the subjects of research. Using one example from Nuala's doctoral research, we have also illustrated how these issues may be worked through in practice. Through this we hope to encourage reflexive practices that foreground relationships in comparable ways. We have also used the production of this chapter to show how similar processes operate within the Western academy, and, while acknowledging that these are generally rather easier to manage, we have again drawn attention to the relevance of a relational understanding of the issues at stake.

Throughout this chapter we have focused on practical rather than theoretical considerations. But our elaboration of a relational perspective operates within the broad framework provided by Carol Gilligan's work *In a Different Voice* and elaborated further in her subsequent work (see Gilligan 1982; Brown and Gilligan 1992). By insisting on what she termed an 'ethics of care', Gilligan drew into question the idea that ethical practice necessarily entails the consistent application of universal principles.[13] We have offered accounts of an 'ethics of care' in two settings: that of cross-cultural ethnographic research and of student–supervisor co-authorship within the Western academy. In so doing it is our intention to illustrate the practical relevance of Gilligan's ideas.

Notes

1 Anonymizing the place is necessary to protect the identities of all those who became subjects of the research.
2 At the same time, and as we elaborate more fully later on, we must acknowledge that the voices adopted in the accounts that follow are complex products of complex interactions rather than transparent reflections of individual experiences (see for example Cosgrove and Domosh 1993).
3 The definitions of the terms mission and missionary are complex, but it is worth clarifying that the mission context of my presence was as a teacher working for a Catholic diocese that was once missionary territory. Volunteers such as myself and my husband were not expected to, nor did we at any stage, proselytize.
4 Sister Matron is a nickname which took account of both her religious and nursing status. The same name is used of other women occupying positions of this kind.
5 Mr Kaba is a pseudonym.
6 It is also significant to note that Michael's name and this narrative are offered with his consent, although he reflects upon this particular day as among the most difficult he spent as a volunteer.

7 I use the term 'development' aware of its loaded meanings. In this context it refers to projects involving elements of local administration and participation supported by external funding and cooperation in the areas of health, education and socio-economic development.

8 I, of course, am not the only person to move between contexts, although I am more mobile than the great majority of those with whom I interacted in Africa (see Massey 1994). Consequently, in Western contexts I am generally positioned as an expert on mission life in rural Africa. It is often difficult to dislodge this positioning, but the scope for movement across contexts does facilitate occasional challenges.

9 There may be circumstances in which the researcher is less powerful than those with whom he or she interacts in the field: for example, when I was unable to refuse 'requests' made by the diocesan bishop (also see Aldridge 1997; Hendry 1992). But what I am interested in here arises from the transfer of knowledge to a domain – the Western academy – in which the researcher is (almost) invariably in the more powerful position.

10 The reasons for, and background to, my involvement are beyond the scope of this chapter.

11 Within the supervisory relationship they can be used to good effect in enabling the student to learn a great deal about writing for publication (which is how Nuala described her experience to me). But the possibility of learning and empowerment is always accompanied by the risk of exploitation and disempowerment because of the unequal positions student and supervisor occupy within the academy.

12 In this to-ing and fro-ing, it becomes clear that neither of us writes from a stable position: we edit our own words as well as editing one another's. The text may therefore be viewed as a product of a web of dynamic relationships involving two individuals whose positions are structured, but not fixed, by relations of power.

13 Because she argued her case with reference to evidence about girls and women, Gilligan's ideas are sometimes taken to imply a straightforward gender dichotomy in moral behaviour and reasoning. However, this interpretation is contested by others (most notably by Hekman 1995) who read Gilligan's work as opening up spaces for multiple voices and for alternatives to universalistic notions of the rational autonomous subject. We ally ourselves with this latter interpretation, although we acknowledge that culturally dominant understandings of gender may be at work in stimulating our interest in these issues.

References

Ake, Claude (1979) *Social Science as Imperialism: The Story of Political Development*, Ibadan: Ibadan University Press.

Aldridge, Theresa (1997) 'Precarious positionings: some notes on working within two LETS systems in West London', paper presented at the Cultural Turns/Geographical Turns Conference, University of Oxford.

Berg, Lawrence (1997) 'Worlding the referee: masculinism, emplacement and positionality in peer review', paper presented to the IICCG, Vancouver.

Brown, Lyn Mikel and Gilligan, Carol (1992) *Meeting at the Crossroads: Women's Psychology and Girl's Development*, Cambridge Mass.: Harvard University Press.

Cosgrove, Denis and Domosh, Mona (1993) 'Author and authority. Writing the new cultural geography' in James Duncan and David Ley (eds) *Place/Culture/Representation*, London and New York: Routledge, 25–38.

Curry, Michael R. (1991) 'On the possibility of ethics in geography: writing, citing and the construction of intellectual property', *Progress in Human Geography* 15, 125–147.

Dyck, Isabel (1997) 'Dialogue with difference: a tale of two studies' in John Paul Jones III, Heidi J. Nast and Susan M. Roberts (eds) *Thresholds in Feminist Geography*, Lanham, MD: Rowman and Littlefield, 183–202.

England, Kim (1994) 'Getting personal: reflexivity, positionality and feminist research', *Professional Geographer* 46, 80–89.

Evans, Mel (1988) 'Participant observation. The researcher as research tool' in John Eyles and David M. Smith (eds) *Qualitative Methods in Human Geography*, Cambridge: Polity, 197–218.

Eyles, John and Smith, David M. (eds) (1991) *Qualitative Methods in Human Geography*, Cambridge: Polity.

Fabian, Johannes (1983) *Time and Another. How Anthropology Makes Its Object*, New York: Columbia University Press.

Facio, Elisa (1993) 'Ethnography as personal experience' in John H. Stanfield and Rutledge M. Dennis (eds) *Race and Ethnicity in Research Methods*, London: Sage, 75–91.

Farrow, Heather (1995) 'Researching popular theater in southern Africa: comments on a methodological implementation', *Antipode* 27, 75–81.

Finch, Janet (1984) ' "It's great to have someone to talk to": ethics and politics of interviewing women' in Colin Bell and Helen Roberts (eds) *Social Researching: Politics, Problems and Practice*, London: Routledge, 70–85.

Gibson-Graham, J.K. (1994) ' "Stuffed if I know!": reflections on post-modern feminist research', *Gender, Place and Culture* 1, 205–224.

Gilbert, Melissa (1994) 'The politics of location: doing feminist research at "home" ", *Professional Geographer* 46, 90–96.

Gilligan, Carol (1982) *In a Different Voice*, Cambridge, Mass.: Harvard University Press.

Hekman, Susan (1995) *Moral Voices, Moral Selves*, Cambridge: Polity.

Hendry, Joy (1992) 'The paradox of friendship in the field: analysis of a long-term Anglo-Japanese relationship' in Judith Okely and Helen Callaway (eds) *Anthropology and Autobiography*, London: Routledge.

Jarvie, I. C. (1982) 'The problem of ethical integrity in participant observation' in Robert G. Burgess (ed.) *Field Research*, London: Allen and Unwin, 68–72.

Katz, Cindi (1994) 'Playing the field: questions of fieldwork in geography', *Professional Geographer* 46, 67–72.

Keith, Michael (1992) 'Angry writing: (re)presenting the unethical world of the ethnographer', *Environment and Planning D: Society and Space* 10, 551–568.

Kobayashi, Audrey (1994) 'Coloring the field: gender, "race", and the politics of fieldwork', *Professional Geographer* 46, 73–80.

Madge, Clare (1997) 'The ethics of research in the "Third World" ' in Elsbeth Robson and Katie Willis (eds) *Postgraduate Fieldwork in Developing Areas: A Rough Guide* (second edn) Monograph No. 9, Developing Areas Research Group, Royal Geographical Society (with The Institute of British Geographers), 113–124.

Massey, Doreen (1994) *Space, Place and Gender*, Cambridge: Polity.

Moss, Pamela (1995) 'Reflections on the "gap" as part of the politics of research design', *Antipode* 27, 82–90.

Nagar, Richa (1997) 'Exploring methodological borderlands through oral narratives' in John Paul Jones III, Heidi J. Nast and Susan M. Roberts (eds) *Thresholds in Feminist Geography*, Lanham, MD: Rowman and Littlefield, 203–224.

Nast, Heidi (1994) 'Opening remarks on "Women in the Field"' *Professional Geographer* 46, 54–66.

Ngugi Wa Thiong'o (1992) *Moving the Centre: The Struggle for Cultural Freedoms*, London: James Currey.

Painter, Nell Irvin (1979) *The Narrative of Hosea Hudson: his Life as a Negro Communist in the South* Cambridge, Mass.: Harvard University Press.

Patai, Daphne (1991) 'US academics and Third World women: is ethical research possible?' in Sherna Berger Gluck and Daphne Patai (eds) *Women's Words: The Feminist Practice of Oral History*, London: Routledge, 137–154.

Personal Narratives Group (1989) *Interpreting Women's Lives: Feminist Theory and Personal Narratives*, Indianapolis: Indiana University Press.

Rose, Gillian (1997) 'Situating knowledges: positionality, reflexivities and other tactics', *Progress in Human Geography* 21, 305–320.

Sayer, Andrew, and Storper, Michael (1997) 'Ethics unbound: for a normative turn in social theory', *Environment and Planning D: Society and Space* 15, 1–17.

Scheper-Hughes, Nancy (1992) *Death Without Weeping: The Violence of Everyday Life in Brazil*, Berkeley: University of California Press.

Seiber, Joan E. (1992) *Planning for Ethically Responsible Research: A Guide for Students and Research Review Bodies*, London: Sage.

Shaw, Barbara (1995) 'Contradictions between action and theory: feminist participatory research in Goa, India', *Antipode* 27, 91–99.

Skeggs, Beverley (1997) *Formations of Class and Gender*, London: Sage.

Slater, David (1997) 'Spatialities of power and postmodern ethics – rethinking geopolitical encounters', *Environment and Planning D: Society and Space* 15, 55–72.

Sparke, Matt (1996) 'Displacing the field in fieldwork' in Nancy Duncan (ed.) *BodySpace*, London and New York: Routledge, 212–233.

Whatmore, Sarah (1997) 'Dissecting the autonomous self: hybrid cartographies for a relational ethics', *Environment and Planning D: Society and Space* 15, 37–53.

19 The end of the Enlightenment?
Moral philosophy and geographical practice

Tim Unwin

This chapter focuses on the reasons why ethics matters to us individually, intellectually and personally. In essence, it seeks to explore in theoretical and philosophical terms what I once held to be clear and straightforward: that geographers have both a right and a duty to be involved in social, economic and political change (see Unwin 1996).

More formally, the chapter examines some of John Locke's philosophical arguments in the context of the challenges to the Enlightenment that have been thrown up by the many stranded geographical postmodernisms that have emerged over the last decade (Duncan 1996; Habermas 1981, 1983, 1987; Hoy and McCarthy 1994; Lyotard 1984; Rorty 1986). In particular, it aims to shed light on ethical questions concerning the 'rights' and 'duties' of academics in general, and of geographers in particular. It thus explores the grounds upon which socially situated geographical practice, concerned with improving the world in which we live, can be justified (see also Unwin 1996).

The main reason for focusing on Locke is that he played a central role in shaping the ways in which *modern* Englishmen thought (Cranston 1957; Dunn 1969, 1984; Chappell 1994; Marshall 1994). Spellman (1997: 1) has emphasized Locke's importance by noting that he was one of those who 'inaugurated the eighteenth-century "Age of Enlightenment" when ... the primacy and dignity of the individual male were advanced in many spheres of human activity'. For geographers, Locke is of crucial significance because of his espousal of a liberal political philosophy and for his emphasis on the importance of the environment in the shaping of knowledge. However, as Spellman notes, Locke was thinking and writing in an essentially male dominated world, and recognition needs to be made of this in interepreting his work in the context of late twentieth-century social and political thought and action. By exploring the writings of one of the founders of the Enlightenment, the intention here is nevertheless to shed light on the reasons why I still believe that academic geographers should engage actively in seeking to improve the living experiences of the poor, underprivileged and exploited. In this context, Locke's concerns with rights, duties and revolution are highly pertinent (Unwin 1998; see also Smith 1998).

Context

The key problematic for this chapter has been succinctly highlighted in Nancy Duncan's (1996: 435) recent review of postmodernism in human geography, where she is tempted to suggest that postmodernism is 'an extreme philosophical position that may trouble some by its logical elegance but few in their practical everyday experience'. I am one of her few, who find its challenges profoundly disconcerting for my everyday lived action (Habermas 1974, 1978; Corbridge 1993, 1998; Olsson 1991; Harvey 1996). What worries me, to use Berman's (1988: 9) argument, is that 'postmodernist social thought pours scorn on all the collective hopes for moral and social progress, for personal freedom and public happiness, that are bequeathed to us by the modernists of the eighteenth century Enlightenment'. Duncan (1996) suggests that at the heart of postmodernism is a radical anti-foundationalism, built upon both epistemological and ontological relativism (see also Bauman 1993). This can be seen as contrasting with a foundationalist view which proposes that guarantees of truth exist independently of our knowing. This distinction is fundamental to the arguments that are developed here, because it draws attention to the following key question: if truth is seen as being socially constructed, how is it possible to ground judgments concerning better or worse conditions and actions?

John Locke, ethics and the enlightenment

In his *An Essay Concerning Human Understanding*, John Locke (1975) divided the sciences into three: those concerned with the knowledge of things, φυσικη; those, focusing on the application of our own actions, πρακτικη; and the doctrine of signs, σημειωτικη (Book IV, Chapter XXI). Locke (1975) included ethics within the second of these, πρακτικη, which he defined as:

> The Skill of Right applying of our own Powers and Actions, for the Attainment of Things good and useful. The most considerable under this Head is *Ethicks*, which is the seeking out those Rules, and Measures of humane Actions, which lead to Happiness, and the Means to practice them. The end of this is not bare Speculation, and the Knowledge of Truth, but Right, and a Conduct suitable to it.
>
> (Locke 1975: 720)

It is with this emphasis on the utility of ethics and its concern with human happiness that this chapter is centrally concerned.

A re-examination of some of Locke's conclusions concerning the grounding of knowledge, truth and action is of considerable relevance to contemporary debates over the parasitic (Duncan 1996) relationship between modernity and postmodernity. It is on the connections that he draws between moral requirements and political action that the core of this chapter is focused. Locke's major works are the *Two Treatises of Government*, published in 1689 (Locke 1967), and *An*

Essay Concerning Human Understanding, published in 1690 (Locke 1975).
However, he wrote a number of other important tracts and essays, including *A
Letter Concerning Toleration* published in 1689 (Locke 1963) and *Some Thoughts
Concerning Education*, published in 1693 (Locke 1989) (see also Locke 1954).
Interestingly, in the last of these Locke made specific reference to the utility of
geography (Yolton and Yolton 1989) noting that it should be the first discipline
to be taught to children. Locke (1989) thus commented that:

> *Geography*, I think, should be begun with: For the learning of the Figure of
> the *Globe*, the Situation and Boundaries of the Four Parts of the World, and
> that of particular Kingdoms and Countries, being only the Exercise of the
> Eyes and Memory, a child with pleasure will learn and retain them: And this
> is so certain, that I now live in the House with a Child, whom his Mother has
> so well instructed this way in *Geography*, that he knew the Limits of the Four
> Parts of the World, could readily point, being asked, to any country upon the
> Globe, or any County in the Map of *England*, knew all the great Rivers,
> Promontories, Straits, and Bays in the World, and could find the Longitude
> and Latitude of any Place, before he was six Years old.
>
> (Locke 1989: 235)

Geography has changed somewhat since Locke wrote these words!

Many interpretations of Locke's philosophy see him as having been 'an optimistic thinker whose optimism was founded on understanding not very well what
we ourselves understand altogether better' (Dunn 1984: vii). However, in contrast, Dunn (ibid.) suggests that 'we should see Locke instead as a tragic thinker,
who understood in advance some of the deep contradictions in the modern conception of human reason, and so saw rather clearly some of the tragedy of our
own lives which we still see very dimly indeed'. It is this concern with the contradictions of modern reason that makes Locke such an interesting philosopher to
consider in the context of this book.

In his *Two Treatises of Government*, Locke suggested that people contract in to
civil society for certain clear benefits. In so doing, they agree to surrender personal power to a ruler and magistracy, which Locke believed would ensure the
maintenance of some kind of natural morality more appropriately than in pre-civil
society. However, if the ruling body fails to adhere to the laws of natural morality,
Locke argued that the population has both a *right* and a *duty* to depose it. It is
here that Locke raises the fundamental questions of political action, and how we
should try to live. The nub of Locke's argument is based upon his somewhat
problematic views concerning the relationships between property rights, the laws
of nature, and the responsibilities of government (for a detailed critique see Tully,
1994). In essence, in the first book of his *Two Treatises of Government* (Chapter
IX: 92), Locke argues that government is designed to ensure that individuals'
rights and property are preserved, and that this is achieved by making the Laws of
Society conformable to the Laws of Nature. Of prime importance is his assertion
that: 'Property . . . is for the benefit and sole Advantage of the Proprietor, so that

he may even destroy the thing, that he has Property in by his use of it, where need requires' (Locke 1967: 227–228).

For Locke, as with many later authorities, notably Marx, labour was also crucial to his argument, because it is only through labour that people acquire the right to what has been worked upon, be this land or materials. In his words '*Labour,* in the Beginning, *gave a Right of Property,* where-ever any one was pleased to imploy it' (*Two Treatises of Government,* Book II, Chapter V: 45) (Locke 1967: 317). Locke then argues that governments are necessary in order to protect human entitlements, most notably life, liberty and material possessions such as property (Book II, Chapter V). But, as Dunn (1984: 43) notes, 'Locke, like Thomas Aquinas, believed that all men had a right to physical subsistence which overrode the property rights of other humans'. Although all people, according to Locke, are born free and equal, they enter civil society and agree to be governed in order that their freedoms and human entitlements can be ensured (see also Marshall 1994: 217). When governments act against the laws of nature and lose the trust of their citizens, it then becomes the duty and right of every individual to work in common to overthrow the legislators.

Much of Locke's argument rests on the concept of a general law of nature which provides the basis for human rights and duties. However, as Dunn has stressed, in his *Two Treatises of Government* Locke:

> chose not to discuss at all the question of how men can naturally know the law of nature, the binding law of God, on which, according to the argument of the book, all human rights rested and from which the great bulk of human duties were more or less directly derived.
>
> (Dunn 1984: 29)

Locke takes as given that God has decreed the law of nature, but he argues that human reason can provide some understanding of it. For Locke, the fundamental right and duty of all individuals is to live according to God's law of nature, and it is through reason that people can gain access to it. In the lectures that he gave at Christ Church in 1664, his *Essays on the Law of Nature,* Locke does, though, provide a clear summary of his earlier views concerning this matter. First, he argues that God has indeed given people a rule of morals, or law of nature, and that 'some principle of good and evil is acknowledged by all men' (Locke 1954: 123). This law of nature, he suggests, can be known by anyone if they make proper use of the faculties with which they are endowed by nature. For Locke (1954: 123), there were four possible ways of knowing this law: by inscription or innate knowledge; by tradition or instruction; by sense-experience; and by divine revelation. He rejects the first two and the last of these, and concludes that it is only by sense-perception that we can attain knowledge of natural law. Locke (1954: 147) thus argues that since 'this light of nature is neither tradition nor some inward moral principle written in our minds by nature, there remains nothing by which it can be defined but reason and sense-perception'.

Locke explored these ideas further in his *Essay Concerning Human*

Understanding, where he makes a clear distinction between the character of nature and of moral ideas. As Dunn (1984) has summarized, in Locke's argument:

> Moral ideas were inventions of the human mind, not copies of bits of nature. This contrast has fundamental implications for the character of moral ideas and for how, if at all, these can be known to be valid. It is the foundation in modern philosophical thinking of the presumption of a stark gap between facts about the world (which can potentially be known) and values for human beings (which can merely be embraced or rejected).

(Dunn 1984: 65)

In defence of the Enlightenment

The justification of socially situated academic practice concerned with improving the world in which we live has been debated at length in the literature on philosophy and social theory (see for examples Habermas 1981, 1893, 1987; Hoy and McCarthy 1994; Lyotard 1984; Rorty 1986; Duncan 1996; Harvey 1996), and it is not the intention here to explore this literature at length. Rather, the aim is much more simply to examine the relevance of Locke's basic principles for contemporary geographical practice.

The need for social change

Locke's views concerning how the law of nature can be known are closely related to the question of whether or not there exists a need for social change. Locke thought that the only way in which knowledge of the law of nature could be achieved was through sense-experience. For Locke, therefore, any consideration that we might know that there is something that needs changing in society must thus be through our own sense-experience. At first sight, this solution might appear to be straightforward: we may have suffered some kind of prejudice or racial discrimination; we may empathize with people begging on city streets; we may see devastating poverty in parts of Asia or Africa; we may hold the hand of a dying child in a refugee camp. In all of these instances, our senses *might* tell us that something is wrong or unequal, and in need of change. However, senses tell different people contrasting messages; they are in part socially, economically and culturally constructed. Not everyone having encountered any of the above experiences need *necessarily* conclude that substantial social change is necessary. Thus, for those seeking to minimize labour costs in their production processes, Asian and African poverty may well be a crucial factor in their decisions as to where to locate their factories. The market economy is fundamentally based on inequality and difference, and yet it is accepted by many as right and just.

A second answer to the question might be that we have read and been persuaded by the arguments of other people that these issues are wrong, or morally indefensible. This, in effect, is Locke's knowledge by tradition, and he develops three arguments as to why this is not a sound way of knowing the law of nature:

first, given the diversity of human views on the subject, he suggests that it is difficult to decide precisely on what is true and false; second, if it 'could be learnt from tradition, it would be a matter of trust rather than of knowledge, since it would depend more on the authority of the giver of information than on the evidence of things themselves' (Locke 1954: 131); and third, he asks where tradition begins, suggesting that all initial ideas must come either from inscription, which he has already rejected, or from 'facts perceived by the senses' (Locke 1954: 131).

While these arguments pertain to the question of how we might know anything, they do not actually address whether or not there is a universal law of nature, or absolute moral standard (see Smith 1998). There is a real dilemma, because if a universal position is rejected, in other words if it is argued that there are no fundamental moral standards, then it needs to be asked why there do appear to be some widely accepted moral standards, and why so many people remain concerned about moral issues at all? Locke touched on this issue, but did not ever really resolve it. At times he seemed to suggest that there were indeed some generally accepted moral principles, and yet at others he seems to have rejected this argument. He thus, for example, commented both that 'some principle of good and evil is acknowledged by all men' (Locke, 1954: 123), and also that 'traditions vary so much the world over and men's opinions are so obviously opposed to each other' (ibid.: 129). In seeking to resolve this dilemma, Smith (1998) has drawn attention to Walzer's (1994) distinction between 'thin' universal concepts such as justice, honour and courage, and their 'thick' expressions in particular contexts.

There is, though, a fundamental distinction between knowing a moral standard, and deciding how to act upon it. To an extent, it matters less whether people actually know a universal truth, than how they act on what they think they know. It is therefore possible to put on one side the issue of whether there is a basic law of nature, or moral principles that people *can* know, and instead consider the ways in which individual people decide to act depending on their own moralities. For Locke, the key point was that individuals had both a *right* and a *duty* to act.

Rights and duties

Dunn (1984) notes that, for Locke, the

> most fundamental right and duty is to judge how the God who created them requires them to live in the world which he has also created. His requirement for all men in the state of nature is that they live according to the law of nature. Through the exercise of his reason every man has the ability to grasp the content of this law.
>
> (Dunn 1984: 47)

In essence therefore, Locke argued that it was the duty and right of everyone to obey the law of nature.

If the idea of an absolute law of nature is accepted, it is relatively straight-forward to argue, as does Locke, that it is the right and duty of people to follow its moral principles. In societies with institutionalized education systems, it might then be argued that the right and duty of university institutions and staff should be to seek to identify the characteristics of the law of nature, and to seek to instil an understanding of it in the whole population. This is based on the assumption that universities are established by societies in order to advance knowledge through research and to disseminate this knowledge within those societies for the betterment of their citizens. This view, however, is highly contested, as reflected recently in Britain in the debates over the Dearing Report on the future of higher education (for comments by geographers on this debate see Chalkley 1998). We are on much more difficult ground if we choose to deny the existence of abso-lutes. If there is no absolute truth, if there are no definitive moral standards, and if everything is fluid and relative, what then does the role of university academics become? At one level, it can be argued that since there are no absolutes, there can be no definitive answer to this question. More worryingly, though, if there are no wider points of referral, it becomes difficult to refute attempts by governments to exert greater control over the research and teaching activities of academics and university institutions.

This is where arguments concerning the role of multiple academic voices within society come to the fore. At one level, it can be argued that all members of society should be considered as 'equal', although even this argument implies some refer-ence to broader moral concepts of social equity and approaches a normative and universal stance. If all voices are equally valued, though, it becomes illogical to suggest that any one group of voices (such as academics) should be privileged, or given more credence than any others. However, if it is accepted that universities exist for the reasons outlined above, then adopting the fitness for purpose type argument so beloved of Plato (1974), it is logical to suggest that universities should seek to employ those most suited to undertaking research and to teaching the new generation of citizens. Some caution is, though, necessary. As Plato (1974) commented, it is

> perfectly plain that in practice people who study philosophy too long, and don't treat it simply as part of their early education and then drop it, become, most of them, very odd birds, not to say thoroughly vicious; while even those who look the best of them are reduced by this study you praise so highly to complete uselessness as members of society.

> (Plato 1974: 281)

If knowledge is of benefit to a society, and that society wishes to optimize its resources, then its government can impose both a duty and a right on universities and the academics within them to pursue such knowledge. The difficulty that arises here is over whether or not governments, and in turn universities, actually serve the interests of the majority of the people, or only those of a particular elite.

Following Locke's arguments it can be suggested that if academics consider that the government is not serving the interests of the knowledge that they are pursuing, then they have a duty and right to seek to overthrow such government (compare Locke 1967: 433). More broadly, if society has established and sanctioned universities as the guardians of knowledge, or as the psychoanalysts of their condition (Unwin 1992, 1996), then it remains logical to argue that if academics indeed believe that society is not operating optimally *for whatever reason*, then they have a right and a duty to persuade all members of that society to change it. This right and duty is both the same as that of everyone else in society, as argued by Locke, but also specific to the particular role of academics as those who seek to explore knowledge.

Action

If academics do therefore have a role to play in changing society, how do they decide what they should do about it? In part, answers to this question must reflect a similar set of arguments to those that have already been addressed over how it is possible to know anything at all. Locke provides few clear suggestions as to the actions that people should take when they believe that their government or legislature has failed to live up to the moral codes of the law of nature, and has lost the trust of the people. He thus asks, 'Who shall be Judge whether the Prince or Legislative act contrary to their Trust?' (Locke 1967: 444). However, he does go on to suggest that:

> If a Controversie arise betwixt a Prince and some of the People, in a matter where the Law is silent, or doubtful, and the thing be of great Consequence, I should think the proper *Umpire*, in such a Case, should be the Body of the *People*.
>
> (Locke 1967: 445)

Such a conclusion has important ramifications. As societies and cultures change, Locke seems to be suggesting that the final repository of knowledge about what is right rests with the body of the people. Locke justifies this choice by noting that it was the people who initially reposed their trust in the prince. However, it is possible to develop this argument further. For those addressing the question from an absolutist stance, the implications are that the people should decide, because it is the mass of the people who hold nearer than the prince to whatever absolute truth there is. From a relativist position, though, it can be concluded that the body of the people is the best judge because the views of the population change through time, and whatever that body considers at any specific time is therefore both right and appropriate.

This has wider implications for academic practice in general, and geography in particular. At the broad level of general university practice, such arguments would suggest that academics have a duty, once they have reached the conclusion that society does indeed require change, to ensure that their arguments reach the widest possible public attention, so that the body public can then determine what action to take. One practical implication of this is to ensure that research findings are not just published in obscure academic journals that few people read, but rather that they are made available in appropriate media and in appropriate languages. More specifically with respect to geography, David Harvey (1996) has recently argued that geographical differences are central to understanding political, economic and ecological alternatives to contemporary life. Indeed, with regard to the environmental justice movement, he argues that 'there is a long and arduous road to travel . . . beyond the phase of rhetorical flourishes, media successes, and symbolic politics, into a world of strong coherent political organizing and practical revolutionary action' (Harvey 1996: 402). Such a conclusion applies not only to environmental considerations, but also to the range of research agendas currently being explored by geographers (see Unwin 1992). Geographers, at least in Britain, have by and large been reluctant to engage productively with the mass media, and a strong case can be made for us to enhance the accessibility of our critical research conclusions to members of the society of which we are a part.

Geographers, individuals and society

In concluding her wide-ranging review of postmodernism in human geography, Nancy Duncan (1996: 454) suggests that 'In the end the study of post-modernity is a very broadly defined project, and having been influenced by some of postmodernism's concerns, we may wish to remain skeptical about the use of such totalizing descriptive categories as *postmodernity* itself'. Abandonment of the certainty of universals, and of a belief in the possibility of making the world a better place, lie at the heart of many postmodern interpretations. However, if reduced to the extreme, the relativist stance can lead to a position where individual expression is all that matters, and to a view of society in which anything is accepted. In such circumstances, there would seem to be little point in doing anything, and we would revert to Plato's (1974) condition of being truly useless to society.

This chapter has tried to highlight some of the problems with such an extreme view, and has sought to explore the moral grounds upon which the rights and duties of practising geographers might be advanced (although to compensate for Locke's essential male perspective, see Benhabib 1992). By understanding and criticizing society, we can be in a position to see and argue how it might better be constructed. Through critique of the aspects of the contemporary world with which, *for whatever reason*, we as individuals feel ill at ease, we may be able to provide positive recommendations for social, economic and political reform. As Habermas (1988) recognized in his description of the work of Herbert Marcuse, there is a fundamental connection between negative critique and positive affirmation:

> We all remember what Herbert Marcuse kept denouncing as the evils of our age: the blind struggle for existence, relentless competition, wasteful productivity, deceitful repression, false virility, and cynical brutality. Whenever he felt that he should speak as teacher and philosopher he encouraged the

negation of the performance principle, of possessive individualism, of alienation in labor – as well as in love relations. . .But the negation of suffering was for him only a start. . .Marcuse moved further ahead. He did not hesitate to advocate in an affirmative mood, the fulfillment of human needs, of the need for undeserved happiness, of the need for beauty, of the need for peace, calm and privacy.

(Habermas 1988: 3)

This task remains central for all those who are concerned with the place of humanity; the task of geography.

I have sought not so much to justify Locke's continued relevance to contemporary society, but rather to illustrate the complexity of debate and ideas relating to the origins of modernist thought. Locke played a crucial role in shaping the Enlightenment ideal that it was both possible and desirable to make society a better, more happy experience. The chapter has, though, also emphasized one of the key themes of this book, that the rights and duties of geographers stem from fundamentally important moral principles (see, for example, Proctor 1998). I find it extremely difficult to say exactly *why* I feel that we do indeed have a right and a duty to seek to change society. I remain unconvinced by Locke's argument that this is only by sense-perception, and am inclined to place more emphasis than he was prepared to give to the idea of tradition. Part of the explanation undoubtedly lies in my own construction as a male product of modernism and the Enlightenment, and by the persuasive logic of writers such as Habermas (1978, 1987). However, the practice of geographical research in specific places also plays a key part in revealing the inequalities of contemporary society. The challenge is to know what to do about them. We can choose to ignore them, and remain focused on the publication of seminal papers in high-status academic journals, or we can also seek to change the society of which we are a part through our teaching and lived practice. The choice forces us to engage in profoundly moral questions.

References

Bauman, Z. (1993) *Postmodern Ethics*, Oxford: Blackwell Publishers.
Benhabib, S. (1992) *Situating the Self: Gender, Community and Postmodernism in Contemporary Ethics*, Oxford: Polity Press.
Chalkley, B. (ed.) (1998) 'Arena Symposium: Dearing and Geography', *Journal of Geography in Higher Education*, 22(1): 55–101.
Chappell, V. (1994) 'Locke on the freedom of the will' in G. A. J. Rogers (ed.) *Locke's Philosophy: Content and Context*, Oxford: Clarendon Press, 101–122.
Corbridge, S. (1993) 'Marxisms, modernities and moralities: development praxis and the claims of distant strangers', *Environment and Planning D: Society and Space*, 11: 449–472.
—— (1998) 'Development ethics: distance, difference, plausibility', *Ethics, Place and Environment*, 1(1): 35–54.

Cranston, M. (1957) *John Locke: a Biography*, London: Longmans, Green and Co.

Duncan, N. (1996) 'Postmodernism in human geography' in C. Earle, K. Mathewson and M. S. Kenzer (eds) *Concepts in Human Geography*, Lanham, MD: Rowman & Littlefield, 429–458.

Dunn, J. (1969) *The Political Thought of John Locke*, Cambridge: Cambridge University Press.

—— (1984) *John Locke*, Oxford: Oxford University Press.

Habermas, J. (1974) *Theory and Practice*, London: Heinemann.

—— (1978) *Knowledge and Human Interests*, London: Heinemann.

—— (1981) 'Modernity versus Postmodernity', *New German Critique*, 22: 3–14.

—— (1983) 'Modernity: an incomplete project' in H. Foster (ed.) *The Anti-aesthetics: Essays on Postmodern Culture*, Seattle: Bay Press, 3–15.

—— (1987) *The Philosophical Discourse of Modernity: Twelve Lectures*, Cambridge and Oxford: Polity Press in association with Basil Blackwell.

—— (1988) 'Psychic thermidor and the rebirth of rebellious subjectivity' in R. Pippin, A. Feenberg and C. P. Webel (eds) *Marcuse: Critical Theory and the Promise of Utopia*, Basingstoke: Macmillan.

Harvey, D. (1996) *Justice, Nature and the Geography of Difference*, Oxford: Blackwell Publishers.

Hoy, D. C. and McCarthy, T. (1994) *Critical Theory*, Oxford: Blackwell Publishers.

Locke, J. (1954) *Essays on the Law of Nature*, W. von Leyden (ed.), Oxford: Clarendon Press.

—— (1963) *A Letter Concerning Toleration*, M. Montuori (ed.), The Hague: Martinus Nijhoff.

—— (1967) *Two Treatises of Government*, P. Laslett (ed.), Cambridge: Cambridge University Press, second edn.

—— (1975) *An Essay Concerning Human Understanding*, P. H. Nidditch (ed.), Oxford: Clarendon Press.

—— (1989) *Some Thoughts Concerning Education*, J. W. Yolton and J. S. Yolton (eds), Oxford: Clarendon Press.

Lyotard, J.-F. (1984) *The Postmodern Condition: a Report on Knowledge*, Minneapolis: University of Minnesota Press.

Marshall, J. (1994) *John Locke: Resistance, Religion and Responsibility*, Cambridge: Cambridge University Press.

Olsson, G. (1991) 'Invisible maps', *Geografiska Annaler*, 73B: 85–92.

Plato (1974) *The Republic*, Harmondsworth: Penguin.

Proctor, J. (1998) 'Ethics in geography: giving moral form to the geographical imagination', *Area*, 30(1): 8–18.

Rorty, R. (1986) 'Habermas and Lyotard on postmodernity' in R. Bernstein (ed.) *Habermas and Modernity*, Cambridge, Mass.: MIT Press.

Smith, D. (1997) 'How far should we care?' Paper prepared for the session on Geography and Ethics at the RGS–IBG Conference held at Exeter, 8 January.

—— (1998) 'Geography and moral philosophy: some common ground', *Ethics, Place and Environment*, 1(1): 7–34.

Spellman, W. M. (1997) *John Locke*, London: Macmillan.

Tully, J. (1994) 'Rediscovering America: the *Two treatises* and aboriginal rights' in G. A. J. Rogers (ed.) *Locke's Philosophy: Content and Context*, Oxford: Clarendon Press, 165–196.

Unwin, T. (1992) *The Place of Geography*, Harlow: Longman.

Unwin, T. (1996) 'Notes on critical geography: from the underground', *Praxis*, 32: 1–9.

—— (1998) 'A revolutionary idea' in T. Unwin (ed.) *A European Geography*, Harlow: Addison, Wesley, Longman, 100–114.

Walzer, M. (1994) *Thick and Thin: Moral Argument at Home and Abroad*, London and Notre Dame, Ind.: University of Notre Dame Press.

Yolton, J. W. and Yolton, J. S. (1989) 'Introduction' in J. W. Yolton and J. S. Yolton (eds) *Locke, J. Some Thoughts Concerning Education*, Oxford: Clarendon Press, 1–75.

20 Conclusion

Towards a context-sensitive ethics

David M. Smith

The individual chapters in this volume have spoken for themselves, and some have made links with others to reveal common themes. Furthermore, this editor has already had opportunities elsewhere to express his own views on the moral terrain which we are exploring (Smith 1997c, 1998a), as well as on specific issues at the interface of geography and ethics (Smith 1997a, 1997b, 1998b, 1999). All that is required by way of conclusion, then, is a reminder of major issues which have surfaced in the text, along with a few connections to other relevant work in the field.

Some of the issues may be expressed in the form of tensions. The term 'tension' (some might prefer dialectic) is used here in the positive and creative sense of provoking resolution, or at least moves in that direction, which better captures the project at hand than the more static 'dualism' with its negative connotation of irreconcilable opposites. Tensions running through and connecting a number the essays include those between general (thin) and specific (thick) moralities, between universalism and particularism, between global and local, space and place, between essentialism and individualism or difference, between the natural and the socially constructed, between ethical thought and moral practice, and between is and ought. These and other issues are summarized, under the four headings adopted to structure our contents, followed by some final comments stressing the importance of context in the relationship between ethics and geography.

Ethics and space

The first main issue concerns the geographical specificity of the ways in which humankind attempts to know the world. The essence of geography is recognition of spatial differentiation, which applies both to the ways in which people live and to how they try to make sense of their existence, yet the attempt to transcend the here (and now) in pursuit of generalization is also an important part of our intellectual tradition. Paul Roebuck contrasts universal and contextual conceptions of knowledge, as a fundamental distinction highlighted by the particularist orientation of the discipline of geography. Geography has a long history of pre-occupation with the uniqueness of places. This has never been far beneath the surface of our discourse, even during the era of spatial science when a residual pre-quantitative geography maintained some challenge to the prevailing search for order and regularity. The resurgence of approaches from humanism and the promotion of a 'new' cultural geography in recent years has reinforced a preoccupation with the contextual. Universalism has thus come to be regarded as one of the sins of Enlightenment thinking, at least among much of the discipline's avant-garde influenced by postmodernism with its risk of moral relativism or nihilism. It is not surprising to find indications, in some of our chapters, of moves back to the more secure moral foundations which universalism was once thought to provide. These are echoed in other contemporary work (e.g. Harvey 1996).

Jürgen Habermas (1990: 208) has described moral universalism as a historical result: 'It arose, with Rousseau and Kant, in the midst of a specific society that possessed corresponding features. The last two or three centuries have witnessed the emergence, after a long seesawing struggle, of a *directed* trend toward the realization of basic rights'. But this also describes a geographical result, in the sense that the society in question had a spatial form, and that the emergence of rights was within a geopolitical context of nation states which provided means of associating rights with citizenship. This was a step towards the recognition of universal human rights, brought to its most prolific expression in the United Nations declaration of 1948. The very notion of a 'united nations' underlines the spatial and temporal specificity of this particular conception of rights.

Joan Tronto (1993, Chapter 2) has written with insight on the spatiality of moralities and how they change. She refers to the relationship between the emergence of a universalistic morality and the expanding spatial scope of eighteenth-century life, contrasting this with the earlier contextual morality exemplified by Aristotle's stress on virtues relevant to a particular community with its shared conception of the good. The prevailing morality changed with the geography:

> While humans grew more distant from one another and the bonds between them became more formal and more formally equal, they also had to expand their gaze beyond the local to the national, and indeed sometimes to a global level . . . these ideas required a change in the nature of moral thought from a type of contextual morality to a morality where human reason could be presumed to be universal.
>
> (Tronto 1993: 31–32)

Similarly, the resurgent particularism (or parochialism) sometimes associated with contemporary communitarianism might be interpreted as a reaction to demands of universalism in the face of spatially prescribed self-interest. As Onora O'Neill (1996: 29) puts it: 'In a post-imperial world, cosmopolitan arrangements threaten rich states with uncontrolled economic forces and immigration and demands for aid for the poor of the world, and autocratic states with demands that human rights be guaranteed across boundaries'.

Nicholas Low and Brendan Gleeson raise some of the difficulties posed by globalization for the nation state, threatening the spatial framework within which rights have usually been implemented. Globalization appears to be promoting socio-economic polarization, rather than the universalization of human rights released from national specificity. Ron Johnston echoes this theme in showing how the division of geographical space matters, with respect to what might be regarded as the most basic right of democratic citizenship: the vote. That the value of a person's vote, or those of a group (perhaps racially defined), can be enhanced or diminished by the manipulation of political jurisdictions is well enough known for 'gerrymandering' to find a place in geographical textbooks. But what Johnston reveals is the extreme difficulty of implementing fairness in the sense of equal voting power in any system of spatially defined constituencies. This has important implications for democratic politics in nation states, especially those in which increasingly strident minorities (usually geographically concentrated) are demanding a share of power not generated by the existing electoral process. It is also important for the new institutions of international regulation associated with globalization, and which require the moral force of political legitimacy if they are to serve their purposes.

Globalization also highlights the issue of the spatial extent of moral responsibility to care for others (Smith 1998b), which is often thought to be highly localized. Some of the implications have been explored in the context of development ethics by Stuart Corbridge (1993, 1998), in fine exemplars of geographical engagement with political philosophy. The impulse towards universal benevolence, if not beneficence, stimulated by the Enlightenment, is currently facing renewed constraints encouraged by some strands of feminism as well as communitarianism. Marilyn Friedman (1991: 818) asserts: 'Hardly any moral philosopher, these days, would deny that we are each entitled to favor our loved ones. Some would say, even more strongly, that we ought to favor them, that it is not simply a moral option'. But as she also suggests, favouring our nearest and dearest can help to promote and sustain inequality, given the spatially uneven capacity to meet need with effective care.

Low and Gleeson's call for new forms of rights is reflected in Seamus Grimes' search for new ways of exercising moral responsibility towards distant others. This raises a fundamental issue of moral motivation: whether the empathetic relations which we seem to be able to establish with close persons (emotionally and spatially) can be extended to different as well as distant others. Friedman (1993: 87–88) crystallizes the problem as follows: 'global moral concern is a rational achievement but not an immediate motivation. It is, furthermore, an achievement only for some selves. It is a result of moral thinking that has no necessary motivational source in the self, so not everyone will find it convincing'. This highlights the fragility of global sympathies, and of moral universalism in practice.

How people live in spatial relationships with others is clearly implicated in such attitudes as care and concern. From our earliest days, we live in specific caring relationships with others, and through this learn the value of caring in general. Some persons, in some contexts, may be deprived of this experience, and in extreme cases may come to treat others with disregard or even cruelty. The extension of more remote means of communication, explored by Jeremy Crampton in the context of the Internet, raises the issue of what kind of human relationships are promoted, and constrained, in this way. The 'virtual communities' of 'cyberspace' may bring together far distant persons, and assist them to engage empathetically (Rheingold 1993). They can have their own formal codes of ethics. But as Crampton shows, the Internet may also be a means of remote control. And it is capable of facilitating evil, distributing unsolicited pornographic material, for example, and perhaps encouraging attitudes and behaviour towards others which is less likely in face-to-face contact. The community without propinquity may lack an important moral dimension.

Ethics and place

Place is a fundamental concept in geography. Michael Curry argues for the normativity of place: places have moral as well as causal power. This notion is central to the framework proposed by Robert Sack (1997), which contains a sustained attempt to incorporate a moral perspective into human geography. Geography is at the foundation of moral judgment: 'Thinking geographically heightens our moral concerns; it makes clear that moral goals must be set and justified by us in places and as inhabitants of a world' (Sack 1997: 24). The moral force of a place, Sack argues, involves its capacity to tie together the particular virtues or moral concerns of truth, justice, and the natural, which exist in different and changing mixes in different places, including home, workplace, city and nation: 'Place is a moral force at any scale' (ibid.: 203).

The most graphic expression of the moral significance of place in Sack's work is in the recurrent image of thick and thin. As boundaries become porous, this thins out the meaning of place, and the virtues therein, changing the thicker places of pre-modern society with their strongly partial moral codes. Thus:

> the local and contextual should be thin and porous enough not to interfere with our ability to attain an expanded view, and the local can be understood and accorded respect only if people attain a more objective perspective, enabling them to see beyond their own partiality and to be held responsible for this larger domain.
>
> (Sack 1997: 248)

Sack's imagery helps to dissolve rigid distinctions or dualisms sometimes associated with universalism and particularism, though it could be argued that his decided preference for transcending place risks reproducing the dualism. He sees the practice of transcending partiality as part of growing up, of expanding horizons, of knowing more about the world and its peoples and the consequences of our actions. Our moral perspective becomes less partial, more impartial, as we move from the self in place locally to a more detached view.

Of course, attributing moral agency to place is by no means unproblematic. When Sack (1997: 209) tells us that 'a place that practices slavery is immoral', there is a risk of dehumanizing evil, even of making a fetish of place after the fashion of some more conventional geographical analyses. Places, as constructions of humankind, are made for a particular purpose, which may be judged good or evil according to some criteria external to the specific project. But the built environment cannot itself have moral standing. The function of places can change with human action. The concentration camp may become a museum, testifying to human endurance as well as to cruelty. A dwelling can change from haven to hell, with its occupants. The changing moral force of a place is part of human life, of the interdependence of its geography and history. Indeed, the very term 'place' as used by geographers means much more than a certain location, and it is in the human threads of place that morality enters.

One of the issues raised by Yi-Fu Tuan, in his discussion of evil, is its temporal as well as spatial specificity. Behaviour once taken for granted, even as natural, is today deplored with the benefit of hindsight. This is what we understand by moral progress. The Romans were wrong to throw Christians to the lions, even if they knew no better then. Or were they? Can we expect persons to transcend the values of their time, or place? Evidently, some are able to, and hence argue or act for a better way of life, otherwise the moral particularities of old eras and communities would never have been challenged and changed.

Notions of moral progress rest uneasily with the contemporary reality of such evident evils as the exploitation of child labour, institutionalized torture and genocide. This is exemplified in Gearóid Ó Tuathail's examination of 'ethnic cleansing' in part of the former Yugoslavia. The image of cleansing place or space has been elaborated by David Sibley (1995), as a powerful moral metaphor involving the exclusion of some kinds of people on grounds of difference. Ó Tuathail makes other geographical points in his observations that the ethical engagement of the United States and United Nations included the aerial action of dropping food from above, rather like the bombing from on high which distanciates those involved from the consequences for people on the ground. The designation of supposedly 'safe areas' required a similarly remote cartography. Such ethical engagement without responsibility contributed to the failure to prevent genocide in hitherto safe places with cohesive if heterogeneous communities. How readily ethnic chauvinism can be made to transcend earlier spatially mediated affinities remains one of the moral mysteries of our times, not least with respect to the treatment of Jews during the Holocaust (see for example Bauman 1989; Geras 1995; Rorty 1989; Vetlesen 1993).

Caroline Nagel, like Gearóid Ó Tuathail, demonstrates something of the abiding strength of the case study. Nagel joins other contemporary voices in geography (e.g. Sack 1997), in challenging the preoccupation with difference characteristic of postmodern thought, and of much of the new cultural geography. She points to problems arising from the politics of difference, which in multicultural societies can lead to undue emphasis on group (and local) specificity in struggles for human rights and social justice. A rediscovery of human affinity or sameness seems a necessary move if the assertion of difference is not to degenerate into further cases of (spatial) exclusion and, at the extreme, of the brutal repression or extermination of those perceived to be different. The tension

280 Ethics and knowledge

between universalism or essentialism and particularity or difference thus surfaces again, in the complex relationship between ethics and place, and between ethics and nature.

Ethics and nature

Environmental ethics has been a subject of considerable attention in these recent years of 'green' activism and politics. This is the focus of the first issue of a series of books on philosophy and geography (Light and Smith 1997). David Harvey (1996) has incorporated nature and environment into his analysis of the process of socio-ecological transformation, while nature is central to the approach to justice elaborated by Nicholas Low and Brendan Gleeson (1998). And a new journal appeared in 1998: *Ethics, Place and Environment*.

There are interesting links with a number of other issues on the interface of geography and ethics. The question of obligation to future generations, which is central to concerns about the use of resources, is sometimes coupled with responsibility to distant peoples in space, with time considered more complex in view of uncertainty about how far ahead to go. Environmental ethics are closely related to development issues, and to international justice. An environmental justice movement has emerged (see Harvey 1996, Chapter 13), reflecting concerns about the uneven distribution of hazards generated by waste disposal, industrial plants and other noxious facilities, by population group and place (Cutter 1995).

One of the fundamental issues raised by James Proctor is the distinction between facts and values, highlighted (and perhaps erased) when what is asserted to be natural is also held to be commendable. A common axiom in philosophy is that assertions concerning what should be done cannot logically be derived from statements as to what is. That this line of reasoning is nevertheless frequently adopted in practice can be illustrated no more clearly than in the discourse of human rights, in which the fact of being human entails a specific moral response.

Being human involves certain natural characteristics, yet any suggestion that there may be such a thing as human nature can attract the charge of essentialism today. 'Any definition of human nature is dangerous because it threatens to devalue or exclude some acceptable individual desires, cultural characteristics, or ways of life', according to Iris Marion Young (1990: 36). However, there are increasing indications of dissatisfaction with this position, as Caroline Nagel's essay showed. Terry Eagleton (1996) exemplifies the critique; he approves of postmodernism in challenging various forms of oppression by race, gender, sexuality and so, but is dismissive of a form of reductionism 'which drastically under-values what men and women have in common as natural, material creatures, foolishly suspects all talk of nature as insidiously mystifying, and overestimates the significance of cultural difference' (Eagleton 1996: 14). A similar position is argued in the response of Norman Geras (1995) to the anti-foundationalism and anti-essentialism of Richard Rorty (1989) and his denial of any human nature. Human beings:

are susceptible to pain and humiliation, have the capacity for language and (in a large sense) poetry, have a sexual instinct, a sense of identity, integral beliefs – and then some other things too, like needs for nourishment and sleep, a capacity for laughter and for play, powers of reasoning and invention that are, by comparison with other terrestrial species, truly formidable, and more shared features yet.

(Geras 1995: 66)

These natural facts have moral consequences. Hence the link into those perspectives on development which argue for the satisfaction of certain 'basic' human needs as of right.

Another dimension of the role of the natural in ethical discourse is exemplified in the essay by Sheila Hones. The United States is by no means unique in having historians deploy some kind of natural destiny to legitimize territorial expansion, at the expense of an indigenous population. Western European imperialism often found justification in links between the natural and the moral. For example, David Livingstone (1991) has identified a moral discourse of climate:

discussions of climatic matters by geographers throughout the nineteenth century and well into the twentieth century were profoundly implicated in the imperial drama and were frequently cast in the diagnostic language of ethnic judgement . . . scientific claims were constituted by, and then made to bear the weight of, moralistic appraisals of both people and places.

(Livingstone 1992: 221)

In the prevailing spirit of environmental determinism, racial constitution, including moral character, was attributed to climate.

A further geographical issue raised in these chapters by James Proctor and Jeremy Tasch is that of the spatial scope of environmental concern. For Proctor, the spatial scope of environmental concern is increasingly (though uncritically) global. Tasch shows how the changing economic orientation of part of the former Soviet Union, from a local periphery in relation to the union core to incorporation into the global circulation of capital, has implications for discourses of environmentalism. The local becomes the global, in such universal physical processes as 'global warming', just as local economic activity can have widespread impacts in our increasingly interdependent world. Here are new arguments for universalism, in both environmental and social concern. Yet focus on the global can deny the moral importance of the local, as not all local processes with environmental impact have significant global ramifications.

The notion of boundaries is a metaphor often used to capture the scope of the moral, sometimes without any spatial reference (e.g. Tronto 1993). That the boundaries of moral concern may extend beyond humankind and into the world of animals is demonstrated by Alice Dawson's chapter. And there is, of course, a local specificity to the treatment of animals, some of which are eaten or tormented for public entertainment in some places, but not in others. In Britain, city is sometimes pitted against countryside on the issue of fox hunting, while in Spain bull fighting is claimed to be part of a national way of life. The contested terrain of animal rights is taking on a new twist in the context of genetic engineering, with its capacity to generate replacement organs for human beings. The question of whether such an organ becomes human in a human being, like the practice itself, raises all manner of ethical issues, the consideration of which seems to lag behind the technology. Thus, nature is opening up all kinds of moral complexities that seem to multiply with the supposed advance of science.

Ethics and knowledge

Aspects of professional ethics in geography have been debated ever since the social relevance or radical geography movement of the late 1960s and early 1970s. An early monograph on some of the issues raised by the (re)turn to humanism was produced by Annette Buttimer (1974), followed by a book on research practice by Bruce Mitchell and Dianne Draper (1982). However, it took a few more years for geographers in significant numbers critically to examine their own ways of doing things, not just from the technical point of view but as involving human relationships with a clear ethical dimension. An important stimulus was the gradual replacement of quantitative techniques by qualitative methods (Eyles and Smith 1988).

The essays in Part 4 of this book have demonstrated various strategies for the compilation of knowledge based on awareness of the unequal power between researcher and researched. The conventional relationship empowers academics to represent the lives of others in particular ways, which may be motivated by considerations other than some conception of authenticity or truth – contested as these concepts themselves may be. Thomas Herman and Doreen Mattingly invoke a communicative ethics in engaging participatory research designed to empower local people. Rob Kitchin argues the moral responsibility to involve disabled persons as collaborators as well as research subjects. Robert Rundstrom and Douglas Deur stress the importance of reciprocity between researchers and researched. Nuala Gormley and Liz Bondi explore problems arising from the conflicting agendas of participants in the process of collecting data and producing research findings.

These essays reveal a common commitment to a relational form of ethics. What is right in the conduct of research is discovered in practice, and negotiated among the various participants. The ethics are situated, in the sense that good practice depends on the specific context. The emergence of a relational ethics is sometimes associated with the feminist challenge to the supposed masculinist orientation of mainstream moral philosophy: a dichotomy first recognized by Carol Gilligan (1982) in her distinction between an ethic of care and an ethic of justice. Sarah Whatmore (1997), among others, has drawn attention to the significance of a relational ethics in geographical and environmental research.

However, Rundstrom and Deur stress that, although they emphasize contextuality, they are not willing to argue against ethical universals. They recognize that all people deserve respect, privacy, equitable treatment, and freedom from intrusion and oppression. How such rights are understood and negotiated will depend on the cultural context of both observers and observed. This is a case of the distinction made by Michael Walzer (1994) between those thin moral concepts or values with universal appeal, like liberty and justice, and their contextual thickening among particular people in particular times and places. The imposition of some universal code of professional ethics, even within the confines of a national institution such as the Association of American Geographers or the Royal Geographical Society, faces the opposition of those whose ethics are necessarily contextual.

These issues of relations among researchers and researched merely scratch the surface of debates on contemporary professional ethics in geography (for a recent review, see Hay 1998). Other issues include how individual scholars should respond to the various criteria of performance foisted on the academy by the contemporary fixation with assessment, a problem to which Jeremy Crampton draws attention in the specific context of use of the Internet for dissemination of research findings. The commodification of knowledge, in an environment of free-market fanaticism where everything has its pecuniary value, poses serious difficulties for the practice of creative scholarship. So does the competitive ethic introduced from the world of commerce, for activities in which mutual collaboration and cooperation tend to have been the norm.

The final issue is that raised by Tim Unwin: the moral right and duty of the geographer to engage in activities contributing to social change. Of course, the fact is that we all do this, whether we recognize it and consciously chose the activist role or not. Seldom if ever is our work politically neutral, or so inconsequential as to make no difference to anything. In this world of rising inequality, mass poverty, brutal exploitation and large-scale oppression, it is not difficult to formulate a general principle of professional practice prioritizing work for the benefit of the worst-off in society, consistent with one of the most influential theories of justice yet devised (Rawls 1971). As the South African politician Jan Hofmeyr stated in 1939, when the seeds of apartheid were being sown in the fertile soil of colonialism, in the conflict between democracy and authoritarianism, 'no University worthy of its great tradition can fail to range itself on the side of democracy' (quoted in Paton 1971: 249).

Most of us face a less stark choice, and one less hazardous than that taken by those courageous academics who worked for the overthrow of a racist regime. But whatever the context, what is required is an informed and principled choice, one that can be defended from a moral point of view. At the very least, this calls for self-conscious consideration of our own ethics – not as a firm and fixed code but as a flexible and expanding response to the world in which we live and work, and to the beings we encounter. This book hopes to encourage readers to explore their own personal moral terrain, and in so doing contribute to the good of others.

Ethics and geography

> [P]hilosophy cannot do its job well unless it is informed by fact and experience: that is why the philosopher, while neither a field-worker nor a politician, should try to get close to the reality she describes.
>
> (Nussbaum 1998: 765)

> We won't learn much from what we see if we do not bring to our fieldwork such theories of justice and human good as we have managed to work out until then.
>
> (Nussbaum 1998: 788)

Finally, some thoughts on the relationship between the disciplines of ethics and geography. These will be brief and selective, as both editors have addressed aspects of the subject elsewhere (Proctor 1998; Smith 1998a). Martha Nussbaum, one of the few moral philosophers to make a sustained contribution to practical matters (in development studies), has drawn attention to the importance of philosophers getting close to reality. Herein lies the most obvious and perhaps significant role for the geographer working at the interface with ethics, exemplified by a number of our contributors: to reveal the facts and experiences of real people in their geographical and historical settings, as they try to make moral sense of the world. In a word, it is context that the geographer can help to provide. For the morality that people actually practise and the theories that ethicists devise are embedded within specific sets of social and physical relationships manifest in geographical space, reflecting the particularity of place as well as time.

This points to the need to recognize and elaborate the kind of historical geography of morality and ethics suggested in the section on ethics and space at the beginning of this concluding chapter (see also Smith 1998a: 12–14). The importance of historical context in moral philosophy has been explained at length by Alasdair MacIntyre (1967, 1981), in work where the coupling of 'place' with 'time' is often suggestive of a geographical perspective which remains implicit. For example, in a passage anticipating the 'thick' and 'thin' distinction made by Michael Walzer (1994), mentioned above, he explains: 'What freedom is in each time and place is defined by the specific limitations of that time and place and by the characteristic goals of that time and place' (MacIntyre 1967: 204). Only Yi-Fu Tuan to date has attempted a sustained examination of how morality is lived and imagined by real people in their geographical setting, seeking to link the geographer's traditional interest in transformed nature or the built environment with the 'moral-ethical systems' that human beings have constructed (Tuan 1989).This follows his exploration of how the meaning of the good life varies from culture to culture (Tuan 1986), and has been followed by elaborations of links between the moral and the aesthetic in the context of culture (Tuan 1993).

In this sensitivity to context, geography shares something with two influential contemporary intellectual movements: feminism and multiculturalism. Feminist approaches to ethics have come to the fore in recent years (e.g. Held 1993), and are featured in the second issue of the journal *Ethics, Place and Environment*. These focused initially on the injustice of gender inequality and discrimination against women in patriarchal social orders. They went on to advocate a relational ethics (including research practice, as illustrated by papers in Part 4 of this book), and especially the ethic of care (mentioned above) posited as a challenge to perspectives grounded in impartiality (e.g. Clement 1996; Bowden 1997). They also involve issues concerned with the body and sexuality, and with traditionally female work associated with the caring professions and child-rearing which are claimed to be undervalued by comparison with predominantly male activities.

Examples of the kind of research in which a feminist perspective combines with geographical sensitivity to context can be found in recent applications of the notion of moral geographies (to which reference was made in the Introduction; see also Smith 1998a: 14–18). In a study of local childcare cultures, Sarah Holloway (1998: 31) identifies 'a moral geography of mothering . . . a localized discourse concerned with what is considered right and wrong in the raising of children', with special reference to pre-school education. She uses interview material to compare two parts of the city of Sheffield (UK): affluent Hallam, where the 'good mother' knows a great deal about the range of available services and ensures that her children receive the best care possible, and less prosperous Southey Green, where the 'good mother' sees that her child's name goes on the list for a popular local authority nursery. The social networks of mothers are crucial in reproducing the Hallam vision of good mothering, whereas in Southey Green it is accomplished through links with family and childcare professionals. In a similar vein, Robyn Dowling (1998) has observed complex differences between and within suburbs around moral codes of mothering and participation in paid work. As Holloway (1998: 47) concludes: 'The moral geographies of local childcare cultures are important both in defining mothers as a social group and in influencing the meaning and experience of motherhood for individual women'. They are part of the local construction of personal and group identities, social relations, and the broader cultures reproducing such attributes within a dynamic process allowing for change with changing context.

However, the emphasis of this kind of work on (local) individuality or difference risks obscuring some essential things: one of the tensions identified at the beginning of this Conclusion. While the practice of mothering and its normative evaluation is clearly subject to differences among cultures, there are common features to consider. In the studies cited above, mothers were responding in different ways to common responsibilities to their children. While some childcare can be taken on by men (who are capable of doing far more than is usually the case), it is a natural fact that children are born to women, and from this some moral conclusions flow, such as that the mother is expected to nurture the newly-born infant and not discard it. No matter how flexible and negotiated gender roles and identities may be to some feminists, there are natural limits. Men cannot bear or suckle children. The condemnation which inevitably follows breaches of the moral imperative of women to nurture their young serves to underline its strength, just as some culturally sanctioned practices of infanticide raise the question of whether the society in question can make claims to the good (which links into debates on multiculturalism). The context-sensitive approach of the geographer is valuable in helping to reveal the complex interplay of factors bearing on local practices of child rearing, and the grounds on which wider normative claims might be made. Similarly, the 'universal cross-cultural norms' (Nussbaum 1998: 770) identified in the capabilities approach to development policy requires local investigations into what is actually needed for human functioning in particular societies.

While stressing the imperative of getting closer to reality, Martha Nussbaum also points to the importance of bringing theory to bear on fieldwork. If there is a weakness in the moral geographies approach, it is that it tends not to be linked to theory in moral philosophy. An exception is to be found in a study of the moral geographies of the recreational area of Broadlands in Norfolk (UK), in which David Matless (1994) works around Michel Foucault's three senses of morality: as moral codes, as the exercise of behaviour in transgression of or obedience to a code, and as the way in which individuals act for themselves as ethical subjects in relation to elements of a code.

Lest we take the recognition of context in moral practice, and how to study it in place as unique to geography, reference may be made to the (re)discovery of moralities in their local specificity by some anthropologists (Howell 1997). This even includes a paper on 'the morality of locality', which describes exclusive attitudes to the other, the outsider, in parts of rural northwest England (Rapport 1997). There is much to be learned from this collection's blend of ethnographic method and theoretical deliberation, to strengthen the thick description of work conducted under the rubric of moral geographies, landscapes or locations.

The reference by Holloway (1998) to local childcare cultures provides a link into the second of the intellectual movements introduced above, to which the geographer's sensitivity to context has strong affinity: multiculturalism. Susan Moller Okin (1998) explains two meanings of multiculturalism. One refers to quests for recognition by groups including women, gays and lesbians, and some ethnic and racial minorities, i.e. the politics of identity on the part of those whose voices have hitherto been silent or subdued but who do not usually claim to have, or are considered to have, their own culture in the sense of 'a way of life'. The other meaning is what is usually understood in the context of debates on group rights to defend a culture which provides its members with a way of life across a range of activities, including social, educational, religious, recreational and economic life, in both public and private spheres. Unlike the first sense, the second is likely to involve a shared language and history, and to be spatially expressed with some sense of territoriality adding to other aspects of collective identity. While the distinction is not rigid (for example, a way of life more limited than that invoked by cultural rights and with some attachment to place might be claimed for gays in some cities), it does serve to distinguish different subject matter with which cultural geographers have engaged in recent years.

Work guided by the first sense of multiculturalism raises very obvious moral issues, including the place (in a literal, geographical sense as well as more generally) of particular gender identities and sexual practices. For example, Phil Hubbard (1998) examines the role of 'immoral geographies' in the emergence of red light districts and the marginalization of female prostitutes on the streets of the city of Birmingham (UK). Space and place are shown to contribute to discourses deployed to establish and challenge what kind of behaviour belongs where, reflecting back on normative notions of community. From this comes a more refined sense of the spatial variability of moral codes: of the facts of moral relativism in a descriptive sense, which philosophers are inclined to leave to others (usually specified as anthropologists rather than geographers).

The second sense of multiculturalism raises issues concerning the evaluation of different ways of life. This is the context within which the problem of normative ethical relativism as the doctrine that what actually is right or wrong differs among cultures, confronts universalism as the doctrine that there are ways of comparing different cultures on the basis of better or worse. Claiming that most cultures have as one of their principle aims the control of women by men, Okin (1998) engages a critique of relativism from a feminist perspective. She stresses the tension between group cultural rights and what actually happens within cultures, especially in the private sphere, where gender relations disadvantage women. She also shares the concern expressed in Caroline Nagel's essay over the contemporary deference to difference: 'focusing only on differences among women and bending over backward out of respect for cultural diversity does great disservice to many women and girls throughout the world' (Okin 1998: 666). Exposing those, often hidden, local practices whereby women are dominated and oppressed may facilitate normative comparisons among cultures on the basis of at least this one dimension of better or worse: another avenue for context-sensitive research.

One of Okin's examples gets to the heart of the ethics and geography interface. It is the way of life of Israel's Ultra-Orthodox Jews (or 'Haredi'), in which a privileged practice of religious study is prescribed for men while women are assigned a subservient supportive role, in both cases with no choice but the difficult route of exit from communities, the boundaries of which are far from porous. Furthermore, as in many fundamentalist religions, girls and women are held responsible for male sexual self-control, which requires 'modest' forms of personal presentation. This way of life is vigorously defended by the communities concerned, and is heavily subsidized financially by the state. However, Okin (1998: 672) argues that 'Ultra-Orthodox culture is more likely than a more open and liberal culture to harm the individual interests of both its male and its female children', and concludes that 'its public support is unacceptable for both liberal and specifically feminist reasons'. The details of her argument are less important than her lack of reticence in being prepared to judge this culture as worse than some alternatives. There is a further, distinctively geographical, dimension to this case. The Haredi wish to impose their way of life on others, at least to the extent of restricting activities on the Sabbath, and are attempting to establish cultural domination and political hegemony in the city of Jerusalem by residential expansion into secular neighbourhoods. So, here is a territorial clash, not only of cultures but of ethics or moral philosophies – of fundamental communitarianism versus liberal pluralism, which is surely not immune to normative evaluation in the light of some basic conceptions of human dignity, equality and freedom. And central to such evaluation is the local knowledge, the fact and experience, provided by geographical research (Hasson 1996; see also Sandercock 1998: 177–182).

We have thus returned to another of the tensions identified at the outset of this Conclusion. The tension between universalism and particularism encourages not the advance (or retreat) to one extreme or the other, but the application of theories of justice and the human good, however tentative, to the facts of the local situation. The conclusions may not convince everyone, but at least they provide a basis for further argument, from which to take another tentative step in the direction of discovering or creating a more convincing sense of justice or of the good. If this dialectical relationship between the theory at our disposal and the practice of fieldwork or local case study seems familiar to the geographical reader, then established professional practice already has something to commend it as we explore the new terrain. So, let us proceed with the confidence that we have something to give as well as to take from an engagement with ethics.

Acknowledgement

Thanks to my co-editor James Proctor for some points made here.

References

Bauman, Z. (1989) *Modernity and the Holocaust*, Cambridge: Polity Press.

Bowden, P. (1997) *Caring: Gender-sensitive Ethics*, London: Routledge.

Buttimer, A. (1974) *Values in Geography*, Washington DC: Association of American Geographers, Commission on College Geography, Resource Paper 24.

Clement, C. (1996) *Care, Autonomy, and Justice: Feminism and the Ethic of Care*, Oxford: Westview Press.

Corbridge, S. (1993) 'Marxisms, modernities, and moralities: development praxis and the claims of distant strangers', *Environment and Planning D: Society and Space* 11, 449–472.

—— (1998) 'Development ethics: distance, difference, plausibility', *Ethics, Place and Environment* 1, 35–53.

Cutter, S. L. (1995) 'Race, class and environmental justice', *Progress in Human Geography* 19, 111–122.

Dowling, R. (1998) 'Suburban stories, gendered lives: thinking through difference', in R. Fincher and J. M. Jacobs (eds) *Cities of Difference*, New York and London: Guilford Press.

Eagleton, T. (1996) *The Illusions of Postmodernism*, Oxford: Blackwell Publishers.

Eyles, J. and Smith, D. M. (eds) (1988) *Qualitative Methods in Human Geography*, Cambridge: Polity Press.

Friedman, M. (1991) 'The practice of partiality', *Ethics* 101, 818–835.

—— (1993) *What are Friends For? Feminist Perspectives on Personal Relationships and Moral Theory*, Ithaca and London: Cornell University Press.

Geras, N. (1995) *Solidarity in the Conversation of Humankind: The Ungroundable Liberalism of Richard Rorty*, London: Verso.

Gilligan, C. (1982) *In a Different Voice: Psychological Theory and Women's Development*, Cambridge, Mass.: Harvard University Press.

Habermas, J. (1990) *Moral Consciousness and Communicative Action*, Cambridge: Polity Press.

Harvey, D. (1996) *Justice, Nature and the Geography of Difference*, Oxford: Blackwell Publishers.

Hasson, S. (1996) *The Cultural Struggle over Jerusalem: Accommodation, Scenarios and Lessons*, Jerusalem: The Floersheimer Institute for Policy Studies.

Hay, I. (1998) 'Making moral imaginations. Research ethics, pedagogy, and professional human geography', *Ethics, Place and Environment* 1, 55–75.

Held, V. (1993) *Feminist Morality: Transforming Culture, Society, and Politics*, Chicago and London: The University of Chicago Press.

Holloway S. H. (1998) 'Local childcare cultures: moral geographies of mothering and the social organisation of pre-school education', *Gender, Place and Culture* 5, 29–53.

Howell, S. (ed.) (1997) *The Ethnography of Moralities*, London: Routledge.

Hubbard, P. (1998) 'Sexuality, immorality and the city: red-light districts and the marginalisation of female street prostitutes', *Gender, Place and Culture* 5, 55–72.

Light, A. and Smith, J. M. (eds) (1997) *Space, Place and Environmental Ethics* (*Philosophy and Geography* I), London: Rowman and Littlefield Publishers.

Livingstone, D. L. (1991) 'The moral discourse of climate: historical considerations on race, place and virtue', *Journal of Historical Geography* 17, 413–434.

—— (1992) *The Geographical Tradition*, Oxford: Blackwell Publishers.

Low, N. and Gleeson, B. (1998) *Justice, Society and Nature: An Exploration of Political Ecology*, London and New York: Routledge.

MacIntyre, A. (1967) *A Short History of Ethics: A History of Moral Philosophy from the Homeric Age to the Twentieth Century* (second edn, 1998), London: Routledge.

—— (1981) *After Virtue: A Study in Moral Theory*, London: Duckworth.

Matless, D. (1994) 'Moral geographies in Broadlands', *Ecumene* 1(2), 127–156.

Mitchell, B. and Draper, D. (1982) *Relevance and Ethics in Geography*, London: Longman.

Nussbaum, M. C. (1998) 'Public philosophy and international feminism', *Ethics* 108, 762–796.

Okin, S. M. (1998) 'Feminism and muticulturalism: some tensions', *Ethics* 108, 661–684.

O'Neill, O. (1996) *Toward Justice and Virtue: A Constructive Account of Practical Reasoning*, Cambridge: Cambridge University Press.

Paton, A. (1971) *Hofmeyr*, Cape Town: Oxford University Press.

Proctor, J. D. (1998) 'Ethics in geography: giving moral form to the geographical imagination', *Area* 30, 8–18.

Rapport, N. (1997) 'The morality of locality: on the absolutism of landownership in an English village' in Howell (1997) op. cit., 74–97.

Rawls, J. (1971) *A Theory of Justice*, Cambridge, Mass.: Harvard University Press.

Rheingold, H. (1993) *The Virtual Community: Homesteading on the Electronic Frontier*, Reading, Mass.: Addison-Wesley.

Rorty, R. (1989) *Contingency, Irony, and Solidarity*, Cambridge: Cambridge University Press.

290   *Ethics and knowledge*

Sack, R. D. (1997) *Homo Geographicus: A Framework for Action, Awareness and Moral Concern*, Baltimore and London: The Johns Hopkins University Press.

Sandercock, L. (1998) *Towards Cosmopolis: Planning for Multicultural Cities*, Chichester: John Wiley.

Sibley, D. (1995) *Geographies of Exclusion: Society and Difference in the West*, London and New York: Routledge.

Smith, D. M. (1997a) 'Las dimensiones morales del desarrollo' (Moral dimensions of development), *Economía, Sociedad y Territorio* 1, 1–39.

—— (1997b) 'Back to the good life: towards an enlarged conception of social justice', *Environment and Planning D: Society and Space* 15, 19–35.

—— (1997c) 'Geography and ethics: a moral turn?', *Progress in Human Geography* 21, 596–603.

—— (1998a) 'Geography and moral philosophy: some common ground', *Ethics, Place and Environment* 1, 7–34.

—— (1998b) 'How far should we care? On the spatial scope of beneficence', *Progress in Human Geography* 22, 15–38.

—— (1999) 'Geography, morality and community', *Environment and Planning A* 31, 19–35.

Tronto, J. (1993) *Moral Boundaries: A Political Argument for an Ethic of Care*, London: Routledge.

Tuan, Y.-F. (1986) *The Good Life*, Madison: University of Wisconsin Press.

—— (1989) *Morality and Imagination: Paradoxes of Progress*, Madison: University of Wisconsin Press.

—— (1993) *Passing Strange and Wonderful: Aesthetics, Nature and Culture*, Washington DC: Island Press.

Vetlesen, A. J. (1993) 'Why does proximity make a moral difference? Coming to terms with lessons learned from the Holocaust', *Praxis International* 12, 371–386.

Walzer, M. (1994) *Thick and Thin: Moral Argument at Home and Abroad*, Notre Dame, Ind. and London: Notre Dame University Press.

Whatmore, S. (1997) 'Dissecting the autonomous self: hybrid cartographies for a relational ethics', *Environment and Planning D: Society and Space* 15, 37–53.

Young, I. M. (1990) *Justice and the Politics of Difference*, Princeton, NJ: Princeton University Press.

Index

Birdsall, Stephen 8, 194, 195
Black, R. 7, 65, 68
Blackstone, Sir William 30, 39
Blake, William 149
Bondi, Liz 136, 141, 194, 207, 208, 209, 224, 229, 251, 257, 282
bonsai 113–14
Bosnia: NATO 129; safe areas 125–6; and UN 124–5, 129; UNPROFOR 124; *see also* Srebrenica
Bosnian Muslim army 125
Bosnian Serb army 93–4, 120, 129–30
Both, N. 120, 125, 127, 128, 129
boundaries 281
Boundary Commission 50–1
Bourdieu, Pierre 99
Boutros Gali, Boutros 128
brain drain 68
Britain: electoral system 47–8, 49, 50–2, 55n1; Muslim communities 133–5, 136; Norfolk 8, 286; Portsmouth 8
Brundtland Commission 33
Budiansky, Stephen 200–1
bullying 110
Bureau of Indian Affairs 240
Bush, George 122, 130
Bushmen 109–10
Buttimer, A. 5, 7, 95, 157, 158, 194, 282
Byron, Margaret 239, 244, 246, 248n5

Cabot, John 97
Callicott, J. B. 150, 151, 155
Canadian troops 128
capitalism: global 30; inequality 138–9; local community 31; and morality 35–6; and other 38–9; and poverty 62; rights 31
care ethics 4, 259, 282
caring 11, 277, 285
Cartesian anxiety of modernism 20–1
Cartesian dualism 26
Castells, Manuel 60, 61, 64, 68, 74
categorization 134
Catholic Church 66
cats 114
children 110, 285–6; access 219; communication 214, 215–16; cruelty of 109; as pets 114–15; photography 218–19
Chinese Research Academy for Environmental Sciences 180
Christianity, disenchantment 24
Churchill, Winston 54
cities 8, 108–9

citizenship 18, 34, 140
City Heights, San Diego 212–13, 215, 216, 217–18
"City Moves" 214, 215, 218–19
civil rights 231–2
civil war 167, 169–70, 172
Clark, J. 7, 9, 158
Clarke, Samuel 102
Clifford, J. 209, 244
climate 8, 281
Clinton, Charles 237, 241, 245
coalitions 56n8
Cold War 122
collaboration 231, 282
Collingwood, R. G. 21
colonialism 32, 64, 237–8, 283
commodification 73, 283
communication 73, 122, 214, 215–16, 219
communicative ethics 207, 216, 282
community: and capitalism 31; ethnic 184; fairness 47–8; inclusive 216; Internet 85; partnership with state 67; social relations 214–15; virtual 73
community arts projects 210, 213–14, 219
compartmentalization 115–17
consequentialism 103
constituencies, electoral 50–1
consumption 34, 108
context 103, 282–3, 284–5
contextual ethics 181–5, 238, 239
contract researchers 241
cookies on Internet 81
Corbridge, Stuart 7, 8, 62, 64, 65, 67, 264, 277
Cortez, Hernando 97
Cosgrove, D. 5, 159
Cottle, S. 137, 143
Council for Exceptional Children 229
"Counter-Enlightenment" (Berlin) 20
Covenant on Civil and Political Rights, UN 66
Covenant on Economic, Social and Cultural Rights, UN 66
Crampton, Jeremy 10, 18, 74, 76, 80, 277, 283
Crawford High School 213, 218
Creek Nation 240
critical geography 165, 226
critical social theory 207, 212
Croft, S. 223, 231
cross-cultural research: code of ethics 244–6; colonization 237; empathy

Printed in the United States
by Baker & Taylor Publisher Services

Printed in the United States
by Baker & Taylor Publisher Services